Adjoint Equations and Analysis of Complex Systems

Mathematics and Its Applications

Managing Editor:

M. HAZEWINKEL

Centre for Mathematics and Computer Science, Amsterdam, The Netherlands

Volume 295

Adjoint Equations and Analysis of Complex Systems

by

Guri I. Marchuk
Institute of Numerical Mathematics,
Russian Academy of Sciences, Moscow, Russia

KLUWER ACADEMIC PUBLISHERS
DORDRECHT / BOSTON / LONDON

Library of Congress Cataloging-in-Publication Data

Marchuk, G. I. (Gurii Ivanovich), 1925-
    Adjoint equations and analysis of complex systems / by Guri
I. Marchuk.
        p.   cm. -- (Mathematics and its applications ; v. 295)
    Includes index.

    1. Adjoint differential equations. 2. Perturbation (Mathematics)
3. Differentiable dynamical systems.   I. Title. II. Series:
Mathematics and its applications (Kluwer Academic Publishers) ; v.
295.
QA372.M3687   1994
515'.352--dc20   94-22318

ISBN 978-90-481-4444-0

Published by Kluwer Academic Publishers,
P.O. Box 17, 3300 AA Dordrecht, The Netherlands.

Kluwer Academic Publishers incorporates
the publishing programmes of
D. Reidel, Martinus Nijhoff, Dr W. Junk and MTP Press.

Sold and distributed in the U.S.A. and Canada
by Kluwer Academic Publishers,
101 Philip Drive, Norwell, MA 02061, U.S.A.

In all other countries, sold and distributed
by Kluwer Academic Publishers Group,
P.O. Box 322, 3300 AH Dordrecht, The Netherlands.

The manuscript was translated from Russian by
Dr. Guennadi Kontarev

Printed on acid-free paper

Table of Contents

Author's Preface to the English Edition

New statements of problems arose recently demanding thorough analysis. Notice, first of all, the statements of problems using adjoint equations which gradually became part of our life. Adjoint equations are capable to bring fresh ideas to various problems of new technology based on linear and nonlinear processes. They became part of golden fund of science through quantum mechanics, theory of nuclear reactors, optimal control, and finally helped in solving many problems on the basis of perturbation method and sensitivity theory. To emphasize the important role of adjoint problems in science one should mention four-dimensional analysis problem and solution of inverse problems.

This range of problems includes first of all problems of global climate changes on our planet, state of environment and protection of environment against pollution, preservation of the biosphere in conditions of vigorous growth of population, intensive development of industry, and many others. All this required complex study of large systems: interaction between the atmosphere and oceans and continents in the theory of climate, cenoses in the biosphere affected by pollution of natural and anthropogenic origin. Problems of local and global perturbations and models sensitivity to input data join into common complex system.

Many of these and other analogous complex systems do not answer yet the most important questions of theory and applications, though the experience accumulated by science is as rich as to give powerful impulse to these problems on the basis of new statements. This author hopes, that the readers will find in the book interesting approaches to their own investigations.

The book was thoroughly revised for the english edition and supplemented with new results. It has two appendices related respectively to applications of the splitting method to complex problems and to a difference analogue of a heat diffusion equation.

The author is thankful to V. Lebedev and V. Agoshkov for a number of valuable comments, to G. Kontarev for translating the book, and to E. Artyomov and V. Shutyaev for editing.

G. Marchuk

Introduction

Adjoint equations are increasingly widespread throughout mathematics and its applications. Originally defined by Lagrange, adjoint operators have since been thoroughly substantiated theoretically and broadly applied in solving many problems in mathematical physics. The true meaning of the adjoint equations theory, though, was duly appreciated for the first time by physicists developing quantum mechanics. Schrödinger's equation required a set of adjoint equations and functions to be developed at least for eigenvalue problems [251]. Adjoint equations became for the first time mathematically indispensable for formulation of the small perturbation theory in spectral problems.

Theorizing on perturbations has since attracted many researchers. The first mathematically strict perturbation theory have probably evolved in F. Rellich [242, 243]. The mathematical perturbation theory was further developed by K. O. Friedrichs [189], T. Kato [199], N. N. Bogolubov and Y. A. Mitropolski [14], A. B. Vasiljeva and V. F. Butusov [24], M. I. Vishik and L. A. Lusternik [25–27], O. A. Oleinik [135], B. De Sz.-Nagy [187], C. A. Lomov [65], N. N. Moiseev [125], V. P. Maslov [118], V. A. Trenogin [150], A. N. Filatov [159–161], R. Bellman [176], M. D. Van Dyke [261], and by many others. Their works went on to apply the perturbation theory to broad ranges of mathematical physics problems. The ideas common for all those works, though, were the decomposability of the solution in powers of the small parameter and justification of convergence of the obtained series to the exact solution of a problem.

The nuclear reactors theory generated the next surge of interest in adjoint equations and the small perturbation theory. The nuclear reactors theory involved complex problems of neutron transport theory in decelerating media and required adequate adjoint equations and solutions of main and adjoint equations as applied to estimation of the first eigenvalue of the spectral problem a sustainable stationary chain reaction involved. The formulation originated in A. Weinberg's and E. Wigner's research [264] for the simplest diffusion models of neutron transport and deceleration, and was further generalized for kinetic reactor equations in L. N. Usachov [152], G. I. Marchuk and V. V. Orlov [110], A. I. Mogilner and V. Y. Pupko et all [142]. The small perturbation theory built on main and adjoint reactor equations was essential for development of nuclear reactor physics and nuclear power projects.

B. B. Kadomtsev [50] formulated a new approach to adjoint problems for point sources in the neutron transport problem. G. I. Marchuk and V. V. Orlov [110] offered general formulation of adjoint problems in relation to selected linear functionals of problems, particularly measurements of physical processes, characteristic of the radiation transport theory functionals. These are no spectral problems any more but mathematical physics' problems, with given sources assigned within equations, or in initial data, or in boundary conditions.

This author went on, in his later works, developing the adjoint problems theory in relation to assigned functionals, for ranges of mathematical physics' problems. The effort proved productively applicable in many other sciences. More or less general approaches in research of complex systems and mathematical models followed up, to constitute the essence of this author's years-long research in different fields of mathematics and its applications to problems of diffusion, environment protection models, theory of climate and its changes, mathematical problems of processing the information provided by satellites, mathematical models in immunology, and others. A rational approach to solution of inverse problems and to mathematical experiment planning was evolving parallel to development of adjoint equations techniques. This author researched planning in collaboration with S. M. Yermakov [100].

A few words should be said about nonlinear problems in mathematical physics. As is well known, adjoint equations techniques, in their classic form, have been built for linear equations only and on Lagrange's identity. This author encountered nonlinear problems dealing with hydrodynamic equations describing processes in the atmosphere and the ocean. The problems required special adjoint equations techniques, and an unexpected solution was found through construction of conventional equations, adjoint in the Lagrangean sense, which required, in their coefficients, main equation's solutions. Consecutive solution of the main, and then adjoint, equations resulted in the small perturbation theory.

Nonlinear problems has become the subject of extensive research in recent years, generating generalizations of the adjoint equations theory originally meaningful for ranges of problems. Approaches formulated by V. P. Maslov [118], V. S. Vladimirov and I. V. Volovich [33, 34] arouse great interest. A theory of nonlinear adjoint equations proposed by M. M. Weinberg [20] is likely to be as productive. This author is also responsible for some research in these areas in collaboration with V. I. Agoshkov [91, 225, 226].

We award so much attention to the small perturbation theory for one serious reason. Variations of functionals obtained by special formulas of the perturbation theory make it in effect possible to link them with variations of coefficients and other inputs of a problem, initial and boundary conditions included. This enables us to develop measurement devices or techniques to obtain maximum informative solutions to inverse problems involving finding unknown parameter variations through known variations of functionals. It is especially important for the analysis of complex systems, where it is quite rarely possible to establish immediate links between variations of functionals and those of parameters.

So broad a scope of problems involving adjoint equations naturally required mathematical justification. A special book the author wrote in collaboration with V. I. Agoshkov and V. P. Shutyaev [94] covers thoroughly these subjects. The book discusses not only adjoint equations and problems but also presents a rather general form of research of the perturbation theory providing a mathematically strict justification of new issues.

Application of adjoint equations in general, and their application to solution of problems in mathematical physics is thoroughly enough substantiated theoretically in special literature. In this monograph, the author will tentatively attract to those new approaches attention of a more or less broad group of scholars and researching engineers in the hope to provide a stimulus to devise experimentation designing/planning techniques for solution of applied problems. These approaches provide in many cases necessary information on physical processes and unknown characteristics of complex systems. With this sort of readership in mind, this author will try to present the subject in the simplest possible form. The techniques discussed may be generalized and mathematically strictly substantiated using special literature (see the end of the book for the titles).

I would like to mark one more feature of the adjoint problems theory. Adjoint problems provide immediately the closest approach to problems of optimal control, as developed by R. Bellman [176], L. S. Pontryagin [140], N. N. Krasovsky [53], J.-L. Lions [214], R. Glowinski [195], A. Balakrishnan [172], and many other researchers, as well as to problems of sensitivity of main problems' solutions, or functionals of those, to problem inputs. This, in turn, provides a deeper insight into the nature of processes modelled, and more precise mathematical statements, which are especially important in solving nonlinear problems notorious for their exceptional resistance to analysis and interpretation. The value of adjoint problems theory cannot be overappreciated in this area.

Finally, a number of important notes. This author's intention was to make the ideas of construction and applications of adjoint equations familiar to a broad circle of researchers, who know just the basic course of mathematics read to the mathematicians and research workers interested in construction of mathematical models realized on computers. Therefore, the author did not pursue the maximum generality and accuracy of presentation, considering, as a rule, just a general scheme. Strictly speaking, all the functions and solutions of problems discussed in the book, must be considered as generalized ones. We assume always that the right-hand parts of equations, initial and boundary conditions provide the existence of problems. In some cases, when input data are sufficiently smooth, the generalized solutions become classic. In the cases of discontinuous functions and operations with them, or δ-functions we do not introduce the generalized solutions inherent to such classes of input data and solutions, being sure that the readers may formulate themselves the generalized statements, if needed, using special literature.

Of course, some statements in this book might be made more general and accurate, but this would narrow the circle of readers who are interested, first of all, in algorithmic side of applications. Therefore this author recommends for more general and strict statements of problems in the class of generalized functions refer to special literature (for example, S. L. Sobolev [255], L. Schwartz [252], J.-L. Lions and E. Magenes [216], V. Vladimirov [30, 32]).

All the considerations at operator level are presented in the book in Hilbert spaces, though it should be noticed, that these considerations as well as results and conclusions are applicable, as a rule, for a more general case of Banach spaces (see [95]).

The author should like to express his profound gratitude to Academician V. S. Pugachev who read this manuscript and contributed important remarks, and to professor J. Fletcher for valuable discussions on applications of adjoint equations to the problem of climate changes.

PART I

Adjoint Equations and Perturbation Theory

Main and Adjoint Equations. Perturbation Theory

Simple Main and Adjoint Equations of Mathematical Physics

Nonlinear Equations

Inverse Problems and Adjoint Equations

This part of the book covers the basics of the adjoint equations theory and perturbation algorithms and respective applications to solution of mathematical physics' problems. The coverage includes subjects linked with application of adjoint equations and algorithms to solving non-homogeneous stationary and non-stationary problems, eigenvalue problems and computation of linear functionals (Chapters 1 and 2). Chapter 3 theorizes on adjoint equations and perturbation algorithms in nonlinear problems. Chapter 4 covers mathematical physics' inverse problems statements based on adjoint equation techniques and the perturbation theory. Specific examples provided to illustrate the matter may prove usable in dealing with more complex practical problems.

CHAPTER 1

Main and Adjoint Equations. Perturbation Theory

This chapter covers simple problems of mathematical physics, gives elements of the self-adjoint and adjoint problems theory as applied to differential equations, and formulates the basic principles of perturbation theory. Lagrange's identity constitutes the principal mathematical tool to construction of adjoint equations.

1.1. Main and Adjoint Operators in Linear Problems. Elements of Theory

Let us consider an example of simple differential operator of the second order

$$A = -\frac{d^2}{dx^2} \qquad (1.1.1)$$

operating on real functions v with the following properties:

a) $v(x)$ are defined in the domain $\Omega = (0,1)$ as continuous and differentiable in all internal points of the domain;

b) $v(x)$ are quadratically summable on Ω together with their derivatives dv/dx and d^2v/dx^2, i. e.,

$$\int_0^1 \left\{ \left(\frac{d^2v}{dx^2}\right)^2 + \left(\frac{dv}{dx}\right)^2 + v^2(x) \right\} dx < \infty \qquad (1.1.2)$$

As is known, the functions $v(x)$ with the property $\int_0^1 v^2(x)dx < \infty$ form the Hilbert space of functions $H = L_2(\Omega)$. It is easily shown [131, 255], that the functions possessing the properties a) and b) may be

9

considered as almost everywhere continuous in Ω together with their first derivatives up to the boundary. Let us assume then, that the functions $v(x)$ have prescribed values on the boundary of the domain Ω at $x = 0$ and $x = 1$, e. g.,

$$v(0) = v(1) = 0 \tag{1.1.3}$$

Denote by $D(A)$ the set of functions $v(x)$ satisfying conditions a), b), and (1.1.3). We shall refer to this set as the definition domain of the operator A.

Let us introduce an inner product for the functions $v(x)$ and $w(x)$ in the Hilbert space H with the definition domain $\Omega = (0, 1)$:

$$(v, w) = \int_0^1 vw \, dx \tag{1.1.4}$$

Let us affect then by the operator A the function $v(x) \in D(A)$. The result will be a new function Av which is also defined in $\Omega = (0, 1)$ and belongs to H. Now consider the inner product of functions Av and w

$$(Av, w) = -\int_0^1 w \frac{d^2v}{dx^2} dx \tag{1.1.5}$$

and integrate expression (1.1.5) by parts, then

$$(Av, w) = -\left. w \frac{dv}{dx} \right|_{x=0}^{x=1} + \int_0^1 \frac{dw}{dx} \frac{dv}{dx} dx \tag{1.1.6}$$

The term outside the integral in the right-hand part of (1.1.6) is equal to zero, since $v \in D(A)$ and $w \in D(A)$ and any function of this set equals zero at the boundaries of intervals according to (1.1.3). Then (1.1.6) results into

$$(Av, w) = \int_0^1 \frac{dw}{dx} \frac{dv}{dx} dx \tag{1.1.7}$$

Let us again integrate the integral in (1.1.7) by parts. We obtain

$$(Av, w) = \left. \frac{dw}{dx} v \right|_{x=0}^{x=1} - \int_0^1 v \frac{d^2w}{dx^2} dx \tag{1.1.8}$$

The term outside the integral is also equal to zero due to condition (1.1.3). Then we obtain

$$(Av, w) = - \int_0^1 v \frac{d^2 w}{dx^2} dx \qquad (1.1.9)$$

Comparing (1.1.5) and (1.1.9), we arrive at the conclusion, that

$$(Av, w) = (v, Aw) \qquad (1.1.10)$$

Expression (1.1.10) is the Lagrange identity for symmetric operators. Or, if (1.1.10) holds for the functions v, $w \in D(A)$, A may also be referred to as a symmetric operator.

Let us turn to spectral problems theory propositions for symmetric operators. Let us assume a spectral problem in the form

$$Av = \lambda v \qquad (1.1.11)$$

where the operator A is defined still in (1.1.1) and the function $v \in D(A)$ and therefore satisfies homogeneous boundary conditions (1.1.3). In the case of this simple problem the operator form (1.1.11) may be replaced by differential one

$$\frac{d^2 v}{dx^2} + \lambda v = 0, \quad v(0) = v(1) = 0 \qquad (1.1.12)$$

All the non-zero functions $v(x)$ and the numbers λ satisfying (1.1.12) must be found. As is known, the solutions are functions

$$v_k(x) = \sin k\pi x, \quad k = 1, 2, \ldots \qquad (1.1.13)$$

with

$$\lambda_k = k^2 \pi^2$$

The system of eigenfunctions (1.1.13) is evidently orthogonal:

$$\int_0^1 v_k(x) v_n(x) dx = \begin{cases} 1/2, & \text{if } k = n, \\ 0, & \text{if } k \neq n \end{cases} \qquad (1.1.14)$$

As is know, again, functions (1.1.13) form a complete system which enables representation of any function of $H = L_2(\Omega)$ as a series.

Let us now examine nonhomogeneous equation

$$Av = f \qquad (1.1.15)$$

where A is still defined as in (1.1.1) and f is the source function in the Hilbert space H with the definition domain $\Omega = (0,1)$. In a differential form, we shall have

$$-\frac{d^2v}{dx^2} = f, \quad v(0) = v(1) = 0 \qquad (1.1.16)$$

R e m a r k. We assume here and forth while examining the un-ordinary equations of (1.1.15), (1.1.16) type that the right-hand part $f(x)$ is sufficiently smooth, so that the solution v of problem (1.1.15) exists and belongs to $D(A)$. For example, if $f(x)$ is continuous on the interval [0,1], then there exists classic solution v of problem (1.1.16) which is double differentiable function on [0,1] and, naturally, belongs to $D(A)$. If $f(x)$ has a discontinuity of the first type, then it is possible to seek a generalized solution in the following sense: find $v \in D(A)$ so that $(Av, w) = (f, w)$ for all $w \in H$ [255]. The notion of generalized solution may be, if needed, extended in this concrete case on the basis of representation (1.1.7). We may seek the generalized solution v of problem (1.1.16) in the Sobolev's class $\overset{\circ}{W}_2^1(0,1)$ of functions quadrati-cally summable on Ω together with their first derivatives and satisfying the condition (1.1.3) with v satisfying the integral identity

$$\int_0^1 \frac{dv}{dx}\frac{dw}{dx}dx = \int_0^1 f(x)w(x)dx$$

for all $w \in \overset{\circ}{W}_2^1(0,1)$. We may take for $f(x)$ in the last case even δ-func-tions, since the right-hand part is a linear continuous functional. But then $D(A)$ must be extended. Here we may consider as $D(A)$ all the class of functions $w \in \overset{\circ}{W}_2^1(0,1)$. $\square$

Let us multiply the equation from (1.1.16) by the function $v_k(x)$ from (1.1.13) and integrate the result over all the definition domain Ω of the solution. We obtain

$$-\int_0^1 v_k \frac{d^2v}{dx^2}dx = \int_0^1 fv_k dx \qquad (1.1.17)$$

Let us introduce the notation

$$\int_0^1 f v_k dx = f_k$$

and represent the function $v(x)$ in the form of the Fourier series

$$v = \sum_{n=1}^{\infty} \alpha_n v_n \qquad (1.1.18)$$

Let us substitute (1.1.18) into (1.1.17) and take into consideration that any function $v_n(x)$ satisfies

$$A v_n = \lambda_n v_n \qquad (1.1.19)$$

Then

$$\sum_{n=1}^{\infty} \alpha_n \lambda_n \int_0^1 v_n(x) v_k(x) dx = f_k$$

Given the orthogonality condition of (1.1.14), we obtain an expression for the Fourier coefficients

$$\frac{\lambda_k \alpha_k}{2} = f_k \quad \text{or} \quad \alpha_k = 2 f_k / \lambda_k$$

Thus (1.1.16) is solved as

$$v = 2 \sum_{n=1}^{\infty} f_n v_n(x) / \lambda_n \qquad (1.1.20)$$

As a specific example, let $f(x) = \sin \pi x$. Then (1.1.15) has the following form:

$$-\frac{d^2 v}{dx^2} = \sin \pi x, \quad x \in (0,1), \quad v(0) = v(1) = 0$$

Using the functions $v_k(x)$ from (1.1.13) we compute f_k:

$$f_k = \int_0^1 f v_k dx = \int_0^1 \sin \pi x \sin \pi k x \, dx = \frac{1}{2} \delta_{1k}$$

where

$$\delta_{1k} = \begin{cases} 0, & k = 1 \\ 1, & k \neq 1 \end{cases}$$

Now, knowing f_k and $\lambda_k = k^2\pi^2$ let us define the solution v using (1.1.20):

$$v = 2\sum_{n=1}^{\infty} f_n v_n(x)/\lambda_n = \sum_{n=1}^{\infty} \frac{\delta_{1n}v_n(x)}{n^2\pi^2} = \frac{1}{\pi^2}\sin\pi x$$

which is immediately verifiable.

Let us take up a particular example of a non-symmetric operator. Assume that

$$A = -\frac{d^2}{dx^2} + \frac{d}{dx}$$

Let us still assume that A affects the functions v which belong to the set $D(A)$ defined above by properties a) and b), and condition (1.1.3).

Let us compose an inner product of the functions $v \in D(A)$ and $w \in D(A^*)$ with the definition domain $\Omega = (0,1)$. Notice, that the set $D(A^*)$ the elements w belong to is not yet defined. Properties of the set will be specified in further transformations which are to lead us to the Lagrange identity. Let us then consider

$$(Av, w) = -\int_0^1 w\frac{d^2v}{dx^2}dx + \int_0^1 w\frac{dv}{dx}dx \qquad (1.1.21)$$

Let us integrate twice by parts the first expression in the right-hand part of (1.1.21) and once the second expression with respect to conditions of (1.1.3)

$$v(0) = v(1) = 0 \qquad (1.1.22)$$

Then let us accept that the analogous conditions

$$w(0) = w(1) = 0 \qquad (1.1.23)$$

are met for $w(x)$. Then the terms outside the integral will equal zero and we obtain (1.1.21) in the form

$$(Av, w) = -\int_0^1 v\frac{d^2w}{dx^2}dx - \int_0^1 v\frac{dw}{dx}dx \qquad (1.1.24)$$

Thus, as we derived (1.1.24) we have made a series of transformations which require not only quadratic summability of the function $w(x)$ at

$\Omega = (0, 1)$ but also that of this function derivatives dw/dx, d^2w/dx^2 as well, that is

$$\int_0^1 \left\{ \left(\frac{d^2w}{dx^2} \right)^2 + \left(\frac{dw}{dx} \right)^2 + w^2(x) \right\} dx < \infty$$

If we add boundary condition (1.1.23) on the function $w(x)$ to these properties, i. e.,

$$w(0) = w(1) = 0$$

we obtain the set $D(A^*)$. If we then introduce

$$A^* = -\frac{d^2}{dx^2} - \frac{d}{dx}$$

the final form of (1.1.24) will be

$$(Av, w) = (v, A^*w), \quad v \in D(A), \quad w \in D(A^*) \qquad (1.1.25)$$

Relationship (1.1.25) is the Lagrange identity. Therefore we have two operators: the main operator A and adjoint to it A^*.

We shall refer to the equation

$$Av = f$$

with the operator A and a certain right-hand part $f \in H$ as the *main equation*, and to the nonhomogeneous equation with the operator A^*

$$A^*w = p$$

as the *adjoint equation*. Here p is still an arbitrary function and its form and properties will be defined depending on the problem to be solved. The equations $Av = f$ and $A^*w = p$ assuming a differential form will be

$$-\frac{d^2v}{dx^2} + \frac{dv}{dx} = f(x), \quad x \in (0, 1), \quad v(0) = v(1) = 0$$

$$-\frac{d^2w}{dx^2} - \frac{dw}{dx} = p(x), \quad x \in (0, 1), \quad w(0) = w(1) = 0$$

and are referred to respectively as the main and adjoint problems.

Let us take up a numerical example: $f(x) = x(1-x)$, $p(x) = 1$. Then $Av = f$ and $A^*w = p$ are

$$-\frac{d^2v}{dx^2} + \frac{dv}{dx} = x(1-x), \quad x \in (0,1), \quad v(0) = v(1) = 0$$

and

$$-\frac{d^2w}{dx^2} - \frac{dw}{dx} = 1, \quad x \in (0,1), \quad w(0) = w(1) = 0$$

The finite difference method on the uniform grid $x_i = ih$, $i = \overline{0,N}$, $h = 1/N$ was applied in solution of these problems, and schemes of the second order of approximation in h were constructed. The three-diagonal systems of linear algebraic equations hereby obtained were solved by the factorization, or sweep, method (see [73]). Fig. 1 shows graphically the solutions of the main and adjoint problems.

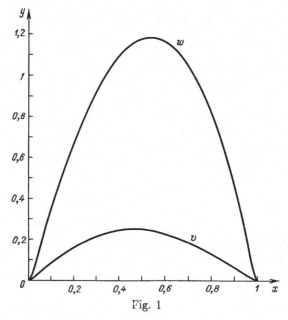

Fig. 1

A non-symmetric operator changes the statement of spectral problem. Indeed, we must consider two problems for real v and w, the main

$$Av = \lambda v \qquad\qquad (1.1.26)$$

and the adjoint

$$A^*w = \lambda w \tag{1.1.27}$$

In the differential representation, the main problem is

$$-\frac{d^2v}{dx^2} + \frac{dv}{dx} = \lambda v, \quad v(0) = v(1) = 0 \tag{1.1.28}$$

and the adjoint

$$-\frac{d^2w}{dx^2} - \frac{dw}{dx} = \lambda w, \quad w(0) = w(1) = 0 \tag{1.1.29}$$

It is easy to find all the solutions of (1.1.28) and (1.1.29):

$$v_k(x) = e^{x/2} \sin k\pi x, \quad k = 1, 2, \ldots \tag{1.1.30}$$

$$w_n(x) = e^{-x/2} \sin n\pi x, \quad n = 1, 2, \ldots \tag{1.1.31}$$

They meet eigenvalues $\lambda_k = 1/4 + \pi^2 k^2$. Obviously, $v_k(x)$ are not orthogonal to each other, as $w_n(x)$ are not either. The functions $v_k(x)$ and $w_n(x)$ are biorthogonal, though:

$$(v_k, w_n) = 0 \quad \text{for} \quad k \neq n \tag{1.1.32}$$

while this inner product differs from zero only if $k = n$. In our simple case, it is equal to

$$(v_n, w_n) = 1/2 \tag{1.1.33}$$

Thus we arrive at biorthogonality condition.

We will require the biorthogonality and completeness of the functions $\{v_k\}$ and $\{w_n\}$ as we expand the functions from H into Fourier series.

Let us discuss the method of determining the set $D(A^*)$ from a given main operator A and the set of functions $D(A)$ the operator A acts on. We will determine these properties of $D(A^*)$ every time bearing in mind the particular statement of a problem we will be researching in differential or integro-differential form.

Let us exemplify the method with the operator

$$A = -\frac{d^2}{dx^2} + \frac{d}{dx} \tag{1.1.34}$$

Let us now assume that A affects the functions $v \in D(A) \subset H$ which satisfy only conditions a) and b). As for the boundary conditions, let us select them as

$$\frac{dv}{dx} + \beta v = 0 \quad \text{at} \quad x = 0$$

$$v = 0 \quad \text{at} \quad x = 1 \tag{1.1.35}$$

where β is a given constant. Let us assume the functions $w \in D(A^*)$ (their properties are unknown yet). We will be determining them consequently, assuming that all the further transformations of the functions w are justified. To conclude, let us formulate a posteriori properties which substantiate the assumption and consider again (1.1.21):

$$(Av, w) = -\int_0^1 w \frac{d^2 v}{dx^2} dx + \int_0^1 w \frac{dv}{dx} dx \tag{1.1.36}$$

Integrating by parts (twice the first term and once the second), the expression in the right-hand part of (1.1.36) we obtain

$$(Av, w) = -\left. w(1)\frac{dv}{dx} \right|_{x=1} + \left. w(0)\frac{dv}{dx} \right|_{x=0} + \left. \frac{dw}{dx} \right|_{x=1} v(1) - \left. \frac{dw}{dx} \right|_{x=0} v(0)$$

$$+ w(1)v(1) - w(0)v(0) + (v, A^*w) \tag{1.1.37}$$

where

$$A^* = -\frac{d^2}{dx^2} - \frac{d}{dx}$$

To meet the Lagrange identity (1.1.24): $(Av, w) = (v, A^*w)$ it is necessary that the terms outside the integrals in (1.1.37) are equal to zero, i. e., the following relationship must be satisfied:

$$-\left. w(1)\frac{dv}{dx} \right|_{x=1} + \left. w(0)\frac{dv}{dx} \right|_{x=0} + \left. \frac{dw}{dx} \right|_{x=1} v(1) - \left. \frac{dw}{dx} \right|_{x=0} v(0)$$

$$+ w(1)v(1) - w(0)v(0) = 0 \tag{1.1.38}$$

Let us simplify (1.1.38) applying conditions (1.1.35) to the functions $D(A)$, i. e.,

$$\left. \frac{dv}{dx} \right|_{x=0} + \beta v(0) = 0, \quad v(1) = 0 \tag{1.1.39}$$

We now exclude $dv/dx(0)$ from (1.1.38), assuming $dv/dx(0) = -\beta v(0)$. Assuming $v(1) = 0$, we rewrite (1.1.38) as

$$- w(1)\frac{dv}{dx}\bigg|_{x=1} - \beta w(0)v(0) - \frac{dw}{dx}\bigg|_{x=0} v(0) - w(0)v(0) = 0 \quad (1.1.40)$$

Reducing the common terms at $dv/dx|_{x=1}$ and $v(0)$ we obtain

$$- w(1)\frac{dv}{dx}\bigg|_{x=1} - \left[(\beta+1)w(0) + \frac{dw}{dx}\bigg|_{x=0}\right]v(0) = 0 \quad (1.1.41)$$

Equality (1.1.41) must hold for arbitrary $dv/dx(1)$ and $v(0)$. Then we must put

$$w(1) = 0, \quad (\beta+1)w(0) + \frac{dw}{dx}\bigg|_{x=0} = 0$$

We now obtain boundary conditions for all functions $w \in D(A^*)$:

$$\frac{dw}{dx} + (\beta+1)w = 0 \quad \text{at} \quad x = 0$$
$$w = 0 \quad \text{at} \quad x = 1 \quad (1.1.42)$$

Now, let us sum it up and formulate all the requirements the functions $w \in D(A^*)$ must meet. As we obtain the Lagrange identity we carry out a string of transformations requiring that the function $w(x)$ and its derivatives dw/dx and d^2w/dx^2 are summable on $\Omega = (0,1)$. Besides, each function $w \in D(A^*)$ must satisfy boundary conditions (1.1.42). If so, the Lagrange identity holds and the operator A^* on $w \in D(A^*)$ is adjoint to A. Therefore nonhomogeneous equation

$$Av = f \quad (1.1.43)$$

with A from (1.1.34) represents a boundary value problem of the type

$$-\frac{d^2v}{dx^2} + \frac{dv}{dx} = f(x), \quad x \in (0,1)$$

$$\frac{dv}{dx} + \beta v = 0 \quad \text{at} \quad x = 0 \quad (1.1.44)$$

$$v = 0 \quad \text{at} \quad x = 1$$

and the adjoint equation

$$A^*w = p \quad (1.1.45)$$

with a right-hand part $p(x)$ respectively puts on the form

$$-\frac{d^2 w}{dx^2} - \frac{dw}{dx} = p(x), \quad x \in (0,1)$$

$$\frac{dw}{dx} + (\beta + 1)w = 0 \quad \text{at} \quad x = 0 \tag{1.1.46}$$

$$w = 0 \quad \text{at} \quad x = 1$$

We illustrate this with two numerical examples. Let us consider

$$\beta = -5, \quad f(x) = \begin{cases} 1, & x \in [0;\ 0.3], \\ 0, & x \notin [0;\ 0.3]; \end{cases} \quad p(x) = \begin{cases} 1, & x \in [0.8;\ 1], \\ 0, & x \notin [0.8;\ 1]. \end{cases}$$

Then (1.1.44) and (1.1.46) take the form

$$-\frac{d^2 v}{dx^2} + \frac{dv}{dx} = f(x), \quad x \in (0,1)$$

$$\frac{dv}{dx} - 5v = 0 \quad \text{at} \quad x = 0 \tag{1.1.47}$$

$$v = 0 \quad \text{at} \quad x = 1$$

$$-\frac{d^2 w}{dx^2} - \frac{dw}{dx} = p(x), \quad x \in (0,1)$$

$$\frac{dw}{dx} - 4w = 0 \quad \text{at} \quad x = 0 \tag{1.1.48}$$

$$w = 0 \quad \text{at} \quad x = 1$$

Simple finite-difference schemes of the second order of approximation were used to solve these problems on the uniform grid $x_i = ih$, where $i = 0, \ldots, N$, $h = 1/N$. The three-diagonal linear equations so obtained were solved by factorization [73]. Fig. 2 shows the diagrams of the solutions of the main and adjoint problems (1.1.47) and (1.1.48).

Let us now assume that

$$\beta = -1, \quad f(x) = \begin{cases} 1, & x \in [0.7;\ 1] \\ 0, & x \notin [0.7;\ 1] \end{cases} \quad p(x) = \begin{cases} 1, & x \in [0;\ 0.3] \\ 0, & x \notin [0;\ 0.3] \end{cases}$$

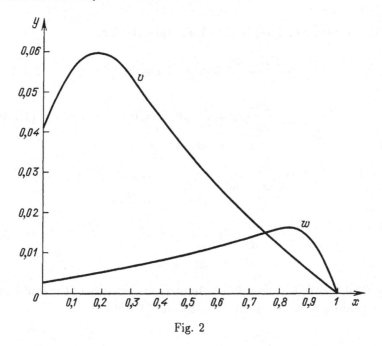

Fig. 2

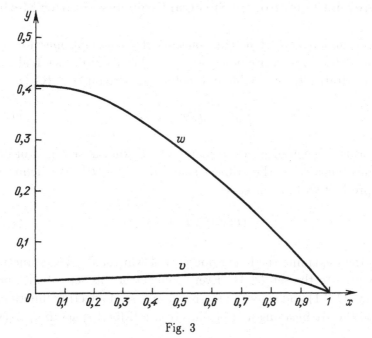

Fig. 3

Then problems (1.1.44) and (1.1.46) take the form

$$-\frac{d^2v}{dx^2} + \frac{dv}{dx} = f(x), \quad x \in (0,1)$$

$$\frac{dv}{dx} - v = 0 \quad \text{at} \quad x = 0 \tag{1.1.49}$$

$$v = 0 \quad \text{at} \quad x = 1$$

$$-\frac{d^2w}{dx^2} - \frac{dw}{dx} = p(x), \quad x \in (0,1)$$

$$\frac{dw}{dx} = 0 \quad \text{at} \quad x = 0 \tag{1.1.50}$$

$$w = 0 \quad \text{at} \quad x = 1$$

Fig. 3 compares the diagrams of numerical solutions of the main and adjoint problems, respectively (1.1.49) and (1.1.50).

1.2. Adjoint Operators in Spectral Problems. Fourier Method

Let us formulate several general concepts of the adjoint operators theory to use for applied problems' analysis. Let us introduce real functions quadratically summable in a definition domain $\Omega \subset \mathbf{R}^n$ $(n \geq 1)$ that is,

$$\int_\Omega u^2(x) \, dx \leq c < \infty \tag{1.2.1}$$

where x is a generalized coordinate. As is known, a range of such functions constitutes the Hilbert space $H = L_2(\Omega)$. We define the inner product within this space

$$(u, w) = \int_\Omega uw \, dx \tag{1.2.2}$$

Let us then examine the linear operator A which affects the functions v belonging to the set $D(A)$. Every element of this set satisfies both condition (1.2.1) and smoothness conditions, so A affecting functions v has also the Av meaning and belongs to the Hilbert space $H = L_2(\Omega)$.

Assume also that the functions satisfy certain boundary conditions and $D(A)$ is dense in H.

Introduce the functions w which belong to a new set, $D(A^*)$, with properties to be established later. Add to the main operator A the adjoint operator A^* satisfying the following Lagrange identity:

$$(Av, w) = (v, A^*w) \qquad (1.2.3)$$

where $v \in D(A)$, $w \in D(A^*)$. The left-hand part of (1.2.3) is usually algorithmized into the right-hand expression through integro-differential transformations of the Green formula type, integration by parts, etc. These transformations will result into extra terms outside (v, A^*w) which will depend on the functions u and v on the boundary of the domain Ω. The aggregate of the extras must be assumed as equal to zero. Using the boundary conditions for the function v we can obtain boundary conditions for the function w to make sure (1.2.3) is satisfied. Transformation of (Av, w) into (v, A^*w) will require application of certain smoothness conditions to the function v, then expression A^*w will have sense and will be a function of the space $H = L_2(\Omega)$.

Therefore, quadratic summability of the function w, together with corresponding smoothness requirements and boundary conditions, makes it possible to determine a posteriori the set $D(A^*)$. This is the general pattern of determination of the adjoint operator A^* and its definition domain $D(A^*)$.

If $A^* = A$ and $D(A^*) = D(A)$, A is a *self-adjoint operator*. The range of these operators is fairly broad and is by itself of much interest.

We have so far discussed properties of main and adjoint operators and those of function classes. Let us now examine a spectral problem. For non-self-adjoint operators, we must consider two homogeneous equations:

$$\begin{aligned} Av &= \lambda v, \quad v \in D(A) \\ A^*w &= \lambda w, \quad w \in D(A^*) \end{aligned} \qquad (1.2.4)$$

We assume that both spectral problems (1.2.4) admit the existence of two complete real systems of eigenfunctions in H: the main $\{v_k\}$ and

the adjoint $\{w_k\}$ corresponding to real eigenvalues $\{\lambda_k\}$ $(k = 1, 2, \ldots)$.[1]
We will need the completeness of eigenfunctions to expand the functions
into Fourier series later on. Meanwhile, let us see that the biorthogonality condition is satisfied. To prove it examine two eigenfunctions v_k
and w_k satisfying respectively

$$\begin{aligned} Av_k &= \lambda_k v_k \\ A^* w_n &= \lambda_n w_n \end{aligned} \qquad (1.2.5)$$

Subtract the inner product of the first of the equations in (1.2.5) and
w_n from the inner product of the second and v_k and obtain

$$(Av_k, w_n) - (v_k, A^* w_n) = (\lambda_k - \lambda_n)(v_k, w_n)$$
$$n = 1, 2, \ldots, \quad k = 1, 2, \ldots \qquad (1.2.6)$$

Since A and A^* are adjoint operators the left-hand side of (1.2.6) is
equal to zero due to the Lagrange identity, so we have

$$(\lambda_k - \lambda_n)(v_k, w_n) = 0 \qquad (1.2.7)$$

If $k \neq 0$ and $\lambda_k \neq \lambda_n$ then for this relation to be valid it is necessary
that

$$(v_k, w_n) = \begin{cases} 0, & k \neq n \\ (v_n, w_n), & k = n \end{cases} \qquad (1.2.8)$$

Conditions (1.2.8) are referred to as *biorthogonality condition*. It is
instrumental in expansion of the function into a Fourier series in eigenfunctions of the main or adjoint problems.

Assume in fact, that we need to represent the function f in H as
the Fourier series

$$f = \sum_{n=1}^{\infty} f_n v_n \qquad (1.2.9)$$

where f_n are Fourier coefficients to be found. To do so, multiply (1.2.9)
by the function w_n, term by term. Then

$$(f, w_k) = \sum_{n=1}^{\infty} f_n(v_n, w_k) \qquad (1.2.10)$$

---

[1]This condition is satisfied by the operator A which possesses a completely continuous inverse
operator with countable set of eigenvalues and, in particular, by all linear differential operators in
boundary value problems of the differential equations theory and in mathematical physics problems
with inverse operators which are Hilbert–Schmidt operators.

All the terms in the series except $n = k$ will be equal to zero with respect to (1.2.8) and the result is

$$(f, w_k) = f_k(v_k, w_k) \qquad (1.2.11)$$

Hence follows

$$f_k = \frac{(f, w_k)}{(v_k, w_k)} \qquad (1.2.12)$$

Expansion of the function f will naturally require a proof, that series (1.2.9) converges to this function.

Any function $g \in H$ is similarly expandable in powers of eigenfunctions w_n of an adjoint operator

$$g = \sum_{n=1}^{\infty} g_n w_n \qquad (1.2.13)$$

This line of reasoning leads us to an expression for the Fourier coefficient g_n:

$$g_n = \frac{(v_n, g)}{(v_n, w_n)} \qquad (1.2.14)$$

We come finally to an equation with the sources. Consider the problem

$$A\varphi = f \qquad (1.2.15)$$

where $\varphi \in D(A)$ and $f \in H$.

Let us solve (1.2.15) using a Fourier series, taking into consideration the functions $\{v_n\}$ and $\{w_n\}$ at our disposal as solutions of spectral problems (1.2.5). Represent the solution φ and the function f as a Fourier series:

$$\varphi = \sum_{n=1}^{\infty} \varphi_n v_n, \quad f = \sum_{n=1}^{\infty} f_n v_n \qquad (1.2.16)$$

The inner product of (1.2.15) and w_k is

$$(A\varphi, w_k) = (f, w_k) \qquad (1.2.17)$$

Substitute (1.2.16) into (1.2.17). Then

$$\sum_{n=1}^{\infty} \varphi_n(Av_n, w_k) = \sum_{n=1}^{\infty} f_n(v_n, w_k) \qquad (1.2.18)$$

According to (1.2.5) obtain

$$Av_n = \lambda_n v_n$$

and biorthogonality of the functions v_n and w_n produces

$$\lambda_k \varphi_k (v_k, w_k) = f_k (v_k, w_k) \qquad (1.2.19)$$

Hence follows

$$\varphi_k = f_k / \lambda_k$$

Therefore the solution of the problem has the form

$$\varphi = \sum_{n=1}^{\infty} \frac{f_n v_n}{\lambda_n} \qquad (1.2.20)$$

Next we need to prove the convergence of the series to the solution of (1.2.15), proceeding from properties of A. Every particular case of proving is specific and requires general or particular results of the functional analysis.

Let us illustrate this solving the problem

$$-\frac{d^2 \varphi}{dx^2} + \frac{d\varphi}{dx} = e^{x/2} \sin 2\pi x, \quad x \in (0,1) \qquad (1.2.21)$$
$$\varphi(0) = \varphi(1) = 0$$

Rewritten in the operator form ((1.2.15)), it has $f(x) = e^{x/2} \sin 2\pi x$ and $A = -d^2/dx^2 + d/dx$ is defined in § 1.1 (see (1.1.34)).

The Fourier method offers the solution of (1.2.21) by using (1.2.20):

$$\varphi = \sum_{n=1}^{\infty} \frac{f_n v_n}{\lambda_n} \qquad (1.2.22)$$

where $f_n = (f, w_n)/(v_n, w_n)$, while v_k and w_k are the eigenvalues of (1.2.5) which in this particular case have the form

$$-\frac{d^2 v_k}{dx^2} + \frac{dv_k}{dx} = \lambda_k v_k, \quad x \in (0,1) \qquad (1.2.23)$$
$$v_k(0) = v_k(1) = 0$$

$$-\frac{d^2 w_n}{dx^2} - \frac{dw_n}{dx} = \lambda_n w_n, \quad x \in (0,1) \qquad (1.2.24)$$
$$w_n(0) = w_n(1) = 0$$

Notice, as it follows from § 1.1 (see (1.1.30)–(1.1.33)) that

$$\lambda_k = 1/4 + \pi^2 k^2, \quad v_k(x) = e^{x/2} \sin \pi k x, \quad w_n(x) = e^{-x/2} \sin \pi n x,$$

$$(v_k, w_n) = \frac{1}{2}\delta_{kn}, \quad \delta_{kn} = \begin{cases} 1, & k = n \\ 0, & k \neq n \end{cases}$$

Use these equalities to compute f_n:

$$f_n = \frac{(f, w_n)}{(v_n, w_n)} = 2\int_0^1 f(x)w_n(x)dx = 2\int_0^1 \sin 2\pi x \sin \pi n x \, dx = \delta_{2n}$$

Then (1.2.22) gives the solution φ of (1.2.21):

$$\varphi = \sum_{n=1}^{\infty} \frac{f_n v_n}{\lambda_n} = \sum_{n=1}^{\infty} \frac{\delta_{2n} v_n}{\lambda_n} = \frac{v_2(x)}{\lambda_2} = \frac{e^{x/2}\sin 2\pi x}{1/4 + 4\pi^2}$$

$$= \frac{4}{1 + 16\pi^2}e^{x/2}\sin 2\pi x \quad (1.2.25)$$

which can be checked immediately substituting φ into (1.2.21).

Let us now consider an adjoint problem with operator A^*:

$$A^*\varphi^* = p \quad (1.2.26)$$

where $\varphi^* \in D(A^*)$, $p \in H$. The Fourier method offers a solution to this one, too. Let us represent the solution φ^* and the function p as a Fourier series in eigenfunctions $\{w_n\}$ of the adjoint operator A^*:

$$\varphi^* = \sum_{n=1}^{\infty} \varphi_n w_n, \quad p = \sum_{n=1}^{\infty} p_n w_n \quad (1.2.27)$$

where φ_n are unknown coefficients and $p_n = \dfrac{(v_n, p)}{(v_n, w_n)}$ (see (1.2.14)).

The inner product of (1.2.26) and v_k is

$$(v_k, A^*\varphi^*) = (v_k, p) \quad (1.2.28)$$

After substitution of series (1.2.27) into (1.2.28) we obtain

$$\sum_{n=1}^{\infty} \varphi_n(v_k, A^* w_n) = \sum_{n=1}^{\infty} p_n(v_k, w_n) \quad (1.2.29)$$

Since (see (1.2.5))
$$A^*w_n = \lambda_n w_n$$
and biorthogonality condition (1.2.8) is satisfied, it follows from (1.2.29) that
$$\lambda_k \varphi_k(v_k, w_k) = p_k(v_k, w_k)$$

Hence
$$\varphi_k = p_k / \lambda_k \tag{1.2.30}$$

and according to (1.2.27) the solution of our adjoint problem is
$$\varphi^* = \sum_{n=1}^{\infty} \frac{p_n w_n}{\lambda_n} \tag{1.2.31}$$

Just like in the case of the main problem, we must then prove the convergence of series (1.2.31) to the solution of problem (1.2.26), which is done in every particular case on the basis of the functional analysis' theorems, proceeding from properties of the operator A^*. As is known, if λ_n happens to be a complex figure, a complex-adjoint figure $\bar{\lambda}_n$ will replace λ_n in an adjoint spectral problem for w_n.

1.3. Adjoint Equations and Functionals. Elements of Theory

Now we come to main and adjoint equations with sources. Let us start with simple examples and consider the problem
$$-\frac{d^2\varphi}{dx^2} = f(x) \quad x \in (0,1),$$
$$\varphi(0) = \varphi(1) = 0 \tag{1.3.1}$$

Let us assume that the solution φ of problem (1.3.1) belongs to the set $D(A)$ as discussed in § 1.1 and $f(x)$ belongs to $H = L_2(0,1)$ i.e., to the space of real square-summable functions.[2] The operator $A = -d^2/dx^2$ on the functions within $D(A)$ is symmetric. It is easy to obtain the solution of the problem in explicit form:
$$\varphi(x) = x \int_0^1 dx' \int_0^{x'} f(x'')dx'' - \int_0^x dx' \int_0^{x'} f(x'')dx' \tag{1.3.2}$$

---

[2]See the Remark at page 12.

Let us suppose that we need to actually know just a particular functional of the solution $u(x)$ in $\Omega = (0,1)$ rather than the solution itself. Let it be

$$J = \int\limits_0^1 p(x)\varphi(x)dx \qquad (1.3.3)$$

where $p(x)$ is an instrumental or measurement characteristic. For example, if

$$p(x) = \begin{cases} 1, & \text{if } 0 \le \alpha \le x \le \beta \le 1 \\ 0, & \text{otherwise} \end{cases} \qquad (1.3.4)$$

then it means, that the instrument registers solution values within $\alpha \le x \le \beta$ and does not respond to the solution in the other part of the interval.

However, we can obtain the same functional J solving a specifically formulated adjoint problem:

$$-\frac{d^2\varphi^*}{dx^2} = p(x)$$
$$\varphi^*(0) = \varphi^*(1) = 0 \qquad (1.3.5)$$

where $p(x)$ is defined in functional (1.3.3.) Let us assume, that $\varphi^*(x)$ belongs to the set $D(A)$ which the solution of the main problem belongs to. Then multiply equation from (1.3.1) by φ^*, the equation from (1.3.5) by φ, subtract the results, and integrate the difference over all the definition domain of the solution:

$$-\int\limits_0^1 \varphi^* \frac{d^2\varphi}{dx^2} dx + \int\limits_0^1 \varphi \frac{d^2\varphi^*}{dx^2} dx = \int\limits_0^1 f\varphi^* dx - \int\limits_0^1 p\varphi dx \qquad (1.3.6)$$

Double integration by parts of one of the terms in the left-hand part of (1.3.6) with regard to boundary condition demonstrably shows that it is equal to zero. We obtain the same directly if we use the Lagrange identity. Then (1.3.6) changes into

$$\int\limits_0^1 f\varphi^* dx - \int\limits_0^1 p\varphi dx = 0 \qquad (1.3.7)$$

But the second term in (1.3.7) is the sought-for functional. Then we have two equivalent formulas to define it:

$$J = \int_0^1 p\varphi dx, \quad J = \int_0^1 f\varphi^* dx \qquad (1.3.8)$$

We thus obtain a dual formula for definition of one functional. The second formula in (1.3.8) happens to be more useful in many complex problems, especially if the source function $f(x)$ changes in the variations of the problem while the operator and boundary conditions do not.

As a numerical example, let us have for the main problem

$$-\frac{d^2\varphi}{dx^2} = f(x), \quad f(x) = \begin{cases} 1, & x \in [0; \ 0.4], \\ 0, & x \notin [0; \ 0.4] \end{cases}$$
$$\varphi(0) = \varphi(1) = 0;$$

where we need to compute functional (1.3.3) where

$$p(x) = \begin{cases} 1, & x \in [0.7; \ 1] \\ 0, & x \notin [0.7; \ 1] \end{cases}$$

We can compute this functional using the formula

$$J = \int_0^1 f(x)\varphi^*(x)dx = \int_0^{0.4} \varphi^*(x)dx$$

where $\varphi^*(x)$ is the solution of the adjoint problem

$$-\frac{d^2\varphi^*}{dx^2} = p(x), \quad x \in (0,1)$$

$$\varphi^*(0) = \varphi^*(1) = 0$$

The main and the adjoint problems received numerical solutions by the finite element method on a uniform grid $x_i = ih$ where $i = 0, \ldots, N$, $h = 1/N$. Fig. 4 shows the diagrams of the solutions of these problems. The Simpson formula produced computed values of the functionals

$$J^{(1)} = \int_{0.7}^1 \varphi(x)dx, \quad J^{(2)} = \int_0^{0.4} \varphi^*(x)dx$$

and it was established, that

$$J^{(1)} = 0.0035997229, \quad J^{(2)} = 0.0035997221$$

The values of the functionals $J^{(1)}$ and $J^{(2)}$ differ in this case only in their tenth digits after the point (as $h = 10^{-2}$).

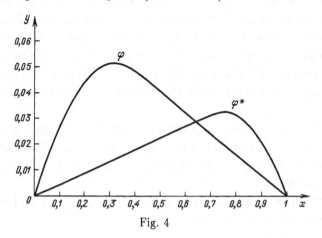

Fig. 4

If the physical measurement changes, so does the functional. Therefore, a new problem (1.3.5) with different instrumental characteristic $p(x)$ wants solution.

Let us now discuss in this simple example, the problem of sensitivity of the functional J to the source $f(x)$. Let us assume that instead of problem (1.3.1), referred to as unperturbed, we have to deal with a new perturbed problem:

$$-\frac{d^2\varphi'}{dx^2} = f'(x)$$
$$\varphi'(0) = \varphi'(1) = 0 \tag{1.3.9}$$

We assume that $f'(x) = f(x) + \delta f$. Since the problem (1.3.9) is linear, its solution has the form

$$\varphi'(x) = \varphi(x) + \delta\varphi(x)$$

where the prime denotes the perturbed function. Then for functional variations in the formula

$$J' = J + \delta J$$

we have two equivalent relationships:

$$\delta J = \int_0^1 p(x)\delta\varphi(x)dx, \quad \delta J = \int_0^1 \delta f(x)\varphi^*(x)dx \qquad (1.3.10)$$

Notice the second formula in (1.3.10) which links a variation of the source in the main problem with a variation of the functional under study. Besides, there is no need to solve perturbed problem (1.3.9) to find $\delta\varphi$. Here $\varphi^*(x)$ is the weight function responsible for the sensitivity of the functional. Therefore the adjoint function φ^* is occasionally referred to as a *function of information importance* or simply *importance*.

What counts is, the initial problems with even self-adjoint operators with the functionals introduced here, generate simply adjoint problems, which will be referred to, for convenience, as adjoint with regard to selected functionals.

As a numerical example, let us assume that $f(x) = x$ and $p(x) \equiv 1$. Then main problem (1.3.1) has the form

$$-\frac{d^2\varphi}{dx^2} = x, \quad x \in (0,1)$$
$$\varphi(0) = \varphi(1) = 0$$

and we have to compute functional (1.3.3):

$$J = \int_0^1 \varphi(x)dx$$

Assume then, that $f'(x) = f(x) + \delta f$ where $\delta f = \varepsilon \sin \pi x$. Perturbed problem (1.3.9) has then the form

$$-\frac{d^2\varphi'}{dx^2} = x + \sin \pi x, \quad x \in (0,1)$$
$$\varphi'(0) = \varphi'(1) = 0$$

To compute the corrections $\delta J = J' - J$ use the formulas

$$\delta J^{(1)} = \int_0^1 \delta\varphi(x)dx, \quad \delta J^{(2)} = \varepsilon \int_0^1 \sin \pi x \varphi^*(x)dx$$

where $\varphi^*(x)$ is the solution of the adjoint problem (1.3.5) which, for this particular case, has the form

$$-\frac{d^2\varphi^*}{dx^2} = 1, \quad x \in (0,1)$$

$$\varphi^*(0) = \varphi^*(1) = 0$$

All the three problems (main, perturbed and adjoint) were solved by the finite-difference method and the values of the solutions φ, φ', φ^* were found at the nodes of a grid $x_i = ih$ where $i = 0, \ldots, N$, $h = 1/N$ with the step $h = 0.01$. Assumed were the values $\varepsilon = 1$, $\varepsilon = 0.1$ and $\varepsilon = 0.01$. Fig. 5 presents the diagrams of the functions φ, φ', and φ^* which show that $\varphi' \to \varphi$ for $\varepsilon \to 0$.

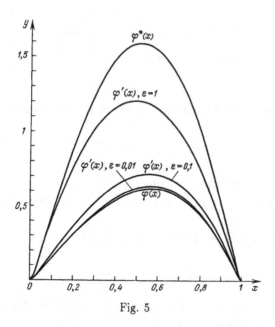

Fig. 5

Table 1 compares values of the functionals $\delta J^{(1)}$ and $\delta J^{(2)}$ for different ε. Notice, that the differences in the values $\delta J^{(1)}$ and $\delta J^{(2)}$ are due to approximation errors of the computation algorithm.[3]

---

[3]In the numerical examples we assume hereinafter φ, φ', φ^* as approximate solutions to respective problems.

Let us now examine, as an example, a non-self-adjoint operator. Assume

$$A = -\frac{d^2}{dx^2} + \frac{d}{dx} \qquad (1.3.11)$$

as one discussed in §1.1 and affecting, within H, functions which belong to $D(A)$. The main problem will have the form

$$-\frac{d^2\varphi}{dx^2} + \frac{d\varphi}{dx} = f(x), \quad x \in (0,1)$$
$$\varphi(0) = \varphi(1) = 0 \qquad (1.3.12)$$

Table 1

	$\varepsilon = 1$	$\varepsilon = 0.1$	$\varepsilon = 0.01$	$\varepsilon = 0$
$\delta J^{(1)}$	$5.59095 \cdot 10^{-2}$	$5.59096 \cdot 10^{-3}$	$5.59094 \cdot 10^{-4}$	0
$\delta J^{(2)}$	$5.59417 \cdot 10^{-2}$	$5.59417 \cdot 10^{-3}$	$5.59418 \cdot 10^{-4}$	0

Let us assume that we need the value of the functional

$$J = \int_0^1 p(x)\varphi(x)dx \qquad (1.3.13)$$

where $p(x)$ is a given function from H.

Let us see how to obtain J with another formula using a solution of related adjoint problem.

In § 1.1 the adjoint to A operator A^* had the form

$$A^* = -\frac{d^2}{dx^2} - \frac{d}{dx} \qquad (1.3.14)$$

It is defined on functions φ^* within the set $D(A^*)$ satisfying the boundary conditions $\varphi^*(0) = \varphi^*(1) = 0$.

Let us now consider a problem adjoint to (1.3.12):

$$-\frac{d^2\varphi^*}{dx^2} - \frac{d\varphi^*}{dx} = p(x)$$
$$\varphi^*(0) = \varphi^*(1) = 0 \qquad (1.3.15)$$

where the right-hand part $p(x)$ determines functional (1.3.13).

Multiply equation from (1.3.12) by φ^* and equation from (1.3.15) by φ, subtract from each other and integrate in x on $[0,1]$. Then

$$\int\limits_0^1 \left(-\varphi^*\frac{d^2\varphi}{dx^2} + \varphi^*\frac{d\varphi}{dx}\right)dx + \int\limits_0^1 \left(\varphi\frac{d^2\varphi^*}{dx^2} + \varphi\frac{d\varphi^*}{dx}\right)dx$$

$$= \int\limits_0^1 f\varphi^*dx - \int\limits_0^1 p\varphi dx \quad (1.3.16)$$

Double integration of one of the terms in the left-hand part of (1.3.16) by parts with regard to boundary conditions shows that the left-hand part of (1.3.16) is equal to zero, as follows from the Lagrange identity which we used in §1.1 to find the adjoint operator A^*. Hence, from (1.3.16)

$$\int\limits_0^1 f\varphi^*dx = \int\limits_0^1 p\varphi dx$$

i. e., together with (1.3.13) the formula to determine the functional J is

$$J = \int\limits_0^1 f\varphi^*dx \quad (1.3.17)$$

As a numerical example, let us examine a main problem in the form

$$-2\frac{d^2\varphi}{dx^2} + \frac{d\varphi}{dx} + 3\varphi = f(x) \quad x \in (0,1)$$

$$\varphi(0) = \varphi(1) = 0$$

where

$$f(x) = \begin{cases} 1, & x \in [1/2,\ 2/3] \\ 0, & x \notin [1/2,\ 2/3] \end{cases}$$

Let us suppose we are seeking the value of the functional

$$J = \int\limits_{1/4}^{1/3} \varphi(x)dx.$$

The respective adjoint problem is

$$-2\frac{d^2\varphi^*}{dx^2} - \frac{d\varphi^*}{dx} + 3\varphi^* = p(x), \quad x \in (0,1)$$

$$\varphi^*(0) = \varphi^*(1) = 0$$

where

$$p(x) = \begin{cases} 1, & x \in [1/4,\ 1/3] \\ 0, & x \notin [1/4,\ 1/3] \end{cases}$$

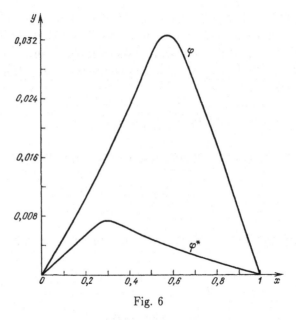

Fig. 6

The main and the adjoint problem were solved by the finite-difference method on a grid with the step $h = 0.01$. Fig. 6 shows the diagrams of the functions φ and φ^*. The value of functional J was computed twice, by formulas (1.3.13) and (1.3.17):

$$J \approx 0.15047 \cdot 10^{-2}, \quad J \approx 0.14999 \cdot 10^{-2}$$

We see, that the difference in the values is of the order of $O(h^2)$. The Simpson quadrature rule was used to compute the integrals with the step $h = 0.01$.

Let us tackle now the problem of sensitivity of the functional J to the source $f(x)$. Assume instead of problem (1.3.12) referred to hereinafter as unperturbed, another, the perturbed:

$$-\frac{d^2\varphi'}{dx^2} + \frac{d\varphi'}{dx} = f'(x) \qquad (1.3.18)$$
$$\varphi'(0) = \varphi'(1) = 0$$

where $f'(x) = f(x) + \delta f(x)$.

Represent φ' in the form

$$\varphi'(x) = \varphi(x) + \delta\varphi(x)$$

where $\varphi(x)$ is the solution of unperturbed problem (1.3.12) and $\delta\varphi = \varphi' - \varphi$. Then, like in (1.3.10), obtain for variations of the functional $\delta J = J' - J$ two equivalent formulas:

$$\delta J = \int_0^1 p(x)\delta\varphi(x)dx \qquad (1.3.19)$$

$$\delta J = \int_0^1 \delta f(x)\varphi^*(x)dx \qquad (1.3.20)$$

Let us take up a numerical example:

$$f(x) = \sin \pi x, \quad \delta f = \varepsilon \sin 2\pi x, \quad p(x) = \begin{cases} 1, & 0 \le x \le 1/2 \\ 2(1-x), & 1/2 < x \le 1 \end{cases}$$

Problems (1.3.12), (1.3.15), (1.3.18) find their solutions by the finite element method on a uniform grid with the step $h = 01$. Values of the functionals $\delta J^{(1)} = \int_0^1 p(x)\delta\varphi(x)dx$, $\delta J^{(2)} = \int_0^1 \delta f(x)\varphi^*(x)dx$ were computed for $\varepsilon = 1$, $\varepsilon = 0.1$, and $\varepsilon = 0.01$. Table 2 compares these values.

Table 2

	$\varepsilon = 1$	$\varepsilon = 0.1$	$\varepsilon = 0.01$	$\varepsilon = 0$
$\delta J^{(1)}$	$0.7227 \cdot 10^{-2}$	$0.7229 \cdot 10^{-3}$	$0.7231 \cdot 10^{-4}$	0
$\delta J^{(2)}$	$0.7241 \cdot 10^{-2}$	$0.7240 \cdot 10^{-3}$	$0.7239 \cdot 10^{-4}$	0

To conclude this Section, notice that the formulas of the type as in (1.3.19), (1.3.20) can be derived also in the general case with an arbitrary non-self-adjoint operator A. We shall discuss this in more detail in § 1.5.

1.4. Adjoint Equations and Importance of Information

Let us now discuss adjoint equations from a general point of view (see [110]). Formulate

$$A\varphi(x) = f(x) \qquad (1.4.1)$$

where A is a linear operator, $\varphi(x)$ belongs to a set of functions $D(A)$ with the definition domain $\Omega \subset \mathbf{R}^n$ ($n \geq 1$). Assume now functions $A\varphi$ and $f(x)$ as square-summable and belonging to the real Hilbert space $H = L_2(\Omega)$. Assume x as a set of all variables of the problem under consideration.

Consider now the inner product of the functions $f(x)$ and $h(x)$ in H

$$(g, h) = \int_\Omega g(x)h(x)dx \qquad (1.4.2)$$

As was said above, solution of many applied problems usually implies obtaining values of a variable which is a functional of $\varphi(x)$. Any variable which is a linear continuous functional of $\varphi(x)$ can be expressed in the form of the inner product according to Riesz theorem [151]

$$J_p[\varphi] = \int_\Omega p(x)\varphi(x)dx = (p, \varphi) \qquad (1.4.3)$$

where $J_p[\varphi]$ is the functional under consideration, $p(x)$ is a characteristic of an instrument employed to measure the physical process.

Therefore, we have to deal with physical values which can be expressed as a linear continuous functional of $\varphi(x)$:

$$J_p[\varphi] = (p, \varphi) \qquad (1.4.4)$$

Now introduce, together with the operator A, operator A^*, adjoint to A and defined by the Lagrange identity

$$(Ah, g) = (h, A^*g) \qquad (1.4.5)$$

for any functions g and h which belong to the sets $D(A)$ and $D(A^*)$, respectively. As was said in § 1.2, properties of functions from the set $D(A)$ are assigned extra following the analysis of initial problem (1.4.1), by smoothness and boundary conditions, while the properties of the set $D(A^*)$ are defined in transformation of the left-hand part of (1.4.5) into the right-hand part so that the Lagrange identity is satisfied. Hence the requirements are obtained for both smoothness and boundary conditions for functions in $D(A^*)$.

Referring to equation (1.4.1) as the main, introduce an adjoint equation

$$A^*\varphi^*(x) = p(x) \tag{1.4.6}$$

where $p(x)$ is the function of measuring instrument as in functional (1.4.4).

Subtract the inner product of equation (1.4.6) and φ from the inner product of equation (1.4.1) and φ^*:

$$(A\varphi, \varphi^*) - (\varphi, A^*\varphi^*) = (f, \varphi^*) - (p, \varphi) \tag{1.4.7}$$

The left-hand part of (1.4.7) is equal to zero, since the Lagrange identity is valid for φ and φ^* as well as for g and h which belong to $D(A^*)$. Then

$$(f, \varphi^*) - (p, \varphi) = 0 \tag{1.4.8}$$

Or, with regard to (1.4.4), (1.4.8) produces two equivalent formulas:

$$J_p[\varphi] = (p, \varphi), \quad J_f[\varphi^*] = (f, \varphi^*) \tag{1.4.9}$$

while

$$J_p[\varphi] = J_f[\varphi^*] \tag{1.4.10}$$

If we seek the value of the functional $J_p[\varphi]$ we can thus obtain it either solving equation (1.4.1) to define the value with the formula

$$J_p[\varphi] = (p, \varphi)$$

or solving equation (1.4.6) to find the same value with the formula

$$J_p[\varphi] = J_f[\varphi^*] = (f, \varphi^*)$$

Since $J_p = J_f$ we shall denote these functionals as J_p.

Therefore, a function $\varphi^*(x)$ can be related to every linear functional $J_p[\varphi] = (p, \varphi)$. This function will satisfy equation (1.4.6), while $p(x)$ should be just used as a free term in this equation since it characterizes a physical process or measurement we are interested in.

Let us assume a "unit-power source" located in a medium at a point $x = x_0$ i. e.,

$$f(x) = \delta(x - x_0) \qquad (1.4.11)$$

Since[4]

$$(\varphi(x), \delta(x - x_0)) = \varphi(x_0)$$

then

$$J_p[\varphi] = J_{f=\delta(x-x_0)}[\varphi^*] = \varphi^*(x_0) \qquad (1.4.12)$$

It follows that the adjoint function φ^* describes the dependence of the functional $J_p[\varphi] = (p, \varphi)$ on the unit power source location point.

Let us imagine a physical system where a variable $J_p[\varphi]$ is measured which is a linear functional of the solution linked, for instance, to the density of particles $\varphi(x)$ in the phase space. If a certain amount of the particles is let into a location of the system or, on the contrary, is removed thence, the value measured of the variable $J_p[\varphi]$ will respectively increase or decrease depending on the location where the particle number is measured. As is seen from our line of reasoning above, this dependence is described by the adjoint function $\varphi^*(x)$ satisfying equation (1.4.6). The adjoint function $\varphi^*(x)$ thus contributes particles located in one or another location within the system, to the functional $J_p[\varphi]$ we are interested in. Therefore the function φ^* is sometimes referred to as the value of the substance at the point x_0 with regard to the functional of the problem. Functionals may naturally differ. Therefore, particular problem (1.4.6) with a given function $p(x)$ should be formulated for every functional.

---

[4]Strictly speaking, the relationship $(\varphi(x), \delta(x - x_0)) = \varphi(x_0)$ should be understood in the sense of the theory of distributions [30]. Omitting details, we shall assume, that the functions $\varphi \in D(A)$ are continuous in x, then $J_p[\varphi] = (\varphi, \delta(x - x_0)) = \varphi(x_0)$ is a linear continuous functional on $D(A)$.

Let us exemplify it assuming that main problem (1.4.1) describes the simplest process of diffusion and has the form

$$-\frac{d^2\varphi}{dx^2} = f(x), \quad x \in (0,1)$$
$$\varphi(0) = \varphi(1) = 0 \qquad (1.4.13)$$

where $f(x)$ is a given function.

Let us now assume an instrument measuring a variable

$$J_p[\varphi] = (p, \varphi) \equiv \int_0^1 p(x)\varphi(x)dx \qquad (1.4.14)$$

which is a linear functional of the solution φ to problem (1.4.13). The function $p(x)$ is linked to a characteristic of the instrument; assume that it is assigned by the formula

$$p(x) = p_0 \sin \pi x, \quad p_0 = \text{const} > 0 \qquad (1.4.15)$$

Then

$$J_p[\varphi] = p_0 \int_0^1 \sin \pi x \varphi(x)dx \qquad (1.4.16)$$

We know that the value of functional (1.4.16) can be computed in another way. To do so, tackle adjoint problem (1.4.6) with the right-hand part $p(x)$ as in (1.4.15). In our example, it takes the form

$$-\frac{d^2\varphi^*}{dx^2} = p_0 \sin \pi x, \quad x \in (0,1)$$
$$\varphi^*(0) = \varphi^*(1) = 0 \qquad (1.4.17)$$

The solution of adjoint problem (1.4.17) in this case is explicit:

$$\varphi^*(x) = \frac{p_0}{\pi^2} \sin \pi x \qquad (1.4.18)$$

Then the value of the functional $J_p[\varphi]$ is offered by the second formula of (1.4.9):

$$J_p[\varphi] = J_f[\varphi^*] = (f, \varphi^*)$$

or, if we use (1.4.18), we obtain

$$J_p[\varphi] = \frac{p_0}{\pi^2} \int_0^1 f(x) \sin \pi x dx \qquad (1.4.19)$$

There is no need to solve (1.4.13) and find φ while computing $J_p[\varphi]$ with formula (1.4.19): substitution of the given function $f(x)$ into (1.4.19) will suffice.

Let us assume a "unit-power source" located in a medium at a point $x = x_0$, that is

$$f(x) = \delta(x - x_0)$$

Then from (1.4.19) we obtain

$$J_p[\varphi] = \frac{p_0}{\pi^2} \sin \pi x_0 \qquad (1.4.20)$$

It follows from (1.4.12) that the right-hand part of (1.4.20) is nothing but the information value of the solution at the point x_0 with regard to the functional $J_p[\varphi] = (p_0 \sin \pi x, \varphi)$. It is easy to deduce from (1.4.20) that, if we want to place a source of particles at a point where their contribution to the functional $J_p[\varphi]$ (or the number of particles which are registered by an instrument with a given characteristic $p = p_0 \sin \pi x$) would be maximum, we need to locate the source at the point $x_0 = 1/2$, since

$$\max_{x_0} J_p[\varphi] = \max_{x_0} \frac{p_0}{\pi^2} \sin \pi x_0 = \frac{p_0}{\pi^2}$$

Notice that though this example is trivial, the principle of calculation of efficient location of sources (or sinks, in control systems, etc.) at specific points (or in specific areas, etc.) on the basis of study of the behavior of solutions to respective adjoint equations, can underly algorithms of solutions to many important applied problems in optimal control, environment protection, etc. It will be discussed in more detail in the following chapters of this book.

Understanding the adjoint function φ^* as the substance value leads to a clear concept of the perturbation theory for any functional $J_p[\varphi]$. If we change in effect the number of particles in a volume unit Δx near a point x by a value δN, a corresponding change in the value J_p will be expressed in an equation as

$$\delta J_p = \delta N \varphi^*(x) \qquad (1.4.21)$$

If small alterations occur in parameters of a system we examine, so that the operator A from (1.4.1) becomes the operator $A + \delta A$, they correspond to changes in the number of particles in every volume unit Δx by a value $\delta N = -\Delta x \delta A \varphi$. Rewrite the total change of the functional J_p due to such alterations as

$$\delta J_p = - \int_{\Omega} \varphi^*(x) \delta A \varphi(x) dx \qquad (1.4.22)$$

The derivation of (1.4.22) will follow in practical applications.

Relationship (1.4.21) makes possible measuring distribution of the value function in a system by altering, by some known method, the number of particles at various points x in a system and measuring corresponding changes in the variable J_p.

The notion of value discussed here can be useful in the theory of measuring instruments. An instrument is in effect usually designed to measure one certain variable J_p. Therefore, a quite definite value function $\varphi^*(x)$ can be applied to each instrument, to be once measured or calculated. If the distribution of a substance and its value are known, (1.4.22) can be used for the measurements in one of two ways. First, measuring the values δJ_p for various medium parameter changes δA, we can use (1.4.22) to determine values δA, i. e. various characteristics of interaction of particles with the matter. It is possible to measure this way, for example, the cross-sections of neutrons' interaction with the matter of different samples in the radiation transport theory. Samples could be put into the instrument and $\delta \Sigma = \delta A$ determined by the change in the variable J_p. Second, (1.4.22) makes it possible to correct the measurements of J_p to allow for various perturbants within the instrument. Finally, the definition of the notion of value provides equations for the function φ^* proceeding immediately from the physical meaning of this variable just like the neutron number conservation law provides the neutron flux equation.

1.5. Adjoint Equations and Perturbation Theory for Linear Functionals

We examined in § 1.3 the problem of sensitivity of the functional $J_p = (p, \varphi)$ to the source $f(x)$ and provided simple examples thereof.

Now let us examine a more common case and formulate basic formulas of the perturbation theory for linear functionals, starting with the following example.

Examine the problem

$$-\frac{d^2\varphi}{dx^2} + \frac{d\varphi}{dx} = f(x), \quad x \in (0,1)$$
$$\varphi(0) = \varphi(1) = 0$$

(1.5.1)

where the solution φ belongs to the set $D(A)$ as introduced in § 1.2, $f(x)$ belongs to the space $H = L_2(0,1)$. The operator $A = -d^2/dx^2 + d/dx$ operates in H and is defined on functions φ from $D(A)$. Problem (1.5.1) will be referred to as unperturbed.

Suppose, our interest is not actually in the value of solution $\varphi(x)$ of problem (1.5.1), but rather in a particular functional from this solution

$$J_p[\varphi] = \int_0^1 p(x)\varphi(x)dx$$

(1.5.2)

where $p(x)$ is a given function from H.

As was shown in § 1.3, the value of the functional (1.5.2) can be found with another formula:

$$J_p[\varphi] = \int_0^1 f(x)\varphi^*(x)dx$$

(1.5.3)

where $\varphi^*(x)$ is the solution of the following adjoint problem:

$$-\frac{d^2\varphi^*}{dx^2} - \frac{d\varphi^*}{dx} = p(x) \quad x \in (0,1)$$
$$\varphi^*(0) = \varphi^*(1) = 0$$

(1.5.4)

Let us now have instead of (1.5.1) a perturbed problem, such as

$$-\frac{d^2\varphi'}{dx^2} + \frac{d\varphi'}{dx} + \delta g(x)\varphi' = f'(x), \quad x \in (0,1)$$
$$\varphi'(0) = \varphi'(1) = 0$$

(1.5.5)

where $f'(x) = f(x) + \delta f$; $\delta g(x)$, $\delta f(x)$ are given perturbant functions.

Represent the solution φ' of perturbed problem (1.5.5) as

$$\varphi'(x) = \varphi(x) + \delta\varphi$$

(1.5.6)

where $\varphi(x)$ is the solution of unperturbed problem (1.5.1), $\delta\varphi = \varphi' - \varphi$. The value of the functional changes, and we obtain

$$J_p' = J_p + \delta J_p \qquad (1.5.7)$$

where

$$\delta J_p = \int_0^1 p(x)\delta\varphi(x)dx$$

Let us establish a link between the changes of the functions $\delta g(x)$, $\delta f(x)$ and the change of the functional δJ_p. To do so, we use the solution of adjoint problem (1.5.4). Multiply the equation of (1.5.5) by $\varphi^*(x)$ and that of (1.5.4) by φ', integrate the results in x on $(0,1)$ and subtract from each other. Then

$$\int_0^1 \left(-\frac{d^2\varphi'}{dx^2}\varphi^* + \frac{d\varphi'}{dx}\varphi^* + \delta g(x)\varphi'\varphi^* \right) dx + \int_0^1 \left(\frac{d^2\varphi^*}{dx^2}\varphi' + \frac{d\varphi^*}{dx}\varphi' \right) dx$$

$$= \int_0^1 f'(x)\varphi^*(x)dx - \int_0^1 p(x)\varphi'(x)dx \quad (1.5.8)$$

Assume now that φ, $\varphi',\varphi^* \in D(A)$. Integrating by parts one of expressions in the left-hand part of (1.5.8), transform it into

$$\int_0^1 \left(-\frac{d^2\varphi'}{dx^2}\varphi^* + \frac{d\varphi'}{dx}\varphi^* + \delta g(x)\varphi'\varphi^* \right) dx + \int_0^1 \left(\frac{d^2\varphi^*}{dx^2}\varphi' + \frac{d\varphi^*}{dx}\varphi' \right) dx$$

$$= \int_0^1 \delta g\varphi^*\varphi'dx \quad (1.5.9)$$

and, doing the same with the right-hand part of (1.5.8) with regard to (1.5.7), obtain

$$\int_0^1 f'(x)\varphi^*(x)dx - \int_0^1 p(x)\varphi'(x)dx = \int_0^1 f(x)\varphi^*(x)dx - \int_0^1 p(x)\varphi(x)dx$$

$$+ \int_0^1 \delta f(x)\varphi^*(x)dx - \int_0^1 p(x)\delta\varphi(x)dx = \int_0^1 \delta f(x)\varphi^*(x)dx - \delta J_p \quad (1.5.10)$$

We have thus used the adjointment relationship obtained in § 1.3

$$\int_0^1 f(x)\varphi^*(x)dx = \int_0^1 p(x)\varphi(x)dx \qquad (1.5.11)$$

From (1.5.8)–(1.5.10), we obtain

$$\int_0^1 \delta g(x)\varphi^*(x)\varphi'(x)dx = \int_0^1 \delta f(x)\varphi^*(x)dx - \delta J_p \qquad (1.5.12)$$

Hence we obtain the formula of the perturbation theory to find changes of the functional δJ_p through δg, δf:

$$\delta J_p = \int_0^1 \varphi^*(x)[\delta f(x) - \delta g(x)\varphi'(x)]dx \qquad (1.5.13)$$

If $\delta g = 0$, (1.5.13) changes into (1.3.20) we obtained in § 1.3.

Now assume perturbations are small and approximately $\varphi'(x) \approx \varphi(x)$ in (1.5.13). Then the formula of small perturbation theory is

$$\delta J_p \approx \int_0^1 \varphi^*(x)[\delta f(x) - \delta g(x)\varphi(x)]dx \qquad (1.5.14)$$

Examine a numerical example. Assume in the problem above that $f(x) = \sin \pi x$, $\delta f(x) \equiv 0$,

$$p(x) = \begin{cases} 1, & x \in [0, 1/2] \\ 2(1 - x), & x \in [1/2, 1] \end{cases}$$

Assume then $\delta g(x) \equiv \varepsilon = const$, varying as

$$\varepsilon = 1, \quad \varepsilon = 0.1, \quad \varepsilon = 0.01$$

Problems (1.5.1), (1.5.4), and (1.5.5) were solved numerically (with the finite element method on a uniform grid with the step $h = 0.01$) for each case. Fig. 7 shows diagrams of the functions $\varphi(x)$, $\varphi^*(x)$, and $\varphi'(x)$ for $\varepsilon = 1$. To compare the values of the functional J_p' for different ε, use equivalent formulas for the perturbed problem (see (1.5.13)):

$$J_p^{(1)} = \int_0^1 p(x)\varphi'(x)dx = J_p'$$

$$J_p^{(2)} = \int_0^1 p(x)\varphi(x)dx - \int_0^1 \varphi^*(x)\delta g(x)\varphi'(x)dx$$

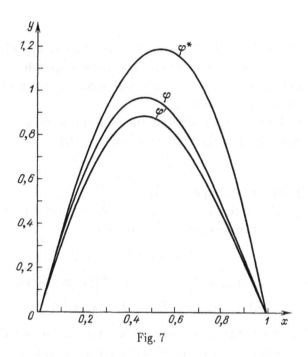

Fig. 7

Substituting the approximate value φ for φ' in $J_p^{(2)}$, we obtain (with regard to small perturbations (see (1.5.14)))

$$J_p^{(3)} = \int\limits_0^1 p(x)\varphi(x)dx - \int\limits_0^1 \varphi^*(x)\delta g(x)\varphi(x)dx$$

Table 3 presents values of $J_p^{(i)}$ ($i = 1, 2, 3$) for different ε.

Table 3

	$\varepsilon = 1$	$\varepsilon = 0.1$	$\varepsilon = 0.01$	$\varepsilon = 0$
$J_p^{(1)}$	0.0550755	0.0597395	0.0602504	0.0603166
$J_p^{(2)}$	0.0551037	0.0597829	0.0602948	0.0603166
$J_p^{(3)}$	0.0546047	0.0597775	0.0602947	0.0603166

It is clear form the table that numerical results (for any ε) confirm that perturbation theory formula (1.5.13) is correct. Besides, it follows from the table that formula (1.5.14) of small perturbation theory works predictably effectively for small ε.

Let us now derive the perturbation theory formula as in (1.5.13) for the general case. Assume the main problem as (1.4.1)

$$A\varphi(x) = f(x) \qquad (1.5.15)$$

which we will refer to as unperturbed.

If the properties of the medium the field of the function φ interacts with change, i. e., if the operator A turns into

$$A' = A + \delta A$$

the solution φ changes as well as the value of the functional

$$\varphi'(x) = \varphi(x) + \delta\varphi(x), \quad J'_p = J_p + \delta J_p$$

Establish a link between the operator's variation δA and the functional's variation δJ_p. The perturbed problem corresponding to (1.5.15) will have the form

$$A'\varphi' = f' \qquad (1.5.16)$$

where

$$f' = f + \delta f$$

Add to problem (1.5.16) an adjoint equation corresponding to the unperturbed problem:

$$A^*\varphi^* = p \qquad (1.5.17)$$

where A^* is the adjoint operator determined by the Lagrange identity (1.4.5). Take an inner product of equation (1.5.16) and φ^* and of equation (1.5.17) and φ', subtract the results from each other and obtain

$$(A'\varphi', \varphi^*) - (\varphi', A^*\varphi^*) = (f', \varphi^*) - (p, \varphi') \qquad (1.5.18)$$

Assume that $D(A) = D(A^*)$ and $\varphi, \varphi' \in D(A)$, $\varphi^* \in D(A^*)$. Transform the left-hand part of (1.5.18), while $A' = A + \delta A$, into

$$(A\varphi', \varphi^*) - (\varphi', A^*\varphi^*) + (\delta A\varphi', \varphi^*) = (\delta A\varphi', \varphi^*) \qquad (1.5.19)$$

We have thus used the Lagrange identity

$$(A\varphi', \varphi^*) - (\varphi', A^*\varphi^*) = 0$$

Transform the right-hand part of (1.5.18) into

$$(f, \varphi^*) - (p, \varphi) + (\delta f, \varphi^*) - (p, \delta\varphi) = (\delta f, \varphi^*) - \delta J_p \qquad (1.5.20)$$

We have used (1.4.8) as we simplified the left-hand part of (1.5.20), i. e.,

$$(f, \varphi^*) - (p, \varphi) = 0$$

Rewrite the expression (1.5.18) with regard to (1.5.19) and (1.5.20) as

$$(\delta A\varphi', \varphi^*) = (\delta f, \varphi^*) - \delta J_p \qquad (1.5.21)$$

Hence we arrive at the formula of perturbation theory

$$\delta J_p = (\delta f - \delta A\varphi', \varphi^*) \qquad (1.5.22)$$

Assuming the perturbations small and approximating $\varphi' = \varphi$, we arrive at the formula of small perturbation theory

$$\delta J_p = (\delta f - \delta A\varphi, \varphi^*) \qquad (1.5.23)$$

We have been until now using the adjoint equations technique in differential statements. However, it is possible first to approximate the main problem and convert it into a system of linear algebraic equations

$$A^h v^h = f^h \qquad (1.5.24)$$

and define the functional[5]

$$J_p^h = (p^h, v^h)_h \qquad (1.5.25)$$

We obtain the adjoint problem in the form

$$(A^h)^* w^h = p^h \qquad (1.5.26)$$

The operators A^h and $(A^h)^*$ must satisfy the Lagrange identity in a finite-dimensional space

$$(A^h v^h, w^h)_h = ((A^h)^* w^h, v^h)_h$$

---

[5] Here $(\cdot, \cdot)_h$ is an inner product in corresponding finite-dimensional space.

As a result,

$$J_f^h = (f^h, w^h)_h \qquad (1.5.27)$$

Then in numerical calculations we will always have

$$J_p^h = J_f^h \qquad (1.5.28)$$

1.6. Simple Nonlinear Problems

In this section, we use very simple examples to outline a way to a generalized adjoint equations theory and perturbation theory for nonlinear problems. General statements of nonlinear problems will be discussed in Chapter 3.

Let us assume that an operator $A(v)$ belongs to the real Hilbert space $H = L_2(0,1)$:

$$A(v) = -\frac{d}{dx} v \frac{d}{dx} \qquad (1.6.1)$$

affecting the functions $v(x)$ which belong to the set $D(A) \subset H$ and are defined on $\Omega = (0,1)$. Assume further that the functions $v \in D(A)$ are sufficiently smooth, so that

$$A(v)v = -\frac{d}{dx} v \frac{dv}{dx} \in H \qquad (1.6.2)$$

and v satisfies certain homogeneous boundary conditions, say

$$\frac{dv}{dx} = 0 \quad \text{at} \quad x = 0$$

$$\qquad (1.6.3)$$

$$\frac{dv}{dx} = 0 \quad \text{at} \quad x = 1$$

Examine formally the inner product

$$(A(v)v, w) = -\int_0^1 w \frac{d}{dx} v \frac{dv}{dx} dx \qquad (1.6.4)$$

Integrate by parts the right-hand part of (1.6.4) and obtain

$$(A(v)v, w) = -\left. wv \frac{dv}{dx} \right|_{x=0}^{x=1} + \int_0^1 v \frac{dw}{dx} \frac{dv}{dx} dx \qquad (1.6.5)$$

Under conditions (1.6.3), the term outside the integral in (1.6.5) turns to zero. Therefore

$$(A(v)v, w) = \int_0^1 \frac{dv}{dx}\left(v\frac{dw}{dx}\right)dx \qquad (1.6.6)$$

Integrate by parts the right-hand part of (1.6.6) once more. Then

$$(A(v)v, w) = v\left(v\frac{dw}{dx}\right)\Big|_{x=0}^{x=1} - \int_0^1 v\frac{d}{dx}v\frac{dw}{dx}dx \qquad (1.6.7)$$

Introduce now a set of functions $D(A^*) \subset H$, assuming that the functions $w \in D(A^*)$ are sufficiently smooth, so that the operator $d/dx\,(vd/dx)$ has a meaning on w and

$$\frac{d}{dx}v\frac{dw}{dx} \in H, \quad v \in D(A), \quad w \in D(A^*)$$

Assume, too, that the functions $w \in D(A^*)$ on the boundary of the domain $\Omega = (0,1)$ satisfy the conditions

$$\frac{dw}{dx} = 0 \quad \text{at} \quad x = 0$$

$$\frac{dw}{dx} = 0 \quad \text{at} \quad x = 1 \qquad (1.6.8)$$

Then if $w \in D(A^*)$, the term outside the integral in the formula (1.6.7) turns to zero, and we obtain

$$(A(v)v, w) = -\int_0^1 v\frac{d}{dx}v\frac{dw}{dx}dx \qquad (1.6.9)$$

Introduce an operator according to the formula

$$A^*(v) = -\frac{d}{dx}v\frac{d}{dx} \qquad (1.6.10)$$

It operates in H affecting the functions $w \in D(A^*)$. Then (1.6.9) turns into the Lagrange identity

$$(A(v)v, w) = (v, A^*(v)w), \quad v \in D(A), \quad w \in D(A^*) \qquad (1.6.11)$$

and $A^*(v)$ happens to be adjoint to $A(v)$. Compared to the linear operator, the adjoint one depends, however, on the function v itself of the set $D(A)$. The method we suggest still allows to apply the adjoint linear operators theory to a range of nonlinear operators in problems with given functionals

$$J_p[\varphi] = (p, \varphi)$$

where p is a measurement characteristic as a function of $x \in (0, 1)$ or a function of x and of the solution φ.

Now examine the main problem in the form

$$-\frac{d}{dx}\varphi\frac{d\varphi}{dx} = f(x)$$

$$\frac{d\varphi}{dx} = 0 \quad \text{at} \quad x = 0 \tag{1.6.12}$$

$$\frac{d\varphi}{dx} = 0 \quad \text{at} \quad x = 1$$

where $f(x)$ is a given function in H.

Assume that problem (1.6.12) does have a solution which is a sufficiently smooth function $\varphi \in D(A)$.[6] We assume that we want the value of the functional

$$J_p[\varphi] = \int_0^1 p(x)\varphi(x)dx, \quad p \in H \tag{1.6.13}$$

As we examine (1.6.12), let us introduce the adjoint problem

$$-\frac{d}{dx}\varphi\frac{d\varphi^*}{dx} = p(x)$$

$$\frac{d\varphi^*}{dx} = 0 \quad \text{at} \quad x = 0 \tag{1.6.14}$$

$$\frac{d\varphi^*}{dx} = 0 \quad \text{at} \quad x = 1$$

where p is the function determining functional (1.6.13).

---

[6]The solution φ of problem (1.6.12), generally speaking, is not unique. We examine one of the solutions, for example, non-negative one, assuming that it exists.

Assume that problem (1.6.14) does have a solution which is a sufficiently smooth function $\varphi^* \in D(A^*)$. Notice, that a solution of adjoint problem (1.6.14) is not unique and we examine below one of solutions φ^*.

Notice an important distinction of an adjoint problem for nonlinear equation from an adjoint problem for linear one: as we formulate the linear adjoint problem, it includes only values present in the main equation, coefficients and parameters. Besides, the right-hand part of the adjoint equation is the instrument function $p(x)$. Therefore the main and adjoint problems are solvable separately as not interdependent.

The situation is different in a nonlinear problem. We must first find a solution to the main problem and only then proceed to solving the adjoint problem, since the adjoint operator now depends on the solution $\varphi(x)$ of the main problem.

Represent now the functional J_p through the solution of adjoint problem (1.6.14). Multiply equation (1.6.12) by φ^* and equation (1.6.14) by φ, integrate in x on $(0,1)$, subtract the results from each other and obtain

$$-\int_0^1 \varphi^* \frac{d}{dx} \varphi \frac{d\varphi}{dx} dx + \int_0^1 \varphi \frac{d}{dx} \varphi \frac{d\varphi^*}{dx} dx = \int_0^1 f(x)\varphi^* dx - \int_0^1 p(x)\varphi dx$$

The left-hand part of this relationship turns to zero under the Lagrange identity (1.6.11). Then the adjointment relationship is

$$\int_0^1 f(x)\varphi^* dx = \int_0^1 p(x)\varphi dx \qquad (1.6.15)$$

Hence, with regard to (1.6.13), we obtain two equivalent formulas to compute the functional J_p:

$$J_p = \int_0^1 p(x)\varphi dx, \quad J_p = \int_0^1 f(x)\varphi^* dx$$

We have regarded the function p as independent of the solution φ until now. However, we can call off the assumption and obtain

$$J_p = \int_0^1 p(\varphi, x)\varphi dx, \quad J_p = \int_0^1 f(x)\varphi^* dx \qquad (1.6.16)$$

If we assume, in particular, that

$$p(\varphi, x) = \varphi$$

then obtain the energy functional

$$J_p = \int_0^1 \varphi^2 dx \qquad (1.6.17)$$

As for the second formula in (1.6.16), J_p will remain as a linear functional of a solution to the adjoint problem

$$J_p = \int_0^1 f\varphi^* dx$$

Examine an example. Let us assume that $f(x) = \cos 2\pi x$. Then the main problem (1.6.12) is

$$-\frac{d}{dx}\varphi\frac{d\varphi}{dx} = \cos 2\pi x, \quad x \in (0,1)$$

$$\frac{d\varphi}{dx} = 0 \quad \text{at} \quad x = 0$$

$$\frac{d\varphi}{dx} = 0 \quad \text{at} \quad x = 1$$

The problem has the solution $\varphi = (1/\pi)\cos \pi x$. If we have to compute energy functional (1.6.17), it will be easy as

$$J_p = \int_0^1 \varphi^2 dx = \frac{1}{\pi^2} \int_0^1 \cos^2 \pi x dx = \frac{1}{2\pi^2}$$

Compute now this functional using the second formula from (1.6.16). Then examine adjoint problem (1.6.14) with $p = \varphi = 1/\pi \cos \pi x$:

$$-\frac{d}{dx}\frac{1}{\pi}\cos \pi x\frac{d\varphi^*}{dx} = \frac{1}{\pi}\cos \pi x, \quad x \in (0,1)$$

$$\frac{d\varphi^*}{dx} = 0 \quad \text{at} \quad x = 0$$

$$\frac{d\varphi^*}{dx} = 0 \quad \text{at} \quad x = 1$$

This problem is linear as concerns φ^* and its solution is evident in $L_2(0,1)$

$$\varphi^*(x) = \frac{1}{\pi^2}\ln|\cos\pi x|, \quad x \neq \frac{1}{2}$$

which is an almost everywhere differentiable function with the continuous "flux" $\varphi d\varphi^*/dx$. From (1.6.16)

$$J_p = \int\limits_0^1 f(x)\varphi^*(x)dx = \frac{1}{\pi^2}\int\limits_0^1 \cos 2\pi x \ln|\cos\pi x|dx$$

Integration by parts provides an easy proof that

$$J_p = \frac{2}{\pi^2}\int\limits_0^{1/2} \cos^2\pi x \ln(\cos\pi x)dx = \frac{1}{2\pi^2}$$

which coincides with the value

$$J_p = \int\limits_0^1 \varphi^2 dx$$

As we noticed before, the solution to adjoint problem (1.6.14) in this example is not unique, and we may take for φ^* any of solutions to (1.6.14).

Let us now derive a perturbation theory formula for the functional J_p.

As we examine (1.6.12) which we will refer to as unperturbed one, introduce the perturbed problem:

$$-\frac{d}{dx}\varphi'\frac{d\varphi'}{dx} = f'(x)$$

$$\frac{d\varphi'}{dx} = 0 \quad \text{at} \quad x = 0 \tag{1.6.18}$$

$$\frac{d\varphi'}{dx} = 0 \quad \text{at} \quad x = 1$$

Assume that

$$\varphi' = \varphi + \delta\varphi, \quad f' = f + \delta f \tag{1.6.19}$$

where $\delta\varphi$ is a variation of the solution of the perturbed problem depending on the perturbation of the source δf. Now assume that the solution φ' to (1.6.18) does exist and is sufficiently smooth. As in the case of the main problem (1.6.12), it is not unique and we consider one of solutions.

Next examine adjoint unperturbed problem (1.6.14) with regard to functional (1.6.13):

$$-\frac{d}{dx}\varphi\frac{d\varphi^*}{dx} = p(x)$$

$$\frac{d\varphi^*}{dx} = 0 \quad \text{at} \quad x = 0 \qquad (1.6.20)$$

$$\frac{d\varphi^*}{dx} = 0 \quad \text{at} \quad x = 1$$

Multiply equation from (1.6.18) by φ^* and equation from (1.6.20) by φ, integrate the results on the interval $(0,1)$, subtract the results from each other and obtain

$$-\int_0^1 \varphi^* \frac{d}{dx}(\varphi + \delta\varphi)\frac{d\varphi'}{dx}dx + \int_0^1 \varphi'\frac{d}{dx}\varphi\frac{d\varphi^*}{dx}dx$$

$$= \int_0^1 f'(x)\varphi^*dx - \int_0^1 p(x)\varphi'dx \quad (1.6.21)$$

Under the Lagrange identity (1.6.11), the left-hand part of the relationship (1.6.21) is reduced into

$$-\int_0^1 \varphi^* \frac{d}{dx}\delta\varphi\frac{d\varphi'}{dx}dx \qquad (1.6.22)$$

Rewrite the right-hand part of (1.6.21), with regard to adjointment relationship (1.6.15) as

$$\int_0^1 \delta f\varphi^*dx - \int_0^1 p\delta\varphi dx \qquad (1.6.23)$$

Present (1.6.21) using (1.6.22) and (1.6.23) as

$$-\int_0^1 \varphi^* \frac{d}{dx}\delta\varphi\frac{d\varphi'}{dx}dx = \int_0^1 \delta f'\varphi^*dx - \delta J_p$$

We assumed above that

$$\delta J_p = \int_0^1 p\delta\varphi dx$$

Hence we have the expression

$$\delta J_p = \int_0^1 \delta f \varphi^* dx + \int_0^1 \varphi^* \frac{d}{dx} \delta\varphi \frac{d\varphi'}{dx} dx \qquad (1.6.24)$$

Integrate by parts the second integral in the right-hand part of (1.6.24) with regard to the fact, that the function φ' satisfies the boundary conditions in (1.6.18). Then the term outside the integral will turn to zero, and we finally obtain

$$\delta J_p = \int_0^1 \delta f \varphi^* dx - \int_0^1 \delta\varphi \frac{d\varphi'}{dx} \frac{d\varphi^*}{dx} dx \qquad (1.6.25)$$

If the perturbations are rather small, or, admissibly $\varphi' \approx \varphi$ with small-value accuracy of the first order, we obtain the formula of the small perturbation theory

$$\delta J_p \cong \int_0^1 \delta f \varphi^* dx - \int_0^1 \delta\varphi \frac{d\varphi}{dx} \frac{d\varphi^*}{dx} dx \qquad (1.6.26)$$

Application of (1.6.26) requires knowledge of a solution to the perturbed problem besides those to the unperturbed one, φ and φ^*, as the term $\delta\varphi = \varphi' - \varphi$ is under the integral sign. We will see that this formula, or its generalized versions, will play an important role in processing experimental data for assessment of the functional δJ_p, since it will require only an approximate value of the variation $\delta\varphi$ obtainable by simple estimations.

In an example, assume $f(x) = \cos 2\pi x$. As seen above, main problem (1.6.12) has a solution $\varphi(x) = \frac{1}{\pi} \cos \pi x$. Examine functional (1.6.13) where $p(x) = \frac{1}{\pi} \cos \pi x$. Then

$$\delta J_p = \int_0^1 p(x)\varphi(x)dx = \frac{1}{2\pi^2}$$

Take for $f'(x)$ the function

$$f'(x) = f(x) + \delta f(x)$$

where $\delta f(x) = \varepsilon \pi \cos \pi x$, $\varepsilon = \text{const} > 0$. Then perturbed problem (1.6.18) assumes the form

$$-\frac{d}{dx}\varphi'\frac{d\varphi'}{dx} = \cos 2\pi x + \varepsilon \pi \cos \pi x, \quad x \in (0,1)$$

$$\frac{d\varphi'}{dx} = 0 \quad \text{at} \quad x = 0$$

$$\frac{d\varphi'}{dx} = 0 \quad \text{at} \quad x = 1$$

How will the functional J_p vary? To compute J_p', use the earlier formula of the small perturbation theory

$$J_p' = J_p + \delta J_p$$

where δJ_p is determined by formula (1.6.26). Finding δJ_p from (1.6.26) requires the knowledge of the solution φ^* of adjoint problem (1.6.20) which then assumes the form

$$-\frac{d}{dx}\frac{1}{\pi}\cos \pi x \frac{d\varphi^*}{dx} = \frac{1}{\pi}\cos \pi x, \quad x \in (0,1)$$

$$\frac{d\varphi^*}{dx} = 0 \quad \text{at} \quad x = 0$$

$$\frac{d\varphi^*}{dx} = 0 \quad \text{at} \quad x = 1$$

From the previous example, the solution to this one will be the function

$$\varphi^*(x) = \frac{1}{\pi^2}\ln|\cos \pi x|, \quad x \neq \frac{1}{2}$$

This function is differentiable almost everywhere, belongs to $L_2(0,1)$, and satisfies (1.6.20) almost everywhere, too.

Besides, formula (1.6.26) for J_p incorporates the value $\delta\varphi = \varphi' - \varphi$. As a rule, it is difficult to find it in explicit form, though it is easy in this example. As we noticed before, a solution of perturbed problem is not unique and we will examine the solution $\varphi' = (1/\pi)\cos \pi x + \varepsilon$, for which the expression $\delta\varphi = \varphi' - \varphi = \varepsilon$ is valid. This restriction separates the unique solution.

Hence, and from (1.6.26),

$$\delta J_p = \int\limits_0^1 \varepsilon\pi\cos\pi x \cdot \frac{1}{\pi^2}\ln|\cos\pi x|dx - \int\limits_0^1 \varepsilon(-\sin\pi x)\left(-\frac{1}{\pi}\frac{\sin\pi x}{\cos\pi x}\right)dx$$

$$= \frac{\varepsilon}{\pi}\left\{\int\limits_0^1 \cos\pi x\ln|\cos\pi x|dx - \int\limits_0^1 \frac{\sin^2\pi x}{\cos\pi x}dx\right\}$$

Integrating by parts, easily obtain

$$\int\limits_0^1 \cos\pi x\ln|\cos\pi x|dx = \int\limits_0^1 \frac{\sin^2\pi x}{\cos\pi x}dx$$

i. e.,

$$\delta J_p \equiv 0.$$

Irrespective of the value of the parameter ε, the value of the functional J_p does not thus vary as we come from the main problem to the perturbed.

The same result is available immediately if we use another representation for δJ_p:

$$\delta J_p = \int\limits_0^1 p\delta\varphi dx = \int\limits_0^1 \frac{1}{\pi}\cos\pi x \cdot \varepsilon dx \equiv 0$$

These simple examples of nonlinear problems show the way to generalizations to be discussed in Chapter 3.

1.7. Adjoint Equations for Non-Stationary Problems

Methods discussed in preceding paragraphs for stationary problems of mathematical physics can be generalized also for non-stationary problems.

Let us begin with the following non-stationary one:

$$\frac{\partial\varphi}{\partial t} - \frac{\partial^2\varphi}{\partial x^2} = f(x,t), \quad 0 < x < 1, \quad 0 < t \le T$$
$$\varphi(0,t) = \varphi(1,t) = 0, \quad \varphi(x,0) = g(x)$$

(1.7.1)

where
$$f \in L_2((0,1) \times (0,T)), \quad g \in L_2(0,1)$$

Assume that the solution φ is sufficiently smooth, continuously differentiable in t on $[0,T]$, and twice continuously differentiable in x on $[0,1]$. Refer to (1.7.1) as the main or unperturbed problem.

Following the procedure in § 1.1, formulate for (1.7.1) a formally related adjoint problem

$$-\frac{\partial \varphi^*}{\partial t} - \frac{\partial^2 \varphi^*}{\partial x^2} = f^*(x,t), \quad 0 < x < 1, \quad 0 \le t < T$$

$$\varphi^*(0,t) = \varphi^*(1,t) = 0, \quad \varphi^*(x,T) = g^*(x) \qquad (1.7.2)$$

$$f^* \in L_2((0,1) \times (0,T)), \quad g^* \in L_2(0,1)$$

where the functions f^*, g^* are yet not defined.

Notice that, unlike in the main problem, (1.7.1), the initial condition for adjoint problem (1.7.2) is given at $t = T > 0$. Assume that the solution φ^* of (1.7.2) exists and is sufficiently smooth.

Now transform: multiply the equation from (1.7.1) by φ^* and the one from (1.7.2) by φ; subtract the results from each other and integrate in t on $[0,T]$ and in x on $[0,1]$, obtaining

$$\int\limits_0^T \frac{\partial}{\partial t} \left(\int\limits_0^1 \varphi^* \varphi dx \right) dt + \int\limits_0^T \left[\int\limits_0^1 \left(-\frac{\partial^2 \varphi}{\partial x^2} \varphi^* + \frac{\partial^2 \varphi^*}{\partial x^2} \varphi \right) dx \right] dt$$

$$= \int\limits_0^T \left[\int\limits_0^1 (f\varphi^* - f^*\varphi) dx \right] dt \quad (1.7.3)$$

The second term in the left-hand part of (1.7.3) is equal to zero (verify this immediately integrating by parts with boundary conditions from (1.7.1), (1.7.2)). Then use the initial conditions to rewrite (1.7.3) as

$$\int\limits_0^1 g^*(x)\varphi_T(x)dx - \int\limits_0^1 \varphi_0^*(x)g(x)dx = \int\limits_0^T \left[\int\limits_0^1 (f\varphi^* - f^*\varphi)dx \right] dt \quad (1.7.4)$$

where
$$\varphi_T(x) = \varphi(x,T), \quad \varphi_0^*(x) = \varphi^*(x,0)$$

Now assume that we seek not actually the value of the solution φ itself, but just a concrete solution's functional, such as

$$J = \int_0^1 g^*(x)\varphi_T(x)dx + \int_0^T \left(\int_0^1 f^*(x,t)\varphi(x,t)dx \right) dt \qquad (1.7.5)$$

where g^*, f^* are given functions.

Use (1.7.4) to obtain another formula and compute J. As in the case of statonary problems (see § 1.3), this formula is to link the value of J and the solution φ^* of the adjoint problem. We have from (1.7.4) and (1.7.5)

$$J = \int_0^1 \varphi_0^*(x)g(x)dx + \int_0^T \left(\int_0^1 f(x,t)\varphi^*(x,t)dx \right) dt \qquad (1.7.6)$$

Like in the stationary case, formula (1.7.6) proves to be more usable in computations than (1.7.5), especially if values of J are sought for a large set of different f.

In certain cases when, for example, functions $f(t,x)$ and $g(x)$ are complicated and it is difficult to find the solution φ of (1.7.1), formula (1.7.6) is also more usable in computing the functional J. As an example, assume

$$f(x,t) = \sqrt[4]{1 + t^{2/3}e^{-t}} \sin 2\pi x, \quad g(x) = 2\sin \pi x + \sin 3\pi x$$

then the main problem, (1.7.1), takes the form

$$\frac{\partial \varphi}{\partial t} - \frac{\partial^2 \varphi}{\partial x^2} = \sqrt[4]{1 + t^{2/3}e^{-t}} \sin 2\pi x, \quad 0 < x < 1, \quad 0 < t \leq T$$

$$\varphi(0,t) = \varphi(1,t) = 0, \quad \varphi(x,0) = 2\sin \pi x + \sin 3\pi x$$

$$(1.7.7)$$

Assume that we seek not the value of the solution φ, but a functional of this solution, of the form

$$J = \int_0^1 \sin \pi x \varphi(x,T)dx \qquad (1.7.8)$$

Computing J with (1.7.8) requires the knowledge of the value of the solution $\varphi(x,t)$ to problem (1.7.7) at $t = T$ which is difficult to find because of the complicated form of the functions f and g. Use then formula (1.7.6). First rewrite (1.7.8) as (1.7.5) where put

$$g^*(x) = \sin \pi x, \quad f^*(x,t) \equiv 0 \qquad (1.7.9)$$

Now build the adjoint problem (1.7.2)

$$-\frac{\partial \varphi^*}{\partial t} - \frac{\partial^2 \varphi^*}{\partial x^2} = 0, \quad 0 < x < 1, \quad 0 \le t < T$$

$$\varphi^*(0,t) = \varphi^*(1,t) = 0, \quad \varphi^*(x,T) = \sin \pi x \qquad (1.7.10)$$

Find easily the solution to adjoint problem (1.7.10), e. g., by the Fourier method. It has the form

$$\varphi^*(x,t) = e^{\pi^2(t-T)} \sin \pi x \qquad (1.7.11)$$

Now, knowing φ^*, use (1.7.6) to compute the functional:

$$J = \int_0^1 e^{-\pi^2 T} \sin \pi x (2 \sin \pi x + \sin 3\pi x) dx$$

$$+ \int_0^T \left(\int_0^1 \sqrt[4]{1 + t^{2/3} e^{-t}} \sin 2\pi x \, e^{\pi^2(t-T)} \sin \pi x dx \right) dt$$

$$= 2 \int_0^1 e^{-\pi^2 T} \sin^2 \pi x dx = \frac{1}{\pi} e^{-\pi^2 T} \qquad (1.7.12)$$

The example above shows how the value of J from (1.7.8) is obtained through adjoint function (1.7.11) explicitly:

$$J = \int_0^1 \sin \pi x \varphi(x,T) dx = \frac{1}{\pi} e^{-\pi^2 T} \qquad (1.7.13)$$

Now, as concerns the sensitivity of the functional J to the source $f(t,x)$, assume that the inputs in problem (1.7.1) are perturbed, i. e., instead of f and g we deal with functions

$$f' = f + \delta f, \quad g' = g + \delta g$$

and face a perturbed problem:

$$\frac{\partial \varphi'}{\partial t} - \frac{\partial^2 \varphi'}{\partial x^2} = f'(x,t), \quad 0 < x < 1, \quad 0 < t \leq T$$

$$\varphi'(0,t) = \varphi'(1,t) = 0, \quad \varphi'(x,0) = g'(x)$$

(1.7.14)

Represent the solution to this one as

$$\varphi'(x,t) = \varphi(x,t) + \delta\varphi(x,t)$$

and obtain for variations of the functional in

$$J' = J + \delta J$$

from (1.7.5) and (1.7.6) two equivalent relationships:

$$\delta J = \int\limits_0^1 g^*(x)\delta\varphi_T \, dx + \int\limits_0^T \left(\int\limits_0^1 f^*\delta\varphi \, dx \right) dt \qquad (1.7.15)$$

$$\delta J = \int\limits_0^1 \varphi_0^*\delta g \, dx + \int\limits_0^T \int\limits_0^1 \delta f \varphi^* \, dx \, dt \qquad (1.7.16)$$

Formula (1.7.16) links directly the functional's variation and variations of input data δf and δg, and is usable in computing immediately δJ, without solving the perturbed problem (1.7.14). We obtained similar formulas in the case of a stationary problem in § 1.3.

As an example, let us assume that

$$f(x,t) = (1 + 4\pi^2 t)\sin 2\pi x, \quad g(x) = 0$$

Then the main problem, (1.7.1), takes the form

$$\frac{\partial \varphi}{\partial t} - \frac{\partial^2 \varphi}{\partial x^2} = (1 + 4\pi^2 t)\sin 2\pi x,$$

$$\varphi(0,t) = \varphi(1,t) = 0, \quad \varphi(x,0) = 0$$

(1.7.17)

Choose, as J from (1.7.5), the following:

$$J = \int\limits_0^1 \sin \pi x \varphi(x,T) \, dx \qquad (1.7.18)$$

where
$$g^*(x) = \sin \pi x, \quad f^*(x,t) \equiv 0$$

Find easily the solution to (1.7.17) in the form

$$\varphi(x,t) = t \sin 2\pi x \tag{1.7.19}$$

Therefore obtain from (1.7.18)

$$J = T \int_0^1 \sin \pi x \sin 2\pi x \, dx = 0 \tag{1.7.20}$$

Let us assume that the initial condition in (1.7.17) changed, and instead of (1.7.17) we have a perturbed problem

$$\frac{\partial \varphi'}{\partial t} - \frac{\partial^2 \varphi'}{\partial x^2} = (1 + 4\pi^2 t) \sin 2\pi x$$

$$\varphi'(0,t) = \varphi'(1,t) = 0, \quad \varphi'(x,0) = e^{\pi^2 T} \sin \pi x \tag{1.7.21}$$

How does functional J vary? Use formula (1.7.16) and assume $\delta f \equiv 0$, $\delta g = e^{\pi^2 T} \sin \pi x$. Then, from (1.7.18) and (1.7.16), obtain

$$J' = J + \delta J = \delta J = \int_0^1 \varphi_0^* \delta g \, dx = \int_0^1 \varphi_0^*(x) e^{\pi^2 T} \sin \pi x \, dx \tag{1.7.22}$$

where $\varphi_0^* = \varphi^*(0,x)$, and $\varphi^*(t,x)$ is the solution of the adjoint problem

$$-\frac{\partial \varphi^*}{\partial t} - \frac{\partial^2 \varphi^*}{\partial x^2} = 0$$

$$\varphi^*(0,t) = \varphi^*(1,t) = 0, \quad \varphi^*(x,T) = \sin \pi x \tag{1.7.23}$$

This problem, as we have seen above, has the solution

$$\varphi^*(x,t) = e^{\pi^2(t-T)} \sin \pi x \tag{1.7.24}$$

Therefore, we have from (1.7.22):

$$J' = \int_0^1 e^{-\pi^2 T} \sin \pi x \, e^{\pi^2 T} \sin \pi x \, dx = \int_0^1 \sin^2 \pi x \, dx = \frac{1}{2} \tag{1.7.25}$$

We have just computed the value of the perturbed J' without actually solving the perturbed problem (1.7.21).

To conclude, let us examine more generalized case. In a non-stationary problem as

$$\frac{\partial \varphi}{\partial t} + A\varphi = f \quad \text{in} \quad \Omega \times (0, T)$$
$$\varphi = g \quad \text{at} \quad t = 0 \tag{1.7.26}$$

A is a linear operator, $\varphi(x, t)$ is assumed as differentiable function in t and at every t belongs to a set of functions $D(A)$ with the definition domain $\Omega \subset \mathbf{R}^n$ ($n \geq 1$). Assume the functions $f(x, t)$, $g(x)$ at every $t \in [0, T]$ squarely summable in x on Ω and belonging to the real Hilbert space $H = L_2(\Omega)$ with a conventional inner product $(\cdot, \cdot)$.

Let us refer to problem (1.7.26) as the main, or unperturbed, and state a corresponding adjoint problem, as

$$-\frac{\partial \varphi^*}{\partial t} + A^* \varphi^* = f^* \quad \text{in} \quad \Omega \times (0, T)$$
$$\varphi^* = g^* \quad \text{at} \quad t = T \tag{1.7.27}$$

where the functions f^*, g^* are yet not defined, A^* is the operator adjoint to A, satisfying the Lagrange identity

$$(A\varphi, \varphi^*) = (\varphi, A^* \varphi^*), \quad \varphi \in D(A), \quad \varphi^* \in D(A^*) \tag{1.7.28}$$

at every t. Notice that problem (1.7.27) is to be solved starting from $t = T$ down to $t = 0$. Calculations will be correct if the operator A^* is positive definite, i. e., $(A^* \varphi^*, \varphi^*) \geq \gamma^2 (\varphi^*, \varphi^*)$ for $\varphi^* \in D(A^*)$ where $\gamma > 0$.

Assume that the solutions φ and φ^* of problems (1.7.26) and (1.7.27) are sufficiently smooth and transform: multiply equation (1.7.26) by φ^* in $L_2(\Omega)$ and equation (1.7.27) by φ; subtract the results from each other and integrate in t on the interval $0 < t < T$ to obtain

$$\int_0^T \frac{\partial}{\partial t}(\varphi^*, \varphi) dt + \int_0^T [(A\varphi, \varphi^*) - (\varphi, A^* \varphi^*)] dt$$

$$= \int_0^T [(f, \varphi^*) - (f^*, \varphi)] dt \tag{1.7.29}$$

From (1.7.28), the second term in the left-hand part of (1.7.29) is equal to zero. Then (1.7.29) with the initial conditions taken into account, can be represented as

$$(g, \varphi_T) - (g, \varphi_0^*) = \int\limits_0^T [(f, \varphi^*) - (f^*, \varphi)]\, dt \qquad (1.7.30)$$

where

$$\varphi_T = \varphi(x, T), \quad \varphi_0^* = \varphi^*(x, 0)$$

If, now, we seek the linear functional of the solution φ which has the form

$$J = (g^*, \varphi_T) + \int\limits_0^T (f^*, \varphi)\, dt \qquad (1.7.31)$$

where f^*, g^* are some given functions, use (1.7.30) to produce another formula to compute J. From (1.7.30) and (1.7.31), we have

$$J = (g, \varphi_0^*) + \int\limits_0^T (f, \varphi^*)\, dt \qquad (1.7.32)$$

This formula links the value of the functional J with the solution φ^* of the adjoint problem, (1.7.27). As above, in some cases this formula proves more usable in computations than (1.7.31).

As concerns sensitivity of J to initial data f and g, let us assume that the initial data (1.7.26) are perturbed, i. e., instead of f and g we have functions

$$f' = f + \delta f, \quad g' = g + \delta g$$

A perturbed problem arises:

$$\frac{\partial \varphi'}{\partial t} + A\varphi' = f' \quad \text{in} \quad \Omega \times (0, T) \qquad (1.7.33)$$
$$\varphi' = g' \quad \text{at} \quad t = 0$$

How does J from (1.7.31) change? Rewrite the solution to (1.7.33) as

$$\varphi'(x, t) = \varphi(x, t) + \delta\varphi(x, t)$$

Then, with regard to (1.7.31), obtain:

$$J = J + \delta J \qquad (1.7.34)$$

where

$$\delta J = (\delta g, \varphi_0^*) + \int_0^T (\delta f, \varphi^*) dt \qquad (1.7.35)$$

or, on the other hand

$$\delta J = (g^*, \delta \varphi_T) + \int_0^T (f^*, \delta \varphi) dt \qquad (1.7.36)$$

Formulas (1.7.35) and (1.7.36) provide two equivalent relationships to determine functional variations δJ. Formula (1.7.35), however, proves more preferable for computations since it links immediately functional variation with initial data variations δf, δg and does not use the value $\delta \varphi$. Therefore, computing deviations of the functional J corresponding to variations in the inputs does not involve solving a large number of perturbed problems of the kind of (1.7.33) for different f' and g', just like in stationary problems. Just solve one adjoint problem (1.7.27) and use formula (1.7.35).

If the operator of the problem varies in perturbed problem (1.7.33), i. e., $A' = A + \delta A$ replaces A, formula (1.7.35) for the variation δJ takes the form

$$\delta J = (\delta g, \varphi_0^*) + \int_0^T (\delta f, \varphi^*) dt - \int_0^T (\delta A \varphi', \varphi^*) dt \qquad (1.7.37)$$

This is proved similarly to (1.5.22) obtained in § 1.5 for a stationary problem.

If perturbations are small and approximately $\varphi' = \varphi$, we obtain a formula of the small perturbation theory

$$\delta J = (\delta g, \varphi_0^*) + \int_0^T (\delta f, \varphi^*) dt - \int_0^T (\delta A \varphi, \varphi^*) dt \qquad (1.7.38)$$

1.8. Adjoint Equations and Simple Inverse Problems

Adjoint equations are particularly applicable in solving the inverse problems where they produce the most correctly formulated algorithms of calculating unknown parameters (e.g. coefficients) of differential operators.

Let us start with a one-dimensional problem,

$$-\frac{d}{dx}k\frac{d\varphi}{dx} = f, \quad x \in (0,1)$$

$$\varphi(0) = \varphi(1) = 0$$

(1.8.1)

where $f \in L_2(0,1)$. Assume, first, that $k = k(x)$ is the sought unknown function. Assume, too, we know a priori, that $k(x)$ is piecewise continuous, i. e.,

$$k(x) = \begin{cases} k_1, & 0 \le x \le 1/2 \\ k_2, & 1/2 < x \le 1 \end{cases}$$

(1.8.2)

where k_1, k_2 are so far unknown values. Assume also that $\varphi(x)$ is a continuous function to be found with a continuous "flow" $k\, d\varphi/dx$, satisfying almost everywhere equation (1.8.1) and boundary conditions $\varphi(0) = \varphi(1) = 0$.

The measurements of two functionals are assumed as

$$J_1 = \int\limits_0^1 p_1\varphi\, dx, \quad J_2 = \int\limits_0^1 p_2\varphi\, dx$$

(1.8.3)

where p_1 and p_2 are the functions of the measurement. Introduce now two adjoint problems corresponding to the functionals J_1 and J_2. The first problem has then the form

$$-\frac{d}{dx}k\frac{d\varphi_1^*}{dx} = p_1, \quad \varphi_1^*(0) = \varphi_1^*(1) = 0$$

(1.8.4)

and the second

$$-\frac{d}{dx}k\frac{d\varphi_2^*}{dx} = p_2, \quad \varphi_2^*(0) = \varphi_2^*(1) = 0$$

(1.8.5)

Assume the functions φ_1^*, φ_2^* as continuous with continuous "flows" $k(d\varphi_1^*/dx)$, $k(d\varphi_2^*/dx)$ and satisfying almost everywhere equations (1.8.4) and (1.8.5) respectively.

To determine $k(x)$ in formula (1.8.2), multiply (1.8.1) by φ_1^* and integrate the result over interval $0 < x < 1$ throughout and obtain

$$-\int_0^1 \varphi_1^* \frac{d}{dx} k \frac{\varphi}{dx} dx = \int_0^1 f\varphi_1^* \, dx \qquad (1.8.6)$$

Integrate by parts the left-hand part of (1.8.6) into

$$-\int_0^1 \varphi_1^* \frac{d}{dx} k \frac{d\varphi}{dx} dx = -\left. k\varphi_1^* \frac{d\varphi}{dx} \right|_{x=0}^{x=1} + \int_0^1 k \frac{d\varphi}{dx} \frac{d\varphi_1^*}{dx} dx \qquad (1.8.7)$$

The term outside the integral in the right-hand part of (1.8.7) is equal to zero under the boundary condition for φ_1^*. Then expression (1.8.6) takes the form

$$\int_0^1 k \frac{d\varphi}{dx} \frac{d\varphi_1^*}{dx} dx = \int_0^1 f\varphi_1^* \, dx \qquad (1.8.8)$$

Similarly, multiply equation (1.8.1) by φ_2^* and integrate it by parts into

$$\int_0^1 k \frac{d\varphi}{dx} \frac{d\varphi_2^*}{dx} dx = \int_0^1 f\varphi_2^* \, dx \qquad (1.8.9)$$

Use (1.8.8) and (1.8.9) to find $k(x)$, remembering that the functionals

$$J_1 = \int_0^1 f\varphi_1^* \, dx, \quad J_2 = \int_0^1 f\varphi_2^* \, dx \qquad (1.8.10)$$

are known from experimental measurements.

With respect to (1.8.10), rewrite (1.8.8) and (1.8.9) as

$$\int_0^1 k \frac{d\varphi}{dx} \frac{d\varphi_1^*}{dx} dx = J_1, \quad \int_0^1 k \frac{d\varphi}{dx} \frac{d\varphi_2^*}{dx} dx = J_2 \qquad (1.8.11)$$

Since (1.8.2) assigns the representation of $k(x)$ as a piecewise continuous function, use (1.8.2) in (1.8.11). Then

$$k_1 \int_0^{1/2} \frac{d\varphi}{dx} \frac{d\varphi_1^*}{dx} dx + k_2 \int_{1/2}^1 \frac{d\varphi}{dx} \frac{d\varphi_1^*}{dx} dx = J_1$$

$$\qquad (1.8.12)$$

$$k_1 \int_0^{1/2} \frac{d\varphi}{dx} \frac{d\varphi_2^*}{dx} dx + k_2 \int_{1/2}^1 \frac{d\varphi}{dx} \frac{d\varphi_2^*}{dx} dx = J_2$$

Assume now that

$$\int\limits_{0}^{1/2} \frac{d\varphi}{dx}\frac{d\varphi_1^*}{dx}dx = a_{11}, \qquad \int\limits_{0}^{1/2} \frac{d\varphi}{dx}\frac{d\varphi_2^*}{dx}dx = a_{21}$$

$$\int\limits_{1/2}^{1} \frac{d\varphi}{dx}\frac{d\varphi_1^*}{dx}dx = a_{12}, \qquad \int\limits_{1/2}^{1} \frac{d\varphi}{dx}\frac{d\varphi_2^*}{dx}dx = a_{22}$$

(1.8.13)

If we admit that the functions φ, φ_1^*, φ_2^* (or at least their derivatives in (1.8.12)) are already known, (1.8.12) gives a two-equation system to determine the unknown k_1 and k_2:

$$a_{11}k_1 + a_{12}k_2 = J_1$$
$$a_{21}k_1 + a_{22}k_2 = J_2$$

(1.8.14)

If the system determinant is

$$\Delta = \begin{vmatrix} a_{11} & a_{12} \\ a_{21} & a_{22} \end{vmatrix} \neq 0$$

then problem (1.8.14) does have a solution. The result will be the more precise, the more predominant are the diagonal terms in (1.8.14). To optimize planning the experiment staged to find k_1 and k_2, choose the functions p_1 and p_2 as

$$p_1(x) = \begin{cases} 1, & 0 \leq x \leq 1/2 \\ 0, & 1/2 < x \leq 1 \end{cases}$$

(1.8.15)

$$p_2(x) = \begin{cases} 0, & 0 \leq x \leq 1/2 \\ 1, & 1/2 < x \leq 1 \end{cases}$$

This is the procedure for solving simple inverse problem where the diffusion coefficient is sought.

Examine an example. Assume the functions $p_1(x)$, $p_2(x)$ assigned in the form (1.8.15) and the functions $\varphi(x)$, $\varphi_1^*(x)$ $\varphi_2^*(x)$ assigned as follows (Fig. 8):

$$\varphi(x) = \sin \pi x,$$

$$\varphi_1^*(x) = \begin{cases} -x^2 + (5/6)x, & 0 \leq x \leq 1/2 \\ -x/3 + 1/3, & 1/2 < x \leq 1 \end{cases}$$

$$\varphi_2^*(x) = \begin{cases} x/3, & 0 \leq x \leq 1/2 \\ -2x^2 + (8/3)x - 2/3, & 1/2 < x \leq 1 \end{cases}$$

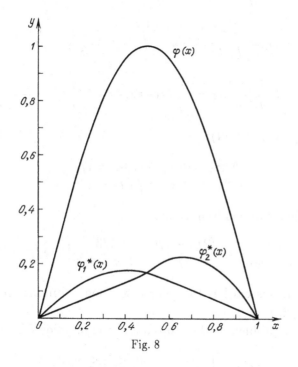

Fig. 8

Assume the function $k(x)$ in (1.8.2) to be sought. Find the values of the functionals J_1, J_2 using (1.8.3):

$$J_1 = \int_0^1 p_1\varphi \, dx = \int_0^{1/2} \varphi(x) \, dx = \int_0^{1/2} \sin \pi x \, dx = \frac{1}{\pi}$$

$$J_2 = \int_0^1 p_2\varphi \, dx = \int_{1/2}^1 \varphi(x) \, dx = \int_{1/2}^1 \sin \pi x \, dx = \frac{1}{\pi}$$

Compute a_{ij} $(i, j = 1, 2)$ from (1.8.13):

$$a_{11} = \int_0^{1/2} \frac{d\varphi}{dx} \frac{d\varphi_1^*}{dx} dx = \int_0^{1/2} \pi \cos \pi x (-2x + \frac{5}{6}) dx = \frac{2}{\pi} - \frac{1}{6}$$

$$a_{12} = \int_{1/2}^1 \frac{d\varphi}{dx} \frac{d\varphi_1^*}{dx} dx = \int_{1/2}^0 \pi \cos \pi x (-\frac{1}{3}) dx = \frac{1}{3}$$

$$a_{21} = \int_0^{1/2} \frac{d\varphi}{dx} \frac{d\varphi_2^*}{dx} dx = \int_0^{1/2} \pi \cos \pi x \frac{1}{3} dx = \frac{1}{3}$$

$$a_{22} = \int_{1/2}^1 \frac{d\varphi}{dx} \frac{d\varphi_2^*}{dx} dx = \int_{1/2}^1 \pi \cos \pi x (-4x + \frac{8}{3}) dx = \frac{4}{\pi} - \frac{2}{3}$$

Now build system (1.8.14):

$$\begin{aligned} (2/\pi - 1/6)k_1 + 1/3k_2 = 1/\pi \\ 1/3k_1 + (4/\pi - 2/3)k_2 = 1/\pi \end{aligned} \tag{1.8.16}$$

The determinant differs from zero:

$$\Delta = \begin{vmatrix} a_{11} & a_{12} \\ a_{21} & a_{22} \end{vmatrix} = \begin{vmatrix} (2/\pi - 1/6) & 1/3 \\ 1/3 & 4/\pi - 2/3 \end{vmatrix} = 8/\pi^2 - 2/\pi$$

therefore, the system has the unique solution. Find from (1.8.16) that $k_1 = 1/2$, $k_2 = 1/4$.

The inverse problem is thus solved and the function $k(x)$ is found:

$$k(x) = \begin{cases} 1/2, & 0 \le x \le 1/2 \\ 1/4, & 1/2 < x \le 1 \end{cases}$$

The function $k(x)$ may naturally have, in other cases, a more complex structure than that assumed in (1.8.2). If continuous on the interval $0 \le x \le 1$ and certain information has been a priori available on this function, it may be usually represented as

$$k(x) = \sum_{s=1}^m \alpha_s w_s(x) \tag{1.8.17}$$

where $w_s(x)$ are given approximating functions. Determining α_k then requires first the solution of m adjoint problems of the type:

$$-\frac{d}{dx}k\frac{d\varphi_s^*}{dx} = p_s(x)$$

$$\varphi_s^*(0) = \varphi_s^*(1) = 0, \quad s = 1, 2, \ldots, m \tag{1.8.18}$$

where $p_s(x)$ is the function linked with the measurement of the functional

$$J_s = \int_0^1 p_s\varphi \, dx \tag{1.8.19}$$

Knowing the solutions of m adjoint problems (1.8.18), we may proceed to determining α_s. Multiply equation (1.8.1) by $\varphi_s^*(x)$, integrate in x on the interval $[0, 1]$ and obtain

$$-\int_0^1 \varphi_s^*\frac{d}{dx}k\frac{d\varphi}{dx}dx = \int_0^1 f\varphi_s^* \, dx \tag{1.8.20}$$

Integrating by parts the left-hand part of (1.8.20) with respect to homogeneous boundary conditions on φ_s^* from (1.8.18), obtain

$$\int_0^1 k\frac{d\varphi_s^*}{dx}\frac{d\varphi}{dx}dx = \int_0^1 f\varphi_s^* \, dx \tag{1.8.21}$$

The functional J_s is evidently represented as

$$J_s = \int_0^1 f\varphi_s^* \, dx \tag{1.8.22}$$

Check it immediately multiplying equation from (1.8.1) by φ_s^* and equation from (1.8.18) by φ. Integrate the results in x on the interval $[0, 1]$, subtract from each other and obtain

$$0 = \int_0^1 f\varphi_s^* \, dx - \int_0^1 p_s\varphi \, dx$$

Whence (and from (1.8.19)) follows our statement.

Return now to (1.8.21). With respect to representation (1.8.17) rewrite it as

$$\sum_{s'=1}^{m} \alpha_{s'} a_{ss'} = J_s, \quad s = 1, 2, \ldots, m \qquad (1.8.23)$$

where

$$a_{ss'} = \int_0^1 w_{s'} \frac{d\varphi_s^* \, d\varphi}{dx \, dx} dx$$

Solving m equations (1.8.23) produces the unknown α_s and thus representation for $k(x)$.

If measurements errors are sizeable, values of m functionals prove insufficient for obtaining α_s with the required accuracy. More than m functionals J_s are usually used for higher accuracy. Then the system of equations

$$\sum_{s'=1}^{m} \alpha_{s'} a_{ss'} = J_s, \quad s = 1, 2, \ldots, m, m+1, \ldots, m+n \qquad (1.8.24)$$

becomes overdefined and the least-square method is used to solve it.

Let us formulate an algorithm for the least-square method. According to it (see, e.g. Strang [258]) choose, instead of exact solution of (1.8.24), values $\alpha_{s'}$ which minimize the sum of squares of deviations

$$E = \sum_{s=1}^{m+n} \left(J_s - \sum_{s'=1}^{m} \alpha_{s'} a_{ss'} \right)^2$$

The sum of squares E represents quadratic polynomial of the variables $\alpha_{s'}$. The polynomial is minimum at the values of $\alpha_1, \ldots, \alpha_m$ when all the first derivatives turn to zero:

$$\frac{\partial E}{\partial \alpha_l} = -2 \sum_{s=1}^{m+n} a_{sl} \left(J_s - \sum_{s'=1}^{m} \alpha_{s'} a_{ss'} \right) = 0, \quad l = 1, 2, \ldots, m$$

Hence the $\alpha_{s'}$ satisfy the system:

$$\sum_{s=1}^{m+n} a_{sl} \sum_{s'=1}^{m} \alpha_{s'} a_{ss'} = \sum_{s=1}^{m+n} a_{sl} J_s, \quad l = 1, 2, \ldots, m$$

This system may be rewritten in a matrix form with the following notations:

$$A = \begin{bmatrix} a_{11} & \cdots & a_{1m} \\ \cdots & \cdots & \cdots \\ a_{m1} & \cdots & a_{mm} \\ \cdots & \cdots & \cdots \\ a_{m+1,1} & \cdots & a_{m+n,m} \end{bmatrix}, \quad \bar{\alpha} = \begin{bmatrix} \alpha_1 \\ \cdots \\ \alpha_m \end{bmatrix}, \quad \bar{\beta} = \begin{bmatrix} J_1 \\ \cdots \\ J_{m+n} \end{bmatrix}$$

Then we have the system

$$A^T A \bar{\alpha} = A^T A \bar{\beta}$$

where A^T is the matrix transposed to A. If the columns of the matrix A are linearly independent, the system has the unique solution (see Strang [258]).

Now examine a general procedure for solution of inverse problems using adjoint equations.

Examine the problem

$$A\varphi = f \qquad (1.8.25)$$

where A is a linear operator operating in the Hilbert space H with the definition domain $D(A)$. Assume an inner product (u, v) defined in H and the function $f \in H$ as known. As concerns A, let us assume it a priori as unknown. Let

$$A = \sum_{s'=1}^{m} \alpha_{s'} A_{s'} \qquad (1.8.26)$$

where the operators A_s are known and the coefficients α_s unknown. The problem amounts to finding α_s $(s = 1, 2, \ldots, m)$ using measurement-assigned functionals of the solution

$$J_s = (p_s, \varphi) \qquad (1.8.27)$$

where p_s is the characteristic of the measurement (or the instrument). Introduce a set of adjoint equations

$$A^* \varphi_s^* = p_s, \quad s = 1, 2, \ldots, m, m+1, \ldots, m+n \qquad (1.8.28)$$

Subtracting inner product in H of equation (1.8.25) and φ_s^* and that of (1.8.28) and φ from each other, obtain

$$(A\varphi, \varphi_s^*) - (\varphi, A^* \varphi_s^*) = (f, \varphi_s^*) - (p_s, \varphi) \qquad (1.8.29)$$

The Lagrange identity reduces the left-hand part of this to zero, and

$$(p_s, \varphi) = (f, \varphi_s^*)$$

or, with regard to (1.8.27), we arrive at a dual representation

$$J_s = (p_s, \varphi), \quad J_s = (f, \varphi_s^*) \tag{1.8.30}$$

Multiply now (1.8.25) in H by φ_s^* and obtain

$$(A\varphi, \varphi_s^*) = (f, \varphi_s^*) \tag{1.8.31}$$

Use the representation (1.8.26) to obtain

$$\sum_{s'=1}^{m} \alpha_{s'} a_{ss'} = J_s, \quad s = 1, 2, \ldots, m, m+1, \ldots, m+n \tag{1.8.32}$$

where

$$a_{ss'} = (A_{s'}\varphi, \varphi_s^*)$$

The system (1.8.32) is overdefined. The least square method applied for its solution produces the values α_s ($s = 1, 2, \ldots, m$) initially unknown. This is the procedure for solution of the inverse problem involving restoration of the coefficients α_s in the operator A in (1.8.26). Notice, the procedure discussed in this section implied that the functions φ, φ_s^* are known, as they determine the coefficients $\alpha_{ss'}$ in system (1.8.32). However, these functions are not always initially known. In certain cases they may be replaced with "unperturbed" functions, i. e., solutions of model problems corresponding to problems (1.8.25) and (1.8.28). The perturbation theory methods, which will be discussed in detail in Chapter 4, makes it possible.

Inverse problems may have other solution modifications. It is known from experience that adjoint equations prove to be effectively applicable in solution of inverse problems.

1.9. Perturbation Theory

We have been discussing so far the small perturbation theory as applied to problems of mathematical physics. In many cases, however, perturbations are not small. Sections 1.9, 1.10 and 1.11 discuss obtaining results just in these conditions. Expansion of the function as a power series in a small parameter and consecutive solution of resulting problems is the most applied mathematical procedure. The general perturbation theory for linear problems and theoretical substantiation of the algorithms are discussed in [94, 199, 243].

Let us begin with a general scheme of perturbation algorithms. Let in the nonhomogeneous equation

$$Au = f \qquad (1.9.1)$$

A is a linear operator operating in the Hilbert space H with the definition domain $D(A) \subset H$. Assume (1.9.1) as solvable, i.e., that there is an inverse operator A^{-1} on elements $f \in H$. Alongside problem (1.9.1), referred to as *unperturbed* one, introduce a *perturbed* problem, where the perturbed operator has the same definition domain $D(A)$, $f_\varepsilon \in H$.

$$A_\varepsilon u_\varepsilon = f_\varepsilon \qquad (1.9.2)$$

Assume, that

$$A_\varepsilon = A + \varepsilon \delta A, \quad f_\varepsilon = f + \varepsilon \delta f \qquad (1.9.3)$$

where ε is a numerical parameter formally introduced into the perturbed problem which may be assumed as equal to 1 only by the end of all transformations. The parameter ε may be determined numerically in a problem; then it should be set equal to the given value after the research.

Proceed to the perturbation theory algorithm: represent the solution of perturbed problem as the series

$$u_\varepsilon = u_0 + \varepsilon u_1 + \varepsilon^2 u_2 + \dots \qquad (1.9.4)$$

Substitute (1.9.4) into (1.9.2) and obtain

$$(A + \varepsilon \delta A)(u_0 + \varepsilon u_1 + \varepsilon^2 u_2 + \dots) = f + \varepsilon \delta f \qquad (1.9.5)$$

Equate conventionally the coefficients at the same powers of ε and obtain a system of equations

$$\begin{aligned}
Au_0 &= f, \\
Au_1 &= \delta f - \delta Au_0, \\
Au_i &= -\delta Au_{i-1}, \quad i = 2, 3, \dots
\end{aligned} \qquad (1.9.6)$$

The consecutive solution of these produces the solution of (1.9.4). Truncate series (1.9.4) after a number of terms, e. g.,

$$u_\varepsilon^{(N)} = u_0 + \varepsilon u_1 + \cdots + \varepsilon^N u_N \qquad (1.9.7)$$

The function $u_\varepsilon^{(N)}$ is known as the N-th approximation to u_ε.

Solvability of equation (1.9.2), representability of the solution u_ε as series (1.9.4), solvability of system (1.9.6), convergence of the series (1.9.4) to the solution of equation (1.9.2) must be examined depending on particular statements of problems. General approaches to such analysis are discussed by the author in coauthorship with Agoshkov and Shutyaev in [94].

To simplify representation, assume hereinafter that our reasoning and algorithmic transformations are substantiated and concentrate on principal points of the theory and evolving the algorithms.

Examine the general scheme of perturbation algorithm as applied to perturbed equation (1.9.2) when A_ε and f_ε have the form

$$A_\varepsilon = \sum_{i=0}^{\infty} \varepsilon^i A_i, \quad A_0 = A$$

$$\qquad (1.9.8)$$

$$f_\varepsilon = \sum_{i=0}^{\infty} \varepsilon^i f_i, \quad f_0 = f$$

where $A_i : H \to H$ are given linear operators, $f_i \in H$.

Assume that the perturbed solution of problem (1.9.2) is representable as (1.9.4). Then, like in the previous case, equate the terms at the same powers of ε and obtain a system of equations

$$
\begin{aligned}
Au_0 &= f \\
Au_1 &= f_1 - A_1 u_0 \\
Au_2 &= f_2 - A_1 u_1 - A_2 u_0 \\
&\cdots\cdots\cdots\cdots\cdots\cdots \\
Au_n &= f_n - \sum_{k=1}^{n} A_k u_{n-k}
\end{aligned}
\qquad (1.9.9)
$$

Consecutively solving the equations, find u_0, u_1, u_2, ... and thus the solution u_ε, if the series (1.9.4) converges.

Examine now a simple unperturbed problem

$$-\frac{d^2u}{dx^2} = f, \quad x \in (0,1)$$
$$u(0) = u(1) = 0$$

(1.9.10)

where $f \in L_2(0,1)$, and a perturbed problem

$$-\frac{d^2u_\varepsilon}{dx^2} + \varepsilon g(x)u_\varepsilon = f_\varepsilon, \quad x \in (0,1)$$
$$u_\varepsilon(0) = u_\varepsilon(1) = 0$$

(1.9.11)

where $f_\epsilon = f + \varepsilon\delta f$, $\varepsilon \geq 0$. Here $\delta f \in L_2(0,1)$, g is a piecewise continuous function satisfying the condition

$$g(x) \geq 0.$$

(1.9.12)

In these conditions the solution of the perturbed problem is evidently unique on the interval $[0,1]$. To find the solution to this problem applying the perturbation algorithm, solve consecutively the system of equations

$$-\frac{d^2u_0}{dx^2} = f, \quad u_0(0) = u_0(1) = 0$$
$$-\frac{d^2u_1}{dx^2} = \delta f - gu_0, \quad u_1(0) = u_1(1) = 0$$
$$-\frac{d^2u_i}{dx^2} = -gu_{i-1}, \quad u_i(0) = u_i(1) = 0, \quad i = 1,2,\ldots$$

(1.9.13)

Find the solution to the perturbed problem:

$$u_\varepsilon = u_0 + \varepsilon u_1 + \varepsilon^2 u_2 + \ldots$$

(1.9.14)

Rewrite the solution of unperturbed problem (1.9.10) explicitly:

$$u = x\int_0^1 dx' \int_0^{x'} f(x'')dx'' - \int_0^x dx' \int_0^{x'} f(x'')dx''$$

(1.9.15)

It will be easy to find all the members of series (1.9.14) up to $n = N$ with the N-th order of accuracy in ε.

As a numerical example, examine unperturbed problem (1.9.10) in the form

$$-\frac{d^2u}{dx^2} = 100x, \quad x \in (0,1)$$
$$u(0) = u(1) = 0 \tag{1.9.16}$$

Assume perturbed problem (1.9.11) as

$$-\frac{d^2u_\varepsilon}{dx^2} + \varepsilon \sin \pi x u_\varepsilon(x) = 100x + \varepsilon 100 \sin \pi x, \quad x \in (0,1) \tag{1.9.17}$$
$$u(0) = u(1) = 0$$

where $f = 100x$, $\delta f = 100 \sin \pi x$, $g(x) = \sin \pi x$. Apply the perturbation algorithm to find the solution to (1.9.17). System (1.9.13) (at $i = 0, 1, 2$) then takes the form

$$-\frac{d^2u_0}{dx^2} = 100x, \quad u_0(0) = u_0(1) = 0$$

$$-\frac{d^2u_1}{dx^2} = 100 \sin \pi x - \sin \pi x u_0(x), \quad u_1(0) = u_1(1) = 0 \tag{1.9.18}$$

$$-\frac{d^2u_2}{dx^2} = -\sin \pi x u_1(x), \quad u_2(0) = u_2(1) = 0$$

Problems (1.9.18) are solved by the finite difference method on the uniform grid $x_i = ih$, $i = 0, \ldots, n$, $h = 1/n$, $n = 100$. The simplest finite-difference schemes of the second order of approximation in h were used. The resulting systems with three-diagonal matrices are solved by the factorization method [73]. The cases with $\varepsilon = 10$, $\varepsilon = 1$ and $\varepsilon = 0,1$ were examined and the results presented respectively in Figs. 9, 10 and 11, where, according to (1.9.7),

$$u_\varepsilon^{(1)} = u_0 + \varepsilon u_1, \quad u_\varepsilon^{(2)} = u_0 + \varepsilon u_1 + \varepsilon^2 u_2$$

To check the solution, we solved perturbed problem (1.9.17) by the same finite-difference method. From Figs. 9, 10 and 11 it is clear that all the cases empirically show convergence of the perturbation algorithm we applied. In Fig. 11, the $u_\varepsilon(x)$ and $u_\varepsilon^{(1)}(x)$ curves almost coincide, which means that finding solutions to perturbed problem (1.9.17)

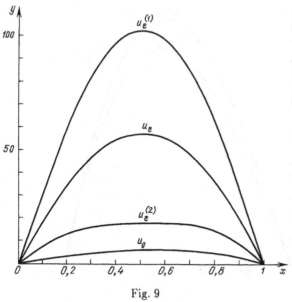

Fig. 9

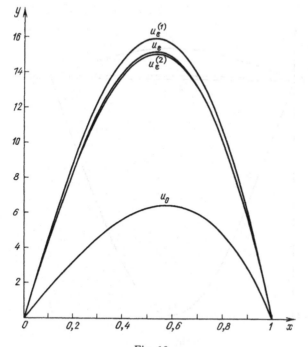

Fig. 10

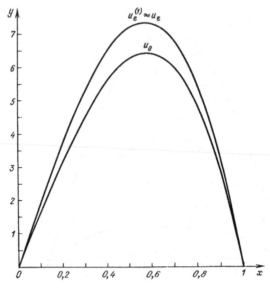

Fig. 11

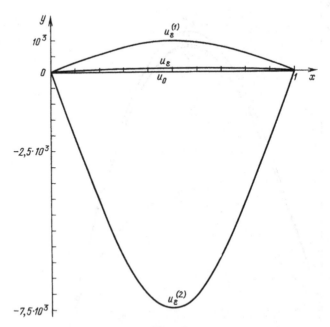

Fig. 12

with the accuracy of $O(h^2)$ would require only finding u_1 from (1.9.18) and constructing $u_\varepsilon^{(1)}(x) = u_0 + \varepsilon u_1$. At $\varepsilon = 1$ and $\varepsilon = 10$, the convergence of perturbation algorithms is slower (see Fig. 9 and Fig. 10).

Fig. 12 shows the case when $\varepsilon = 100$ and the perturbation algorithm does not converge, for ε is too large.

It is easy to find the condition on the parameter ε that could guarantee the convergence of perturbation series (1.9.14) in this case. To obtain such a condition, estimate the norms of functions u_i determined by system (1.9.13). At $u_0 \equiv u$, formula (1.9.15) is valid. Prove the validity of the estimate

$$\|u\| \le \frac{1}{c}\|f\| \tag{1.9.19}$$

where

$$\|u\| = \left(\int_0^1 u^2(x)dx\right)^{1/2}, \quad \|f\| = \left(\int_0^1 f^2(x)dx\right)^{1/2}, \quad c = \sqrt{\frac{189}{43}}$$

From (1.9.15), obtain

$$\|u\|^2 = \int_0^1 \left\{ x \int_0^1 dx' \int_0^{x'} f(x'')dx'' - \int_0^x dx' \int_0^{x'} f(x'')dx'' \right\}^2 dx$$

$$= I_1 + I_2 + I_3 \tag{1.9.20}$$

where

$$I_1 = \int_0^1 x^2 \left(\int_0^1 dx' \int_0^{x'} f(x'')dx''\right)^2 dx, \quad I_2 = \int_0^1 \left(\int_0^x dx' \int_0^{x'} f(x'')dx''\right)^2 dx$$

$$I_3 = -2 \left(\int_0^1 dx' \int_0^{x'} f(x'')dx''\right) \int_0^1 x \left(\int_0^x dx' \int_0^{x'} f(x'')dx''\right) dx$$

For I_1:

$$I_1 = \frac{1}{3} \left(\int_0^1 dx' \int_0^{x'} f(x'')dx''\right)^2 dx \tag{1.9.21}$$

For I_2, valid is inequality

$$I_2 \le \frac{2}{3} \left(\int_0^1 dx' \int_0^{x'} f(x'')dx''\right)^2 dx + \frac{1}{3}\int_0^1 \left(\int_0^x dx' \int_0^{x'} f(x'')dx''\right)^2 dx$$

Since

$$\int\limits_0^1 \left(\int\limits_0^x dx' \int\limits_0^{x'} f(x'')dx'' \right)^2 dx$$

$$\leq \int\limits_0^1 dx \left[\int\limits_0^x \left(\int\limits_0^{x'} dx'' \right)^{1/2} \left(\int\limits_0^{x'} f^2(x'')dx'' \right)^{1/2} dx' \right]^2 \leq \frac{\|f\|^2}{9}$$

then

$$I_2 \leq \frac{2}{3} \left(\int\limits_0^1 dx' \int\limits_0^{x'} f(x'')dx'' \right)^2 dx + \frac{\|f\|^2}{27} \qquad (1.9.22)$$

Integrating by parts, obtain for I_3

$$I_3 = -2 \left(\int\limits_0^1 dx' \int\limits_0^{x'} f(x'')dx'' \right) \left[\frac{1}{2} \int\limits_0^1 dx' \int\limits_0^{x''} f(x'')dx'' \right.$$

$$\left. - \int\limits_0^1 \frac{x^2}{2} \left(\int\limits_0^x f(x'')dx'' \right) dx \right] = - \left(\int\limits_0^1 dx' \int\limits_0^{x'} f(x)dx'' \right)^2$$

$$+ \int\limits_0^1 x^2 \left(\int\limits_0^x f(x'')dx'' \right) dx \cdot \int\limits_0^1 dx' \int\limits_0^{x'} f(x'')dx''$$

Since from the Cauchy–Bunyakovski inequality for the integrals

$$\int\limits_0^1 x^2 \left(\int\limits_0^x f(x'')dx'' \right) dx \leq \int\limits_0^1 x^2 \left(\int\limits_0^x dx'' \right)^{1/2} \left(\int\limits_0^x f^2(x'')dx'' \right)^{1/2} dx$$

$$\leq \|f\| \int\limits_0^1 x^{5/2}dx = \frac{2}{7}\|f\|$$

and

$$\int\limits_0^1 dx' \int\limits_0^{x'} f(x'')dx'' \leq \int\limits_0^1 \left(\int\limits_0^{x'} dx'' \right)^{1/2} \left(\int\limits_0^{x'} f^2(x'')dx'' \right)^{1/2} dx'$$

$$\leq \|f\| \int\limits_0^1 (x')^{1/2}dx' = \frac{2}{3}\|f\|$$

then

$$I_3 \leq -\left(\int_0^1 dx' \int_0^{x'} f(x'')dx''\right)^2 dx + \frac{4}{21}\|f\|^2 \qquad (1.9.23)$$

Substituting estimates (1.9.21) to (1.9.23) for I_1, I_2, and I_3 in (1.9.20), obtain

$$\|u\|^2 \leq \left(\frac{1}{27} + \frac{4}{21}\right)\|f\|^2 = \frac{43}{189}\|f\|^2$$

whence inequality (1.9.19) follows.

Similar estimates may be derived for $u_1, u_2, \ldots$ from (1.9.13), too:

$$\|u_1\| \leq \frac{1}{c}\|\delta f - gu_0\|,$$

$$(1.9.24)$$

$$\|u_i\| \leq \frac{1}{c}\|u_{i-1}\|, \quad i = 2, 3, \ldots$$

Since we have $g = \sin \pi x \leq 1$, it takes the form

$$\|u_i\| \leq \frac{1}{c}\|u_{i-1}\|, \quad i = 2, 3, \ldots$$

Whence consecutively

$$\|u_2\| \leq \frac{1}{c}\|u_1\|, \quad \|u_3\| \leq \frac{1}{c}\|u_2\| \leq \frac{1}{c^2}\|u_1\|, \ldots, \|u_n\| \leq \frac{1}{c^{n-1}}\|u_1\|, \ldots$$

Then series (1.9.14) is majorized above by the numerical series

$$\|u_\varepsilon\| \leq \|u_0\| + |\varepsilon|\|u_1\| + c\|u_1\|\sum_{i=2}^{\infty}\left(\frac{|\varepsilon|}{c}\right)^2 \qquad (1.9.25)$$

which converges for $|\varepsilon| < c$.

Then the condition

$$|\varepsilon| < c \qquad (1.9.26)$$

where $c = \sqrt{189/43}$, is a sufficient condition for convergence of perturbation series (1.9.14).

1.10. Adjoint Equations. Perturbation Algorithms

Examine now an adjoint unperturbed problem (see [94])

$$A^* u^* = p \qquad (1.10.1)$$

related to the main problem (1.9.1). A^* is an operator adjoint to A with the definition domain $D(A^*) \subset H$, $p \in H$. Assume problem (1.10.1) as a solvable.

Like in the main problem, introduce a perturbed adjoint problem

$$A_\varepsilon^* u_\varepsilon^* = p_\varepsilon \qquad (1.10.2)$$

$$A_\varepsilon^* = A^* + \varepsilon \delta A^*, \quad p_\varepsilon = p + \varepsilon \delta p \qquad (1.10.3)$$

with δA^* being a given operator operating in H with the definition domain $D(A^*)$, $\delta p \in H$. Assume, to make the case general, that both A^* and the function of measurement p are perturbed. Physical characteristics of an instrument do not usually change, and in most cases of practical interest $\delta p = 0$, i. e., $p_\varepsilon = p$, though not necessarily always.

Seek the solution to the perturbed problem in the form

$$u_\varepsilon^* = u_0^* + \varepsilon u_1^* + \varepsilon^2 u_2^* + \dots \qquad (1.10.4)$$

Substituting (1.10.4) and (1.10.3) in (1.10.2) and equating the terms at the same power of ε, obtain a system of equations to determine u_i^*:

$$\begin{aligned}
A^* u_0^* &= p \\
A^* u_1^* &= \delta p - \delta A^* u_0^* \\
A^* u_i^* &= -\delta A^* u_{i-1}^*, \quad i = 2, 3, \dots
\end{aligned} \qquad (1.10.5)$$

If just the functions $\{u_i^*\}_{i=1}^N$ are only to be found, the function

$$u_\varepsilon^{*(N)} = u_0^* + \varepsilon u_1^* + \dots + \varepsilon^N u_N^* \qquad (1.10.6)$$

may be assumed as the approximation of Nth order to the solution of perturbed problem (1.10.2).

Examine now the adjoint perturbed equation in general form

$$A_\varepsilon^* u_\varepsilon^* = p_\varepsilon \qquad (1.10.7)$$

where

$$A_\varepsilon^* = \sum_{i=0}^{\infty} \varepsilon^i A_i^*, \quad A_0^* = A^*$$

$$(1.10.8)$$

$$p_\varepsilon = \sum_{i=0}^{\infty} \varepsilon^i p_i, \quad p_0 = p$$

where A_i^* are assigned operators and $p_i \in H$. Find the solution of problem (1.10.7) also as a series

$$u_\varepsilon^* = u_0^* + \varepsilon u_1^* + \varepsilon^2 u_2^* + \ldots$$

Now use conventional technique to obtain the system of equations

$$A^* u_0^* = p$$
$$A^* u_1^* = p_1 - A_1^* u_0^*$$
$$A^* u_2^* = p_2 - A_1^* u_1^* - A_2^* u_0^*$$

$$(1.10.9)$$

$$\ldots\ldots\ldots\ldots\ldots\ldots\ldots$$

$$A^* u_n^* = p_n - \sum_{k=1}^{n} A_k^* u_{n-k}^*$$

Solve these consecutively to find $u_\varepsilon^{*(N)}$ through formula (1.10.6).

Justification of perturbation algorithms for adjoint problems is discussed in [94].

In an example, assume an operator A in the space $H = L_2(0,1)$, defined by the differential expression

$$Au = -\frac{d^2 u}{dx^2} + \frac{du}{dx}$$

with the definition domain $D(A)$ as discussed in § 1.1. Then problem (1.9.1) takes the form

$$-\frac{d^2 u}{dx^2} + \frac{du}{dx} = f, \quad x \in (0,1)$$

$$(1.10.10)$$

$$u(0) = u(1) = 0$$

Write, according to § 1.1, problem (1.10.1) adjoint to (1.10.10), as

$$-\frac{d^2 u^*}{dx^2} - \frac{du^*}{dx} = p, \quad x \in (0,1)$$

$$(1.10.11)$$

$$u^*(0) = u^*(1) = 0$$

Assume knowing the solution to problem (1.10.11) at a given function $p \in H = L_2(0,1)$. Alongside (1.10.11), examine perturbed problem (1.10.2) in the form

$$-\frac{d^2 u_\varepsilon^*}{dx^2} - (1 + \varepsilon g(x))\frac{du_\varepsilon^*}{dx} = p, \quad u_\varepsilon^*(0) = u_\varepsilon^*(1) = 0 \qquad (1.10.12)$$

where $p_\varepsilon = p$, $\delta A^* u_\varepsilon^* = -g(x)du_\varepsilon^*/dx$, $g(x)$ is a sufficiently smooth function.

To find the solution u_ε^* to (1.10.12) with an accuracy of $O(\varepsilon^2)$, use the perturbation algorithm discussed earlier. Expanding as (1.10.4), obtain system (1.10.5) in the form

$$-\frac{d^2 u_0^*}{dx^2} - \frac{du_0^*}{dx} = p, \quad u_0^*(0) = u_0^*(1) = 0$$

$$-\frac{d^2 u_1^*}{dx^2} - \frac{du_1^*}{dx} = g(x)\frac{du_0^*}{dx}, \quad u_1^*(0) = u_1^*(1) = 0 \qquad (1.10.13)$$

$$-\frac{d^2 u_i^*}{dx^2} - \frac{du_i^*}{dx} = g(x)\frac{du_{i-1}^*}{dx}, \quad u_i^*(0) = u_i^*(1) = 0, \quad i = 2, 3, \ldots$$

The first equation in (1.10.13) is just unperturbed adjoint problem (1.10.11) we know the solution to; it is $u_0^* = u^*$. Solving the second equation in (1.10.13) produces u_1^*. Then, according to (1.10.6), we can construct $u_\varepsilon^{*(1)}$:

$$u_\varepsilon^{*(1)} = u_0^* + \varepsilon u_1^*$$

This is in fact the first-order approximation to u_ε, the solution to perturbed problem (1.10.12).

1.11. Perturbation Algorithms for Eigenvalue Problems

Let us examine algorithms for solution of linear eigenvalue problems. Assume A as a linear operator operating in a real Hilbert space H with the definition domain $D(A) \subset H$. Assume also that functions of the set $D(A)$ satisfy certain homogeneous boundary conditions. Other properties of the functions $\varphi \in D(A)$ follow from the requirement that the equation

$$A\varphi = \lambda\varphi \qquad (1.11.1)$$

allows the complete set of eigenfunctions $\{\varphi_i\}$ (the basis in H) corresponding to the eigenvalues $\{\lambda_i\}$.

If A is not self-adjoint, introduce an adjoint operator A^*, according to the Lagrange identity

$$(A\varphi, \ \varphi^*) = (\varphi, A^*\varphi^*) \tag{1.11.2}$$

and obtain adjoint eigenvalue problem

$$A^*\varphi^* = \lambda\varphi^* \tag{1.11.3}$$

Assuming that φ^* belongs to $D(A^*)$, which is the definition domain for the operator A^*, find boundary conditions on $\varphi^* \in D(A^*)$ from another assumption, that the Lagrange identity is valid, then while transforming the left-hand part of the identity into the right-hand part all the terms outside the integrals are assumed equal to zero. Whence obtain the boundary conditions on φ^*. Other properties of the functions from the set $D(A^*)$ follow from the validity of the transformations applied to the Lagrange identity.

Let us refer to equations (1.11.1) and (1.11.3) as *unperturbed*. Introduce a perturbed equation corresponding to (1.11.1):

$$A_\varepsilon\varphi_\varepsilon = \lambda_\varepsilon\varphi_\varepsilon \tag{1.11.4}$$

where A_ε is a linear operator operating in H. Let us derive a formal scheme of the algorithm for calculation of eigenvectors and eigenvalues φ_ε and λ_ε by assigned perturbations in the operator A_ε.

Examine first a non-homogeneous equation corresponding to homogeneous unperturbed equation (1.11.1):

$$A\varphi = \lambda\varphi + f \tag{1.11.5}$$

and a homogeneous adjoint problem also corresponding to unperturbed problem (1.11.1):

$$A^*\varphi^* = \lambda\varphi^* \tag{1.11.6}$$

where the functions φ, φ^*, as well as λ are assumed as real.

The inner product of (1.11.5) and φ^* and of (1.11.6) and φ produce, through the Lagrange identity a condition

$$(f, \varphi^*) = 0 \tag{1.11.7}$$

This means that solution of problem (1.11.5) requires that eigenfunctions of the adjoint problem are orthogonal to the right-hand part of equation (1.11.5). Assume that these are orthogonal.

Assume now that the perturbation of the operator A is determined by a parameter and the representation

$$A_\varepsilon = \sum_{i=1}^{\infty} \varepsilon^i A^{(i)} \qquad (1.11.8)$$

is valid, where $A^{(0)} = A$, and $A^{(i)}$ $(i = 1, 2, \ldots)$ are linear operators within H with the definition domain $D(A)$.

Assume that λ_ε and φ_ε are representable as a series in parameter ε:

$$\lambda_\varepsilon = \sum_{i=0}^{\infty} \varepsilon^i \lambda^{(i)}, \quad \lambda^{(0)} = \lambda$$

$$\varphi_\varepsilon = \sum_{i=0}^{\infty} \varepsilon^i \varphi^{(i)}, \quad \varphi^{(0)} = \varphi \qquad (1.11.9)$$

converging in a neighborhood of $\varepsilon = 0$.

Here λ is the eigenvalue, and φ is the corresponding eigenfunction of unperturbed problem (1.11.1).

Substituting (1.11.9) into equation (1.11.4) and equating the terms at the same powers of ε, obtain a system

$$(A - \lambda^{(0)} E)\varphi^{(0)} = 0, \quad \lambda^{(0)} = \lambda, \quad \varphi^{(0)} = \varphi$$
$$(A - \lambda^{(0)} E)\varphi^{(1)} = \lambda^{(1)} \varphi^{(0)} - A^{(1)} \varphi^{(0)}$$

$$\cdots\cdots\cdots\cdots\cdots\cdots\cdots\cdots\cdots\cdots\cdots\cdots\cdots \qquad (1.11.10)$$

$$(A - \lambda^{(0)} E)\varphi^{(n)} = \sum_{i=1}^{n} \lambda^{(i)} \varphi^{(n-i)} - \sum_{i=1}^{n} A^{(i)} \varphi^{(n-i)}$$

$$\cdots\cdots\cdots\cdots\cdots\cdots\cdots\cdots\cdots\cdots\cdots\cdots$$

where E is the identity operator, i. e., $E\varphi = \varphi$. Rewrite system (1.11.10) as

$$(A - \lambda^{(0)} E)\varphi^{(0)} = 0$$
$$(A - \lambda^{(0)} E)\varphi^{(1)} = f_1$$

$$\cdots\cdots\cdots\cdots\cdots\cdots \qquad (1.11.11)$$

$$(A - \lambda^{(0)} E)\varphi^{(n)} = f_n$$

$$\cdots\cdots\cdots\cdots\cdots\cdots$$

Here f_n $(n = 1, 2, \ldots)$ is determined by the formula

$$f_n = \sum_{i=1}^{n} \lambda^{(i)} \varphi^{(n-i)} - \sum_{i=1}^{n} A^{(i)} \varphi^{(n-i)} \qquad (1.11.12)$$

As is known, solvability of the second and following equations of the system (1.11.11) requires that the right-hand parts f_n are orthogonal to the element φ^*, which is a solution of adjoint problem (1.11.6): $(f_n, \varphi^*) = 0$, $n = 1, 2, \ldots$.

Obtain consecutively the inner product of equations of system (1.11.10) and the function φ^*, which is the solution of adjoint equation (1.11.6), and subtract from each the expressions resulting from the inner product of equation (1.11.6) and respectively $\varphi^{(0)}, \varphi^{(1)}, \varphi^{(2)}, \ldots$. Use the Lagrange identity to obtain

$$\lambda^{(1)}(\varphi^{(0)}, \varphi^*) - (A^{(1)}\varphi^{(0)}, \varphi^*) = 0$$
$$\lambda^{(2)}(\varphi^{(0)}, \varphi^*) + \lambda^{(1)}(\varphi^{(1)}, \varphi^*) - (A^{(2)}\varphi^{(0)} + A^{(1)}\varphi^{(1)}, \varphi^*) = 0$$

$$(1.11.13)$$

. . . .

This leads us to the following algorithm. First, two unperturbed problems are to be solved:

$$A\varphi^{(0)} = \lambda^{(0)}\varphi^{(0)}, \quad A^*\varphi^* = \lambda^{(0)}\varphi^* \qquad (1.11.14)$$

Then using the found $\varphi^{(0)}$ and φ^*, and the first relationship in (1.11.13), find

$$\lambda^{(1)} = (A^{(1)}\varphi^{(0)}, \varphi^*)/(\varphi^{(0)}, \varphi^*) \qquad (1.11.15)$$

Now compute the right-hand part of the second equation in (1.11.10) or (1.11.11):

$$f_1 = \lambda^{(1)}\varphi^{(0)} - A^{(1)}\varphi^{(0)} \qquad (1.11.16)$$

Solve the second equation in (1.11.1) and find $\varphi^{(1)}$. Now $\varphi^{(0)}$, $\varphi^{(1)}$, φ^* and the values $\lambda^{(0)}$, $\lambda^{(1)}$ are unknown. Using the second equation in (1.11.13), find $\lambda^{(2)}$:

$$\lambda^{(2)} = \frac{(A^{(2)}\varphi^{(0)} + A^{(1)}\varphi^{(1)}, \varphi^*) - \lambda^{(1)}(\varphi^{(1)}, \varphi^*)}{(\varphi^{(0)}, \varphi^*)} \qquad (1.11.17)$$

With $\lambda^{(2)}$ we can calculate

$$f_2 = \lambda^{(2)}\varphi^{(0)} + \lambda^{(1)}\varphi^{(1)} - A^{(2)}\varphi^{(0)} - A^{(1)}\varphi^{(1)} \qquad (1.11.18)$$

and solve the following problem:

$$(A - \lambda^{(0)}E)\varphi^{(2)} = f_2 \qquad (1.11.19)$$

Whence find $\varphi^{(2)}$, and so on. All the corrections $\varphi^{(n)}$, $\lambda^{(n)}$ are thus consecutively found and hence φ_ε, which is the eigenfunction of the perturbed problem and corresponds to the perturbed eigenvalue λ_ε. Any eigenvalue may be made more precise this way.

Examine an example. Assume $H = L_2(0,1)$ as a Hilbert space of real functions $u(x)$ defined on the interval $0 \leq x \leq 1$ and integrable with the square, with the inner product

$$(u, v) = \int_0^1 u(x)v(x)dx, \quad u, v \in H \qquad (1.11.20)$$

Assume A as a simple differential operator

$$A\varphi = -2\frac{d^2\varphi}{dx^2} + \varphi, \quad \varphi \in D(A) \qquad (1.11.21)$$

with the definition domain $D(A)$ (as discussed in § 1.1).

Rewrite problem (1.11.1) for a given operator A as a boundary value problem

$$-2\frac{d^2\varphi}{dx^2} + \varphi = \lambda\varphi, \quad x \in (0,1)$$

$$\varphi(0) = \varphi(1) = 0 \qquad (1.11.22)$$

which happens to be the simplest Sturm–Liouville problem. The operator A is know to have a countable set of simple eigenvalues λ_i:

$$\lambda_i = 2\pi^2 i^2 + 1 \qquad (1.11.23)$$

each having a corresponding eigenfunction $\varphi_i = \sqrt{2}\sin\pi ix$, while the elements $\{\varphi_i\}$ ($i = 1, 2, \ldots$) make a complete orthonormal system in $H = L_2(0,1)$.

Alongside A, examine the perturbed operator A_ε within H with the definition domain $D(A)$ defined by the formula

$$A_\varepsilon = A + \varepsilon A^{(1)} \qquad (1.11.24)$$

where $-1 \le \varepsilon \le 1$, $A^{(1)}$ is the linear operator

$$A^{(1)}\varphi = -\frac{d}{dx}p(x)\frac{d\varphi}{dx} + q(x)\varphi(x), \quad \varphi \in D(A) \qquad (1.11.25)$$

and the functions $p(x), q(x)$ are assumed as

$$p(x) = 1 + \sin \pi x, \quad q(x) = -2\pi^2 \sin \pi x$$

Then according to formula (1.11.3), $A_i \equiv 0$ $(i = 2, 3, \ldots)$.

Rewrite eigenvalue problem (1.11.4) for the perturbed operator A_ε as the Sturm–Liouville boundary problem:

$$-\frac{d}{dx}(2 + \varepsilon p(x))\frac{d\varphi_\varepsilon}{dx} + (1 + \varepsilon q(x))\varphi_\varepsilon(x) = \lambda u, \quad x \in (0,1) \qquad (1.11.26)$$
$$\varphi_\varepsilon(0) = \varphi_\varepsilon(1) = 0$$

It is known from such equations' general theory (see [30]), that this problem has a countable set of simple eigenvalues $\lambda_\varepsilon^{(1)}, \lambda_\varepsilon^{(2)}, \ldots$. But eigenvalues cannot be found in an explicit form. However, the perturbation algorithm may be applied, e. g. in calculation of the first eigenvalue $\lambda_\varepsilon = \lambda_\varepsilon^{(1)}$ with an accuracy of $O(\varepsilon^3)$. Assume, that expansions (1.11.9) do take place in this case.

Since the operator A is symmetric, the unperturbed problems in (1.11.14) coincide and have the form of (1.11.22). Then

$$\lambda^{(0)} = 2\pi^2 + 1, \quad \varphi^{(0)} = \varphi^* = \sqrt{2}\sin \pi x \qquad (1.11.27)$$

Using $\varphi^{(0)}$ and φ^*, find $\lambda^{(1)}$ by formula (1.11.15)

$$\lambda^{(1)} = (A^{(1)}\varphi^{(0)}, \varphi^*) = \int_0^1 p(x)\left(\frac{d\varphi^{(0)}}{dx}\right)^2 dx + \int_0^1 q(x)(\varphi^{(0)})^2 dx$$

$$= 2\pi^2 \int_0^1 (1 + \sin \pi x)\cos^2 \pi x \, dx + 2\int_0^1 (-2\pi^2 \sin \pi x)\sin^2 \pi x \, dx$$

$$= \pi^2 - 4\pi \qquad (1.11.28)$$

(Consider that $(\varphi^{(0)}, \varphi^*) = 1$). Calculate f_1 in formula (1.11.16):

$$f_1 = \lambda^{(1)}\varphi^{(0)} - A^{(1)}\varphi^{(0)} = \lambda^{(1)}\sqrt{2}\sin \pi x + \sqrt{2}\pi\frac{d}{dx}(p(x)\cos \pi x)$$

$$- \sqrt{2}q(x)\sin \pi x = \sqrt{2}\pi^2 - 4\sqrt{2}\pi \sin \pi x \quad (1.11.29)$$

and solve the second equation in (1.11.11), rewritten as a boundary value problem

$$-2\frac{d^2\varphi^{(1)}}{dx^2} + \varphi^{(1)} - (2\pi^2 + 1)\varphi^{(1)} = f_1, \quad \varphi^{(1)}(0) = \varphi^{(1)}(1) = 0 \quad (1.11.30)$$

The solution to this one will evidently be the function

$$\varphi^{(1)} = \frac{1}{2\pi}\int_0^x f_1(\tau)\sin \pi(\tau - x)d\tau = \frac{\sqrt{2}}{\pi}\sin \pi x - \frac{\sqrt{2}}{2} - \left(\sqrt{2}x - \frac{\sqrt{2}}{2}\right)\cos \pi x$$

which satisfies the condition

$$(\varphi^{(1)}, \varphi^{(0)}) = \int_0^1 \varphi^{(1)}(x)\varphi^{(0)}(x)dx = 0 \quad (1.11.31)$$

Using $\varphi^{(0)}, \varphi^{(1)}$, and condition (1.11.31), find $\lambda^{(2)}$ with formula (1.11.17):

$$\lambda^{(2)} = (A^{(1)}\varphi^{(1)}, \varphi^*) = \int_0^1 p(x)\frac{d\varphi^{(1)}}{dx}\frac{d\varphi^{(0)}}{dx}dx + \int_0^1 q(x)\varphi^{(1)}(x)\varphi^{(0)}(x)dx$$

$$= \int_0^1 (1 + \sin \pi x)\left(\frac{\pi}{2}(2\sqrt{2}x - \sqrt{2})\sin \pi x\right)(\sqrt{2}\pi \cos \pi x)dx$$

$$+ \int_0^1 (-2\pi^2 \sin \pi x)\left(-\frac{\sqrt{2}}{2} + \frac{\sqrt{2}}{\pi}\sin \pi x - (\sqrt{2}x - \frac{\sqrt{2}}{2})\cos \pi x\right)$$

$$\times (\sqrt{2}\sin \pi x)dx = -(\pi^2 + 8)$$

Write with an accuracy of up to $O(\varepsilon^3)$

$$\lambda_\varepsilon \approx \lambda^{(0)} + \varepsilon\lambda^{(1)} + \varepsilon^2\lambda^{(2)} = 2\pi^2 + 1 + \varepsilon(\pi^2 - 4\pi) - \varepsilon^2(\pi^2 + 8) \quad (1.11.32)$$

Knowing $\lambda^{(2)}$, calculate f_2 with formula (1.11.18), solve problem (1.11.19) and find $\varphi^{(2)}$, then $\lambda^{(3)}$, etc.

The perturbation algorithm which this Section discusses is thus applied to find the perturbed eigenvalue $\lambda_\varepsilon = \lambda_\varepsilon^{(1)}$ with any pre-assigned accuracy ε. Any eigenvalue $\lambda_\varepsilon^{(k)}$ $(k = 2, 3, \ldots)$ in problem (1.11.26) may be computed following this procedure.

CHAPTER 2

Simple Main and Adjoint Equations

of Mathematical Physics

Main and adjoint equations are widely applicable, primarily so in the perturbation theory as applied to functionals and sensitivity to problem's input parameters variations. Such problems are particularly important for complex systems where numerous details may obscure direct and feedback relationships between different factors making difficult assessment of these relationships. Assessing functional sensitivity to parameters' variations, however, offers the right answer in many cases and helps in planning correct experimentation or optimize process control.

This monograph discusses just this type of problems. Starting with simple mathematical physics problems will, however, make understanding the basic ideas easier.

2.1. Diffusion Equation

Substance diffusion problems are perhaps the most common in mathematical physics and its applications. These problems happen to be fairly simple to analyze. Let us therefore start applying adjoint equations just to diffusion problems, the simplest one-dimensional ones, and eventually generalize the theory for multi-dimensionals in specific applications.

Examine a diffusion equation

$$-\frac{d}{dx}k\frac{d\varphi}{dx} + q\varphi = f, \quad x \in (0,1) \tag{2.1.1}$$

where $k(x) \geq 0$, $q(x) \geq 0$, $f(x)$ are piecewise continuous functions in $\Omega = (0,1)$, $\varphi \in D(A)$. All the elements of the set $D(A)$ are continuous, and each has a continuous and differentiable derivative $k\,d\varphi/dx$ so that the derivative $d/dx(k\,d\varphi/dx)$ is piecewise continuous in Ω. Assume that all the functions of the set $D(A)$ satisfy homogeneous boundary conditions

$$\begin{aligned} \alpha\frac{d\varphi}{dx} + \beta\varphi = 0 \quad &\text{at} \quad x = 0 \\ \gamma\frac{d\varphi}{dx} + \sigma\varphi = 0 \quad &\text{at} \quad x = 1 \end{aligned} \tag{2.1.2}$$

where α, β, γ, and σ are given numbers with $\alpha\gamma \neq 0$.

Let us deal with problem (2.1.1) and (2.1.2) in a real Hilbert space $H = L_2(\Omega)$ with an inner product

$$(u,w) = \int_0^1 v(x)w(x)dx$$

where v, $w \in H$. Rewrite this problem in the operator form as

$$A\varphi = f \tag{2.1.3}$$

where A is a linear operator within H with the definition domain $D(A)$ and

$$A\varphi = -\frac{d}{dx}k\frac{d\varphi}{dx} + q\varphi$$

Notice that if boundary conditions (2.1.2) are non-homogeneous, they may be converted into homogeneous in a respective linear transformation of the solutions φ replacing the function $f(x)$ with another.

Referring to problem (2.1.1) and (2.1.2) as the main, formulate an adjoint problem with regard to the functional as

$$J_p = \int_0^1 p\varphi dx \tag{2.1.4}$$

where $p(x)$ is a given function linked to the measurement of the field φ. In certain cases, as was noted above, it is referred to as the function of the instrument resolution.

Therefore introduce an adjoint function φ^*, multiply equation (2.1.1) by φ^* and integrate the result all over the domain $\Omega = (0,1)$ to obtain

$$-\int_0^1 \varphi^* \frac{d}{dx} k \frac{d\varphi}{dx} dx + \int_0^1 q\varphi^* \varphi \, dx = \int_0^1 f\varphi^* dx \qquad (2.1.5)$$

Transform the first of the integrals in (2.1.5) to make the function φ the multiplier: integrate the expression by parts twice and obtain

$$-\int_0^1 \varphi^* \frac{d}{dx} k \frac{d\varphi}{dx} dx = -k\varphi^* \frac{d\varphi}{dx}\Big|_{x=0}^{x=1} + \int_0^1 k \frac{d\varphi^*}{dx} \frac{d\varphi}{dx} dx$$

$$= -k\varphi^* \frac{d\varphi}{dx}\Big|_{x=0}^{x=1} + k\varphi \frac{d\varphi^*}{dx}\Big|_{x=0}^{x=1} - \int_0^1 \varphi \frac{d}{dx} k \frac{d\varphi^*}{dx} dx \qquad (2.1.6)$$

Put (2.1.6) into (2.1.5) and obtain

$$\int_0^1 \varphi \left(-\frac{d}{dx} k \frac{d\varphi^*}{dx} + q\varphi^* \right) dx - k\varphi^* \frac{d\varphi}{dx}\Big|_{x=0}^{x=1} + k\varphi \frac{d\varphi^*}{dx}\Big|_{x=0}^{x=1}$$

$$= \int_0^1 f\varphi^* dx \qquad (2.1.7)$$

The sets of functions $\varphi^* \in D(A^*)$ still remain undefined. Assume now that the function φ^* is continuous and has a continuous and differentiable "flow" $k \, d\varphi^*/dx$ which makes the function $d/dx(k \, d\varphi^*/dx)$ piecewise continuous. Require that the expression outside the integral in the left-hand part of (2.1.7) turns to zero:

$$-k \left(\varphi^* \frac{d\varphi}{dx} - \varphi \frac{d\varphi^*}{dx} \right)\Big|_{x=1} + k \left(\varphi^* \frac{d\varphi}{dx} - \varphi \frac{d\varphi^*}{dx} \right)\Big|_{x=0} = 0 \qquad (2.1.8)$$

The sufficient conditions for this equality will be

$$\varphi^* \frac{d\varphi}{dx} - \varphi \frac{d\varphi^*}{dx} = 0 \quad \text{at} \quad x = 0$$

$$\qquad (2.1.9)$$

$$\varphi^* \frac{d\varphi}{dx} - \varphi \frac{d\varphi^*}{dx} = 0 \quad \text{at} \quad x = 1$$

Exclude the derivatives $d\varphi/dx$ in (2.1.9) expressing them in terms of φ through the boundary conditions of (2.1.2), i. e.,

$$\frac{d\varphi}{dx} = -\frac{\beta}{\alpha}\varphi \quad \text{at} \quad x = 0$$

$$\frac{d\varphi}{dx} = -\frac{\sigma}{\gamma}\varphi \quad \text{at} \quad x = 1$$

(2.1.10)

Then conditions (2.1.9) will transform into

$$\varphi\left(-\frac{\beta}{\alpha}\varphi^* - \frac{d\varphi^*}{dx}\right) = 0 \quad \text{at} \quad x = 0$$

$$\varphi\left(-\frac{\sigma}{\gamma}\varphi^* - \frac{d\varphi^*}{dx}\right) = 0 \quad \text{at} \quad x = 1$$

(2.1.11)

Hence follows that (2.1.8) holds if, together with conditions (2.1.2) for the function φ conditions for the function φ^* are

$$\alpha\frac{d\varphi^*}{dx} + \beta\varphi^* = 0 \quad \text{at} \quad x = 0$$

$$\gamma\frac{d\varphi^*}{dx} + \sigma\varphi^* = 0 \quad \text{at} \quad x = 1$$

(2.1.12)

With regard to conditions (2.1.12), the terms outside the integrals in (2.1.7) turn to zero, then obtain

$$\int_0^1 \varphi\left(-\frac{d}{dx}k\frac{d\varphi^*}{dx} + q\varphi^*\right) dx = \int_0^1 f\varphi^* dx$$

(2.1.13)

Assume now that the function φ^* satisfies the equation

$$-\frac{d}{dx}k\frac{d\varphi^*}{dx} + q\varphi^* = p$$

(2.1.14)

Then (2.1.13) takes the form

$$\int_0^1 p\varphi dx = \int_0^1 f\varphi^* dx,$$

and the functional J_p acquires a dual representation:

$$J_p = \int_0^1 p\varphi dx, \quad J_p = \int_0^1 f\varphi^* dx \qquad (2.1.15)$$

The problem adjoint to the main diffusion problem (2.1.1) and (2.1.2) thus takes the final form

$$-\frac{d}{dx}k\frac{d\varphi^*}{dx} + q\varphi^* = p \qquad (2.1.16)$$

$$\alpha\frac{d\varphi^*}{dx} + \beta\varphi^* = 0 \quad \text{at} \quad x = 0$$

$$\qquad (2.1.17)$$

$$\gamma\frac{d\varphi^*}{dx} + \sigma\varphi^* = 0 \quad \text{at} \quad x = 1$$

This means that the sought functional J_p is obtainable by solving either the main problem (2.1.11) and (2.1.2) or the adjoint problem (2.1.16) and (2.1.17) with the use of one of equivalent representations in (2.1.15). Then we seek the solution of problem (2.1.16) and (2.1.17) among functions of the set $D(A^*)$ which coincides with $D(A)$.

Now formulate a sensitivity theory. Examine the perturbed diffusion problem corresponding to (2.1.1):

$$-\frac{d}{dx}k'\frac{d\varphi'}{dx} + q'\varphi' = f' \qquad (2.1.18)$$

$$\alpha'\frac{d\varphi'}{dx} + \beta'\varphi' = 0 \quad \text{at} \quad x = 0$$

$$\qquad (2.1.19)$$

$$\gamma'\frac{d\varphi'}{dx} + \sigma'\varphi' = 0 \quad \text{at} \quad x = 1$$

where the primed variables may be represented as the sums of main values and variations:

$$\varphi' = \varphi + \delta\varphi, \quad f' = f + \delta f, \quad k' = k + \delta k$$

$$q' = q + \delta q, \quad \alpha' = \alpha + \delta\alpha, \quad \beta' = \beta + \delta\beta$$

$$\gamma' = \gamma + \delta\gamma, \quad \sigma' = \sigma + \delta\sigma$$

with $\alpha'\gamma' \neq 0$. Select as the problem's functional

$$J' = \int_0^1 p\varphi' dx = J_p + \delta J, \quad \delta J = \int_0^1 p\delta\varphi dx \qquad (2.1.20)$$

Multiply equation (2.1.18) by φ^*, which is the solution of adjoint problem (2.1.16) and (2.1.17); equation (2.1.16) by φ', which is the solution of problem (2.1.18) and (2.1.19); integrate the results in x within the limits from $x = 0$ to $x = 1$ and subtract from each other to obtain

$$\int_0^1 \varphi^* \left(-\frac{d}{dx} k' \frac{d\varphi'}{dx} + q'\varphi' \right) dx - \int_0^1 \varphi' \left(-\frac{d}{dx} k \frac{d\varphi^*}{dx} + q\varphi^* \right) dx$$

$$= \int_0^1 f'\varphi^* dx - \int_0^1 p\varphi' dx \quad (2.1.21)$$

Integrate by parts the first and second integrals in the left-hand part of expression (2.1.21) into

$$\int_0^1 \varphi^* \left(-\frac{d}{dx} k' \frac{d\varphi'}{dx} + q'\varphi' \right) dx$$

$$= -\left. k'\varphi^* \frac{d\varphi'}{dx} \right|_{x=0}^{x=1} + \int_0^1 \left(k' \frac{d\varphi'}{dx} \frac{d\varphi^*}{dx} + q'\varphi'\varphi^* \right) dx$$

$$\qquad\qquad (2.1.22)$$

$$\int_0^1 \varphi' \left(-\frac{d}{dx} k \frac{d\varphi^*}{dx} + q\varphi^* \right) dx$$

$$= -\left. k\varphi' \frac{d\varphi^*}{dx} \right|_{x=0}^{x=1} + \int_0^1 \left(k \frac{d\varphi'}{dx} \frac{d\varphi^*}{dx} + q\varphi'\varphi^* \right) dx$$

Use the conditions (2.1.17) and (2.1.19) and put (2.1.22) into (2.1.21) to obtain

$$\int_0^1 \delta k \frac{d\varphi'}{dx} \frac{d\varphi^*}{dx} dx + \int_0^1 \delta q\varphi'\varphi^* dx + \delta\left(\frac{k\sigma}{\gamma}\right)\bigg|_{x=1} \varphi'(1)\varphi^*(1)$$

$$- \delta\left(\frac{k\beta}{\alpha}\right)\bigg|_{x=1} \varphi'(0)\varphi^*(0) = \int_0^1 f'\varphi^* dx - \int_0^1 p\varphi' dx \quad (2.1.23)$$

Here

$$\delta\left(\frac{k\sigma}{\gamma}\right) = \frac{k'\sigma'}{\gamma'} - \frac{k\sigma}{\gamma}, \quad \delta\left(\frac{k\beta}{\alpha}\right) = \frac{k'\beta'}{\alpha'} - \frac{k\beta}{\alpha}$$

Then transform the right-hand part of (2.1.23) with regard to

$$\int_0^1 f'\varphi^* \, dx = \int_0^1 f\varphi^* \, dx + \int_0^1 \delta f\varphi^* \, dx = J_p + \int_0^1 \delta f\varphi^* \, dx$$

$$\int_0^1 p\varphi' \, dx = \int_0^1 p\varphi \, dx + \int_0^1 p\delta\varphi \, dx = J_p + \delta J \tag{2.1.24}$$

where $\delta J = \int_0^1 p\delta\varphi \, dx$, and obtain, as a result, the perturbation theory formula:

$$\delta J = -\,\delta\left(\frac{k\sigma}{\gamma}\right)\bigg|_{x=1} \varphi'(1)\varphi^*(1) + \delta\left(\frac{k\beta}{\alpha}\right)\bigg|_{x=0} \varphi'(0)\varphi^*(0)$$

$$-\int_0^1 \delta q\varphi'\varphi^* + \int_0^1 \delta f\varphi^* dx - \int_0^1 \delta k\frac{d\varphi'}{dx}\frac{d\varphi^*}{dx}dx \tag{2.1.25}$$

Substituting approximately the unperturbed solution φ for the perturbed φ', obtain the small perturbation theory formula

$$\delta J = -\,\delta\left(\frac{k\sigma}{\gamma}\right)\bigg|_{x=1} \varphi(1)\varphi^*(1) + \delta\left(\frac{k\beta}{\alpha}\right)\bigg|_{x=0} \varphi(0)\varphi^*(0)$$

$$+\int_0^1 \delta f\varphi^* - \int_0^1 \delta q\varphi^*\varphi dx - \int_0^1 \delta k\frac{d\varphi}{dx}\frac{d\varphi^*}{dx}dx \tag{2.1.26}$$

Remember that

$$\delta\left(\frac{k\sigma}{\gamma}\right) = \frac{k'\sigma'}{\gamma'} - \frac{k\sigma}{\gamma}, \quad \delta\left(\frac{k\beta}{\alpha}\right) = \frac{k'\beta'}{\alpha'} - \frac{k\beta}{\alpha}$$

Formula (2.1.26) gives an estimate of sensitivity of the functional variation δJ to input variations.

Notice that if boundary conditions in problem (2.1.18) and (2.1.19) are not perturbed, then $\alpha' = \alpha$, $\beta' = \beta$, $\gamma' = \gamma$, $\sigma' = \sigma$, and formula (2.1.26) takes the form

$$\delta J = -\frac{\sigma}{\gamma}\delta k(1)\varphi(1)\varphi^*(1) + \frac{\beta}{\alpha}\delta k(0)\varphi(0)\varphi^*(0)$$

$$+ \int_0^1 \delta f \varphi^* dx - \int_0^1 \delta q \varphi^* \varphi dx - \int_0^1 \delta k \frac{d\varphi}{dx}\frac{d\varphi^*}{dx} dx \quad (2.1.27)$$

Examine an example, assuming unperturbed problem (2.1.1) and (2.1.2) as

$$-d^2\varphi/dx^2 + \varphi = e^x, \quad x \in (0,1) \quad (2.1.28)$$

$$-\frac{d\varphi}{dx} + \varphi = 0 \quad \text{at} \quad x = 0$$

$$\frac{d\varphi}{dx} + \varphi = 0 \quad \text{at} \quad x = 1 \quad (2.1.29)$$

Here $k = q = 1$, $f(x) = e^x$, $\alpha = -1$, $\beta = \gamma = \sigma = 1$. The solution φ of problem (2.1.28) and (2.1.29) is easy to find in explicit form:

$$\varphi(x) = \frac{3}{4}e^x - \frac{1}{4}e^{-x} - \frac{1}{2}xe^x \quad (2.1.30)$$

Assume functional (2.1.4) in the form

$$J_p = \int_0^1 \varphi(x)\, dx \quad (2.1.31)$$

In the given case, $p(x) \equiv 1$ and

$$J_p = \int_0^1 \left(\frac{3}{4}e^x - \frac{1}{4}e^{-x} - \frac{1}{2}xe^x\right) dx = \frac{3}{4}e + \frac{1}{4}e^{-1} - \frac{3}{2}$$

Write problem (2.1.16) and (2.1.17) adjoint to the main problem (2.1.28) and (2.1.29) with regard to functional (2.1.31):

$$-\frac{d^2\varphi^*}{dx^2} + \varphi^* = 1, \quad x \in (0,1) \quad (2.1.32)$$

$$-\frac{d\varphi^*}{dx} + \varphi^* = 0 \quad \text{at} \quad x = 0$$

$$\frac{d\varphi^*}{dx} + \varphi^* = 0 \quad \text{at} \quad x = 1 \quad (2.1.33)$$

The problem's solution is

$$\varphi^*(x) = 1 - \frac{1}{2}e^{x-1} - \frac{1}{2}e^{-x} \qquad (2.1.34)$$

Examine perturbed problem (2.1.18) and (2.1.19) corresponding to (2.1.28) and (2.1.29) in the form

$$-\frac{d}{dx}(1 + \varepsilon e^x)\frac{d\varphi'}{dx} + \varphi' = e^x + \varepsilon e^x + \varepsilon(1 + \varepsilon)xe^{2x}, \quad x \in (0,1) \quad (2.1.35)$$

$$-\frac{d\varphi'}{dx} + \varphi' = 0 \quad \text{at} \quad x = 0$$
$$\frac{d\varphi'}{dx} + \varphi' = 0 \quad \text{at} \quad x = 1 \qquad (2.1.36)$$

It was assumed here that $k' = 1 + \delta k$, $\delta k = \varepsilon e^x$, $f'(x) = f(x) + \delta f$, $\delta f = \varepsilon e^x + \varepsilon(1 + \varepsilon)xe^{2x}$, $\delta q = 0$, $\delta \alpha = \delta \beta = \delta \gamma = \delta \sigma = 0$.

To find the value of functional (2.1.20) use formula (2.1.27)

$$J' = \int_0^1 \varphi'(x)dx = J_p + \delta J$$

of the small perturbation theory and obtain

$$J' \approx J'_{(1)} \equiv J_p - \delta k(1)\varphi(1)\varphi^*(1) - \delta k(0)\varphi(0)\varphi^*(0)$$

$$+ \int_0^1 \delta f \varphi^* \, dx - \int_0^1 \delta k \frac{d\varphi}{dx}\frac{d\varphi^*}{dx} dx \qquad (2.1.37)$$

All the functions in the right-hand part of expression (2.1.37) are known. Therefore formula (2.1.37) serves to find an approximate value of J' at small ε without solving perturbed problem (2.1.35) and (2.1.36). In this particular case it is possible to estimate the error of this formula's application, since the accurate solution of perturbed problem (2.1.35) and (2.1.36) is known; it is

$$\varphi'(x) = \varphi(x) + \varepsilon\varphi_1(x), \qquad (2.1.38)$$

where $\varphi_1 = \frac{3}{4}e^x - \frac{1}{4}e^{-x} - \frac{1}{2}xe^x$. Use formulas (2.1.25) and (2.1.37) to obtain, for the difference $J' - J'_{(1)}$,

$$J' - J'_{(1)} = -\varepsilon^2 k_1(1)\varphi_1(1)\varphi^*(1) - \varepsilon^2 k_1(0)\varphi_1(0)\varphi^*(0)$$

$$- \varepsilon^2 \int_0^1 k_1(x)\frac{d\varphi_1}{dx}\frac{d\varphi^*}{dx}dx \quad (2.1.39)$$

where $k_1(x) = e^x$.

Formula (2.1.37) makes it thus possible to find a value of the functional J' with an accuracy up to $O(\varepsilon^2)$. It is easy to compute the difference $J' - J'_{(1)}$ in an explicit form, using the functions $\varphi_1(x)$, $\varphi^*(x)$, $k_1(x)$, through formula (2.1.39):

$$J' - J'_{(1)} = -\varepsilon^2(10e^2 - 13e^{-1} - 18)/72 \approx -0,71\varepsilon^2$$

2.2. Heat Conductivity Equation

The problem of heat conductivity is a typical nonstationary problem in mathematical physics. Historically, the heat conductivity equation helped state and solve many principal problems of the computation theory and build first-rate algorithms for solution of problems in mathematical physics. We will examine in this Chapter the simple problem concerning propagation of heat in a finite homogeneous bar heated by internal sources:

$$\frac{\partial\varphi}{\partial t} - \frac{\partial}{\partial x}k\frac{\partial\varphi}{\partial x} + q\varphi = f(x,t), \quad x \in (0,1), \quad t \in (0,T] \quad (2.2.1)$$

$$\varphi(0,t) = \varphi(1,t) = 0 \quad (2.2.2)$$

$$\varphi(x,0) = g(x) \quad (2.2.3)$$

Assume $k(x,t)$, $q(x,t)$, $f(x,t)$ to be non-negative piecewise continuous functions of the coordinate x and time t, and $g(x)$ to be a function of x. Statements like this prevail in applications.[1]

Assume that $\Omega = (0,1)$, $\Omega_t = (0,T)$ and the solution φ of problem (2.2.1)–(2.2.3) belongs at each t to a set of functions $D(A)$ with the

---

[1]In fact $q \equiv 0$ as a rule in heat conductivity equations, but in nonstationary particle diffusion processes the coefficient q plays an important role as it describes absorbtion by the medium.

definition domain $\overline{\Omega} = [0,1]$. Every element of the set $D(A)$ is continuous and has a continuous differentiable derivative $k \, \partial\varphi/\partial x$, so that the derivative $\partial/\partial x (k \, \partial\varphi/\partial x)$ is quadratically summable on $\Omega \times \Omega_t$.[2] Besides, all the functions of the set $D(A)$ satisfy the homogeneous boundary conditions (2.2.2). Assume also, that the solution φ of problem (2.2.1)–(2.2.3) has a derivative $\partial\varphi/\partial t$ quadratically summable on $\Omega \times \Omega_t$.

Establish conditions in which the solution φ of problem (2.2.1)–(2.2.3) is unique. Suppose this problem has two solutions φ_1 and φ_2. Examine the difference

$$w = \varphi_1 - \varphi_2$$

Then there is a homogeneous problem for the function w:

$$\frac{\partial w}{\partial t} - \frac{\partial}{\partial x} k \frac{\partial w}{\partial x} + qw = 0 \qquad (2.2.4)$$

$$w(0,t) = w(1,t) = 0 \qquad (2.2.5)$$

$$w(x,0) = 0 \qquad (2.2.6)$$

Multiply equation (2.2.4) by w and integrate on $\Omega \times \Omega_t$ to obtain

$$\int_0^T dt \int_0^1 w \left(\frac{\partial w}{\partial t} - \frac{\partial}{\partial x} k \frac{\partial w}{\partial x} + qw \right) dx = 0 \qquad (2.2.7)$$

Integrate (2.2.7) by parts with regard to boundary conditions (2.2.5) and initial data (2.2.6) into

$$\int_0^T dt \int_0^1 \left[\frac{1}{2} \frac{\partial w^2}{\partial t} + k \left(\frac{\partial w}{\partial x} \right)^2 + qw^2 \right] dx = 0 \qquad (2.2.8)$$

Integrate the first term in (2.2.8) in t to obtain finally

$$\frac{1}{2} \int_0^1 w^2(x,T) dx + \int_0^T dt \int_0^1 \left[k \left(\frac{\partial w}{\partial x} \right)^2 + qw^2 \right] dx = 0 \qquad (2.2.9)$$

Since $k \geq 0$, $q \geq 0$ by the assumption, (2.2.9) holds only if

$$w \equiv 0 \qquad (2.2.10)$$

---

[2] Assume hereinafter that $\Omega \times \Omega_t$ is understood as a set of points x, t with $x \in \Omega$, $t \in \Omega_t$.

This means that $\varphi_1 \equiv \varphi_2$, i. e., the solution of problem (2.2.1)–(2.2.3) is in effect unique.

Formulate now the adjoint problem. Multiply equation (2.2.1) by the function φ^* referred to as the adjoint and integrate the result on $\Omega \times \Omega_t$. Then

$$\int\limits_0^T dt \int\limits_0^1 \varphi^* \left(\frac{\partial \varphi}{\partial t} - \frac{\partial}{\partial x} k \frac{\partial \varphi}{\partial x} + q\varphi \right) dx = \int\limits_0^T dt \int\limits_0^1 f\varphi^* dx \qquad (2.2.11)$$

Transform

$$\int\limits_0^T dt \int\limits_0^1 \varphi^* \frac{\partial \varphi}{\partial t} dx = \int\limits_0^1 \varphi^*(x,T)\varphi(x,T)dx - \int\limits_0^1 \varphi^*(x,0)\varphi(x,0)dx$$

$$- \int\limits_0^T dt \int\limits_0^1 \varphi \frac{\partial \varphi^*}{\partial t} dx \quad (2.2.12)$$

$$\int\limits_0^T dt \int\limits_0^1 \varphi^* \frac{\partial}{\partial x} k \frac{\partial \varphi}{\partial x} dx = \int\limits_0^T \left(k \frac{\partial \varphi}{\partial x} \varphi^* - k \frac{\partial \varphi^*}{\partial x} \varphi \right)\Bigg|_{x=0}^{x=1} dt$$

$$= \int\limits_0^T dt \int\limits_0^1 \varphi \frac{\partial}{\partial x} k \frac{\partial \varphi^*}{\partial x} dx \quad (2.2.13)$$

to bring (2.2.11) to

$$\int\limits_0^T dt \int\limits_0^1 \varphi \left(-\frac{\partial \varphi^*}{\partial t} - \frac{\partial}{\partial x} k \frac{\partial \varphi^*}{\partial x} + q\varphi^* \right) dx$$

$$= \int\limits_0^T dt \int\limits_0^1 f\varphi^* dx - \int\limits_0^1 \varphi^*(x,T)\varphi(x,T)dx + \int\limits_0^1 \varphi^*(x,0)\varphi(x,0)dx$$

$$+ \int\limits_0^T k \frac{\partial \varphi}{\partial x} \varphi^* \Bigg|_{x=0}^{x=1} dt - \int\limits_0^T k \frac{\partial \varphi^*}{\partial x} \varphi \Bigg|_{x=0}^{x=1} dt \quad (2.2.14)$$

Assume

$$-\frac{\partial \varphi^*}{\partial t} - \frac{\partial}{\partial x} k \frac{\partial \varphi^*}{\partial x} + q\varphi^* = p(x,t) \qquad (2.2.15)$$

$$\varphi^*(0,t) = \varphi^*(1,t) = 0 \qquad (2.2.16)$$

$$\varphi^*(x,T) = h(x) \qquad (2.2.17)$$

where $p(x,t)$ and $h(x)$ are still arbitrary functions. Assume also, that the solution φ^* of problem (2.2.15)–(2.2.17) belongs at every t to a set $D(A^*)$ coinciding with $D(A)$. With regard to (2.2.15)–(2.2.17) and boundary and initial conditions for the function φ in (2.2.2), (2.2.3) transform (2.2.14) into

$$\int_0^T dt \int_0^1 p\varphi dx = \int_0^T dt \int_0^1 f\varphi^* dx - \int_0^1 h(x)\varphi(x,T)dx$$

or

$$+ \int_0^1 g(x)\varphi^*(x,0)dx \quad (2.2.18)$$

$$\int_0^T dt \int_0^1 p\varphi dx + \int_0^1 h(x)\varphi(x,T)dx$$

$$= \int_0^T dt \int_0^1 f\varphi^* dx + \int_0^1 g(x)\varphi^*(x,0)dx \quad (2.2.19)$$

Introduce the problem's functional

$$J = \int_0^T dt \int_0^1 p\varphi dx + \int_0^1 h(x)\varphi(x,T)dx \qquad (2.2.20)$$

Then, using (2.2.19) obtain the same value of the functional (2.2.20) though expressed in terms of the solution of the adjoint problem:

$$J = \int_0^T dt \int_0^1 f\varphi^* dx + \int_0^1 g(x)\varphi^*(x,0)dx \qquad (2.2.21)$$

Notice, that the functions $p(x,t)$ and $h(x)$ in the functional J defined by (2.2.20) are determined by measurement characteristic of the instrument. The dual formula of the functional J is noticeably important also because instead of repeated solutions of problem (2.2.1)–(2.2.3) to obtain the values of the functional J as in form (2.2.20) at different $f(x,t)$ and $g(x)$ we may solve once problem (2.2.15)–(2.2.17) to obtain φ^* and then use the functional J expression as in (2.2.21) by changing the input functions f and g.

As concerns the perturbation theory at fixed coefficients $k(x,t)$ and $q(x,t)$ of the problem, assume instead of the functions f and g functions $f + \delta f$ and $g + \delta g$. Then the value of the functional J changes by the value of δJ which is easy to find through formulas (2.2.20) and (2.2.21):

$$\delta J = \int_0^T dt \int_0^1 p\delta\varphi dx + \int_0^1 h(x)\delta\varphi(x,T)dx$$

$$\delta J = \int_0^T dt \int_0^1 \delta f \varphi^* dx + \int_0^1 \delta g(x)\varphi^*(x,0)dx \qquad (2.2.22)$$

where $\delta\varphi$ is the value the solution of the problem (2.2.1)–(2.2.3) changes by. The estimate of the functional δJ sensitivity to variations δf and δg evolves from the second formula in (2.2.22), where the functions $\varphi^*(x,t)$ and $\varphi^*(x,0)$ appear as the weights in the formulas of the perturbation theory.

Now assume as varying not only the input data $f(x,t)$ and $g(x)$, but also the equation coefficients $k(x,t)$ and $q(x,t)$. This makes the problem a perturbed problem analogous to (2.2.1)–(2.2.3):

$$\frac{\partial\varphi'}{\partial t} - \frac{\partial}{\partial x}k'\frac{\partial\varphi'}{\partial x} + q'\varphi' = f' \qquad (2.2.23)$$

$$\varphi'(0,t) = \varphi'(1,t) = 0 \qquad (2.2.24)$$

$$\varphi'(x,0) = g' \qquad (2.2.25)$$

where $k' = k + \delta k$, $q' = q + \delta q$, $f' = f + \delta f$, $g' = g + \delta g$.

Seeking a formula for δJ in this case, remember (see Chapter 1), that adjoint unperturbed problem (2.2.15)–(2.2.17) must join to equations system (2.2.23)–(2.2.25). Multiply equation (2.2.23) by φ^* and integrate all over the domain $\Omega \times \Omega_t$. Multiply also adjoint equation (2.2.15) by φ' and integrate over the domain $\Omega \times \Omega_t$, then subtract the results from each other to obtain

$$\int_0^T dt \int_0^1 \left(\varphi^*\frac{\partial\varphi'}{\partial t} + \varphi'\frac{\partial\varphi^*}{\partial t}\right) dx - \int_0^T dt \int_0^1 \left(\varphi^*\frac{\partial}{\partial x}k'\frac{\partial\varphi'}{\partial x} - \varphi'\frac{\partial}{\partial x}k\frac{\partial\varphi^*}{\partial x}\right) dx$$

$$+ \int_0^T dt \int_0^1 (q'\varphi^*\varphi' - q\varphi^*\varphi')dx = \int_0^T dt \int_0^1 f'\varphi^* dx - \int_0^T dt \int_0^1 p\varphi' dx \quad (2.2.26)$$

Consecutively transform the integrals in (2.2.26) with regard to the boundary conditions and initial data for the functions φ' and φ^*. Then

$$\int_0^T dt \int_0^1 \left(\varphi^* \frac{\partial \varphi'}{\partial t} + \varphi' \frac{\partial \varphi^*}{\partial t} \right) dx = \int_0^1 \varphi' \varphi^* dx \bigg|_{t=0}^{t=T}$$

$$= \int_0^1 h\varphi'(x,T)dx - \int_0^1 g'\varphi^*(x,0)dx, \quad (2.2.27)$$

$$\int_0^T dt \int_0^1 \left(\varphi^* \frac{\partial}{\partial x} k' \frac{\partial \varphi'}{\partial x} - \varphi' \frac{\partial}{\partial x} k \frac{\partial \varphi^*}{\partial x} \right) dx$$

$$= \int_0^T \left[\varphi^* \left(k' \frac{\partial \varphi'}{\partial x} \right) - \varphi' \left(k \frac{\partial \varphi^*}{\partial x} \right) \right] \bigg|_{x=0}^{x=1} dt - \int_0^T dt \int_0^1 \delta k \frac{\partial \varphi'}{\partial x} \frac{\partial \varphi^*}{\partial x} dx$$

$$= - \int_0^T dt \int_0^1 \delta k \frac{\partial \varphi'}{\partial x} \frac{\partial \varphi^*}{\partial x} dx \quad (2.2.28)$$

Substituting (2.2.27) and (2.2.28) into (2.2.26), obtain

$$\int_0^1 h\varphi'(x,T)dx - \int_0^1 g'\varphi^*(x,0)dx + \int_0^T dt \int_0^1 \delta k \frac{\partial \varphi'}{\partial x} \frac{\partial \varphi^*}{\partial x} dx$$

$$+ \int_0^T dt \int_0^1 \delta q \varphi' \varphi^* dx = \int_0^T dt \int_0^1 f' \varphi^* dx - \int_0^T dt \int_0^1 p\varphi' dx \quad (2.2.29)$$

Rewrite this as

$$\int_0^T dt \int_0^1 \delta k \frac{\partial \varphi'}{\partial x} \frac{\partial \varphi^*}{\partial x} dx + \int_0^T dt \int_0^1 \delta q \varphi' \varphi^* dx$$

$$= \left(\int_0^T dt \int_0^1 f' \varphi^* dx + \int_0^1 g' \varphi^*(x,0)dx \right)$$

$$- \left(\int_0^T dt \int_0^1 p\varphi' dx + \int_0^1 h\varphi'(x,T)dx \right) \quad (2.2.30)$$

Transform the bracketed expression in the right-hand part of (2.2.30) into

$$\int\limits_0^T dt \int\limits_0^1 f'\varphi^* dx + \int\limits_0^1 g'\varphi^*(x,0)dx$$

$$= J + \int\limits_0^T dt \int\limits_0^1 \delta f\varphi^* dx + \int\limits_0^1 \delta g\varphi^*(x,0)dx \quad (2.2.31)$$

where J is the examined functional determined by formula (2.2.21), while

$$\int\limits_0^T dt \int\limits_0^1 p\varphi' dx + \int\limits_0^1 h\varphi'(x,T)dx = J' = J + \delta J \quad (2.2.32)$$

where J is determined by formula (2.2.20) and

$$\delta J = \int\limits_0^T dt \int\limits_0^1 p\delta\varphi dx + \int\limits_0^1 h\delta\varphi(x,T)dx$$

With regards to (2.2.31) and (2.2.32), reduce expression (2.2.30) to the formula of the perturbation theory for the functional δJ:

$$\delta J = -\int\limits_0^T dt \int\limits_0^1 \left(\delta k \frac{\partial\varphi'}{\partial x}\frac{\partial\varphi^*}{\partial x}dx + \delta q\varphi'\varphi^* - \delta f\varphi^* \right) dx$$

$$+ \int\limits_0^1 \delta g\varphi^*(x,0)dx \quad (2.2.33)$$

In case of small perturbations, substitute φ for φ' in (2.2.33) and obtain the formula of the small perturbation theory:

$$\delta J = -\int\limits_0^T dt \int\limits_0^1 \left(\delta k \frac{\partial\varphi}{\partial x}\frac{\partial\varphi^*}{\partial x}dx + \delta q\varphi\varphi^* - \delta f\varphi^* \right) dx$$

$$+ \int\limits_0^1 \delta g\varphi^*(x,0)dx \quad (2.2.34)$$

As concerns the choice of the functional J, it depends on particular variations and is fixed by the choice of the functions $p(x,t)$ and $h(x)$. Formula (2.2.34) provides the estimate of sensitivity of the functional to small input data variations.

Examine an example. If $k = 1$, $q = 0$, $f(x,t) = 0$, and $g(x,t) = \sin \pi x$, problem (2.2.1)–(2.2.3) has the form

$$\frac{\partial \varphi}{\partial t} - \frac{\partial^2 \varphi}{\partial x^2} = 0, \quad x \in (0,1), \quad t \in (0,T] \qquad (2.2.35)$$

$$\varphi(0,t) = \varphi(1,t) = 0 \qquad (2.2.36)$$

$$\varphi(x,0) = \sin \pi x \qquad (2.2.37)$$

The solution is

$$\varphi(x,t) = \sin \pi x \cdot e^{-\pi^2 t} \qquad (2.2.38)$$

Examine functional (2.2.20) in the form

$$J = \int_0^1 \sin \pi x \cdot \varphi(x,T) dx \qquad (2.2.39)$$

where $p(x,t) \equiv 0$, and $h(x) = \sin \pi x$. From (2.2.38) and (2.2.39),

$$J = e^{-\pi^2 T} \int_0^1 \sin^2 \pi x dx = \frac{1}{2} e^{-\pi^2 T} \qquad (2.2.40)$$

This functional may be computed by formula (2.2.21) for $f(x,t) \equiv 0$, $g(x) = \sin \pi x$, too:

$$J = \int_0^1 \sin \pi x \cdot \varphi^*(x,0) dx \qquad (2.2.41)$$

where $\varphi^*(x,t)$ is the solution of adjoint problem (2.2.15)–(2.2.17) which then has the form

$$-\frac{\partial \varphi^*}{\partial t} - \frac{\partial^2 \varphi^*}{\partial x^2} = 0, \quad x \in (0,1), \quad t \in [0,T) \qquad (2.2.42)$$

$$\varphi^*(0,t) = \varphi^*(1,t) = 0 \qquad (2.2.43)$$

$$\varphi^*(x,T) = \sin \pi x \qquad (2.2.44)$$

The solution of problem (2.2.42)–(2.2.44) in effect is

$$\varphi^*(x,t) = \sin \pi x \cdot e^{\pi^2(t-T)} \tag{2.2.45}$$

therefore from (2.2.41)

$$J = \int_0^1 \sin^2 \pi x \cdot e^{-\pi^2 T} = \frac{1}{2} e^{-\pi^2 T} \tag{2.2.46}$$

which coincides with (2.2.40).

Assume now that the input data of problem (2.2.35)–(2.2.37) have changed and we are examining perturbed problem (2.2.23)–(2.2.25) at $k' = k = 1$, $q' = q = 0$, $f' = f + \delta f$, $\delta f = \sin \pi x \cos^2 \pi x$, $g'(x) = g(x) + \delta g = \sin \pi x + \delta g$, and $\delta g = x(1-x)$. Then the perturbed problem has the form

$$\frac{\partial \varphi'}{\partial t} - \frac{\partial^2 \varphi'}{\partial x^2} = \sin \pi x \cos^3 \pi x, \quad x \in (0,1), \quad t \in (0,T] \tag{2.2.47}$$

$$\varphi'(0,t) = \varphi'(1,t) = 0 \tag{2.2.48}$$

$$\varphi'(x,0) = \sin \pi x + x(1-x) \tag{2.2.49}$$

The functional

$$J' = \int_0^1 \sin \pi x \cdot \varphi'(x,T) dx \tag{2.2.50}$$

is not easy to find in an explicit form, since $\varphi'(x,t)$ is unknown, but we may use formula (2.2.33) of the perturbation theory at $\delta k = \delta q = 0$, $\delta f = \sin \pi x \cos^3 \pi x$, and $\delta g = x(1-x)$. Then, for J'

$$J' = J + \delta J \tag{2.2.51}$$

where

$$\delta J = \int_0^T dt \int_0^1 \delta f \varphi^*(x,t) dx + \int_0^1 \delta g \varphi^*(x,0) dx \tag{2.2.52}$$

and $\varphi^*(x,t)$ is the solution of unperturbed problem (2.2.42)–(2.2.44). From (2.2.45) and (2.2.52) obtain for δJ

$$\delta J = \int_0^T dt \int_0^1 \sin \pi x \cos^2 \pi x \sin \pi x \cdot e^{\pi^2(t-T)} dx$$

$$+ \int_0^1 x(1-x) \sin \pi x \cdot e^{-\pi^2 T} dx = \left(\int_0^T e^{\pi^2(t-T)} dt \right) \left(\int_0^1 \sin^2 \pi x \cos^2 \pi x dx \right)$$

$$+ e^{-\pi^2 T} \int_0^1 x(1-x) \sin \pi x dx$$

Since

$$\int_0^1 \sin^2 \pi x \cos^2 \pi x dx = \frac{1}{\pi} \int_0^\pi \sin^2 y \cos^3 y dy = \frac{1}{\pi} \left(\frac{\sin^3 y}{3} - \frac{\sin^5 y}{5} \right) \Big|_0^\pi = 0$$

$$\int_0^1 x(1-x) \sin \pi x dx = \frac{2}{\pi^2} \int_0^1 \sin \pi x dx = \frac{4}{\pi^2}$$

then

$$\delta J = \frac{4}{\pi^2} e^{-\pi^2 T}, \tag{2.2.53}$$

$$J' = J + \delta J = \frac{1}{2} e^{-\pi^2 T} + \frac{4}{\pi^2} e^{-\pi^2 T} \tag{2.2.54}$$

Formula (2.2.33) of the perturbation theory thus produced immediately the value of perturbed functional J' through the solution of an unperturbed adjoint problem.

2.3. Oscillation Equation

Equations of hyperbolic type play an important role in applications. Their solution definition domain is noticeably limited by the characteristics cone, so that the part of the domain $\Omega \times \Omega_t$ beyond the cone does not affect a solution at point examined.

Many problems in mechanics (relating, e. g. to oscillations of strings, bars, etc.) are described by the oscillation equation

$$\frac{1}{c^2} \frac{\partial^2 \varphi}{\partial t^2} = \frac{\partial}{\partial x} k \frac{\partial \varphi}{\partial x} - qu + f(x,t), \quad x \in (0,1), \quad t \in (0,T] \tag{2.3.1}$$

$$\varphi(0,t) = \varphi(1,t) = 0 \tag{2.3.2}$$

$$\varphi(x,0) = 0, \quad \frac{\partial \varphi}{\partial t}(x,0) = g(x) \tag{2.3.3}$$

where $c = $ const, the coefficients k and q are determined by medium properties; the free member $f(x,t)$ expresses the intensity of external impact and $g(x)$ is the initial displacement velocity.

Assume $k(x,t)$, $q(x,t)$, $f(x,t)$, and $g(x)$ as non-negative piecewise smooth functions.

The one-dimensional oscillation theory have been thoroughly studied. We will then discuss applications of adjoint equations techniques and perturbation theory to problem (2.3.1)–(2.3.3).

Assume that $\Omega = (0,1)$, $\Omega_t = (0,T)$, and that the solution φ of problem (2.3.1)–(2.3.3) at every t belongs to a set of functions $D(A)$ with a definition domain $\bar{\Omega} = [0,1]$. Every element of the set $D(A)$ is continuous and has continuous differentiable derivative $k\,\partial\varphi/\partial x$, so that the derivative $\partial/\partial x(k\,\partial\varphi/\partial x)$ is quadratically summable on $\Omega \times \Omega_t$. Besides, all the functions of the set $D(A)$ satisfy boundary conditions (2.3.2). Assume also, that the solution φ of problem (2.3.1)–(2.3.3) has derivatives $\partial\varphi/\partial t$, $\partial^2\varphi/\partial t^2$, quadratically summable on $\Omega \times \Omega_t$.

To formulate equations adjoint to problem (2.3.1)–(2.3.3) multiply (2.3.1) by a function $\varphi^*(x,t)$ and integrate the result over $\Omega \times \Omega_t$ to obtain

$$\int_0^T dt \int_0^1 \varphi^* \left(\frac{1}{c^2} \frac{\partial^2 \varphi}{\partial t^2} - \frac{\partial}{\partial x} k \frac{\partial \varphi}{\partial x} + q\varphi \right) dx = \int_0^T dt \int_0^1 f\varphi^* dx \qquad (2.3.4)$$

Integrating by parts, obtain

$$\int_0^T dt \int_0^1 \varphi^* \frac{\partial^2 \varphi}{\partial t^2} dx = \int_0^1 \left(\varphi^* \frac{\partial \varphi}{\partial t} - \varphi \frac{\partial \varphi^*}{\partial t} \right) \Bigg|_{t=0}^{t=T} dx - \int_0^T dt \int_0^1 \varphi \frac{\partial^2 \varphi^*}{\partial t^2} dx \quad (2.3.5)$$

$$\int_0^T dt \int_0^1 \varphi^* \left(\frac{\partial}{\partial x} k \frac{\partial \varphi}{\partial x} \right) dx = \int_0^T \left(k\varphi^* \frac{\partial \varphi}{\partial x} - k\varphi \frac{\partial \varphi^*}{\partial x} \right) \Bigg|_{x=0}^{x=1} dt$$

$$+ \int_0^T dt \int_0^1 \varphi \frac{\partial}{\partial x} k \frac{\partial \varphi^*}{\partial x} dx \quad (2.3.6)$$

Reduce (2.3.4) to

$$\int_0^T dt \int_0^1 \varphi \left(\frac{1}{c^2} \frac{\partial^2 \varphi^*}{\partial t^2} - \frac{\partial}{\partial x} k \frac{\partial \varphi^*}{\partial x} + q\varphi^* \right) dx$$

$$= \int_0^T dt \int_0^1 f\varphi^* dx - \frac{1}{c^2} \int_0^1 \varphi^* \frac{\partial \varphi}{\partial t} \bigg|_{t=0}^{t=T} dx + \frac{1}{c^2} \int_0^1 \varphi \frac{\partial \varphi^*}{\partial t} \bigg|_{t=0}^{t=T} dx$$

$$+ \int_0^T k\varphi^* \frac{\partial \varphi}{\partial x} \bigg|_{x=0}^{x=1} dt - \int_0^T k\varphi \frac{\partial \varphi^*}{\partial x} \bigg|_{x=0}^{x=1} dt \quad (2.3.7)$$

Assume

$$\frac{1}{c^2} \frac{\partial^2 \varphi^*}{\partial t^2} = \frac{\partial}{\partial x} k \frac{\partial \varphi^*}{\partial x} - q\varphi^* + p(x,t), \quad x \in (0,1), \quad t \in [0,T) \quad (2.3.8)$$

$$\varphi^*(0,t) = \varphi^*(1,t) = 0 \quad (2.3.9)$$

$$\varphi^*(x,T) = 0, \quad \frac{\partial \varphi^*}{\partial t}(x,T) = h(x) \quad (2.3.10)$$

where $p(x,t)$ and $h(x)$ are yet arbitrarily assigned functions. With regard to (2.3.8)–(2.3.10) and conditions (2.3.2), (2.3.3) from (2.3.7), obtain

$$\int_0^T dt \int_0^1 p\varphi dx = \int_0^T dt \int_0^1 f\varphi^* dx + \frac{1}{c^2} \int_0^1 \varphi^*(x,0)g(x)dx + \frac{1}{c^2} \int_0^1 \varphi(x,T)h(x)dx$$

or

$$\int_0^T dt \int_0^1 p\varphi dx - \frac{1}{c^2} \int_0^1 \varphi(x,T)h(x)dx$$

$$= \int_0^T dt \int_0^1 f\varphi^* dx + \frac{1}{c^2} \int_0^1 \varphi^*(x,0)g(x)dx \quad (2.3.11)$$

Introduce a functional J of the solution of (2.3.1)–(2.3.3) as

$$J = \int_0^T dt \int_0^1 p\varphi dx - \frac{1}{c^2} \int_0^1 \varphi(x,T)h(x)dx \quad (2.3.12)$$

Use the solution of adjoint problem (2.3.8)–(2.3.10) to derive the same value of the functional: from (2.3.11) and (2.3.12) obtain

$$J = \int\limits_0^T dt \int\limits_0^1 f\varphi^* dx + \frac{1}{c^2} \int\limits_0^1 \varphi^*(x,0)g(x)dx \qquad (2.3.13)$$

where $\varphi^* = \varphi^*(x,t)$ is the solution of adjoint problem (2.3.8)–(2.3.10) which depends on the functions $p(x,t)$ and $h(x)$ determining functional (2.3.12).

As already proven, formula (2.3.13) is more preferable for computations in certain cases (e. g. if repeated solution of problem (2.3.1)–(2.3.3) is required for obtaining values of J as in (2.3.12)).

Examine now the sensitivity of the functional J to input data f and g in problem (2.3.1)–(2.3.3). Assume that f and g are perturbed, and instead of these examine

$$f' = f + \delta f, \quad g' = g + \delta g$$

There is in effect a perturbed problem:

$$\frac{1}{c^2}\frac{\partial^2 \varphi'}{\partial t^2} = \frac{\partial}{\partial x}k\frac{\partial \varphi'}{\partial x} - q\varphi' + f', \quad x \in (0,1), \quad t \in (0,T] \qquad (2.3.14)$$

$$\varphi'(0,t) = \varphi'(1,t) = 0 \qquad (2.3.15)$$

$$\varphi'(x,0) = 0, \quad \frac{\partial \varphi'}{\partial t}(x,0) = g' \qquad (2.3.16)$$

Represent its solution as

$$\varphi'(x,t) = \varphi(x,t) + \delta\varphi(x,t)$$

Then, for variations of the functional in

$$J' = J + \delta J$$

obtain from (2.3.12) and (2.3.13) two equivalent relatioships:

$$\delta J = \int\limits_0^T dt \int\limits_0^1 p\delta\varphi dx - \frac{1}{c^2} \int\limits_0^1 h(x)\delta\varphi(x,T)dx \qquad (2.3.17)$$

$$\delta J = \int\limits_0^T dt \int\limits_0^1 \delta f\varphi^* dx + \frac{1}{c^2} \int\limits_0^1 \delta g\varphi^*(x,0)dx \qquad (2.3.18)$$

Formula (2.3.18) directly links variation of the functional with variations of the input data δf and δg. The variation δJ is computable by this formula without solving perturbed problem (2.3.14)–(2.3.16).

Examine a more general instance of perturbation. Assume not only the input data $f(x,t)$ and $g(x)$ as varying, but also the coefficients $k(x,t)$ and $q(x,t)$, which leads to another still perturbed problem

$$\frac{1}{c^2}\frac{\partial^2 \varphi'}{\partial t^2} = \frac{\partial}{\partial x}k'\frac{\partial \varphi'}{\partial x} - q'\varphi' + f', \quad x \in (0,1), \quad t \in (0,T] \qquad (2.3.19)$$

$$\varphi'(0,t) = \varphi'(1,t) = 0 \qquad (2.3.20)$$

$$\varphi'(x,0) = 0, \quad \frac{\partial \varphi'}{\partial t}(x,0) = g' \qquad (2.3.21)$$

where $k' = k + \delta k$, $q' = q + \delta q$, $f' = f + \delta f$, and $g' = g + \delta g$.

Find a formula for δJ for this case. As already done in § 2.3, join adjoint perturbed problem (2.3.8)–(2.3.10) to equations system (2.3.19)–(2.3.21). Multiply equation (2.3.19) by φ^* and integrate it over the domain $\Omega \times \Omega_t$. Then multiply equation (2.3.8) by φ' and integrate it over the domain $\Omega \times \Omega_t$ too. Subtract the results from each other and obtain

$$\int_0^T dt \int_0^1 \left(\varphi^*\frac{1}{c^2}\frac{\partial^2 \varphi'}{\partial t^2} - \varphi'\frac{1}{c^2}\frac{\partial^2 \varphi^*}{\partial t^2}\right) dx$$

$$-\int_0^T dt \int_0^1 \left(\varphi^*\frac{\partial}{\partial x}k'\frac{\partial \varphi'}{\partial x} - \varphi'\frac{\partial}{\partial x}k\frac{\partial \varphi^*}{\partial x}\right) dx + \int_0^T dt \int_0^1 (q'\varphi^*\varphi' - q\varphi^*\varphi')dx$$

$$= \int_0^T dt \int_0^1 f'\varphi^* dx - \int_0^T dt \int_0^1 p\varphi' dx. \qquad (2.3.22)$$

Transform the integrals in (2.3.22) with regard to boundary conditions and initial data for the functions φ' and φ^* and obtain

$$\int_0^T dt \int_0^1 \left(\varphi^*\frac{1}{c^2}\frac{\partial^2 \varphi'}{\partial t^2} - \varphi'\frac{1}{c^2}\frac{\partial^2 \varphi^*}{\partial t^2}\right) dx = \frac{1}{c^2}\int_0^1 \left(\varphi^*\frac{\partial \varphi'}{\partial t} - \varphi'\frac{\partial \varphi^*}{\partial t}\right)\Bigg|_{t=0}^{t=T} dx$$

$$= -\frac{1}{c^2}\int_0^1 \varphi^*(x,0)g'(x)dx - \frac{1}{c^2}\int_0^1 \varphi'(x,T)h(x)dx \qquad (2.3.23)$$

$$\int\limits_0^T dt \int\limits_0^1 \left(\varphi^* \frac{\partial}{\partial x} k' \frac{\partial \varphi'}{\partial x} - \varphi' \frac{\partial}{\partial x} k \frac{\partial \varphi^*}{\partial x} \right) dx$$

$$= \int\limits_0^T \left\{ \varphi^* \left(k' \frac{\partial \varphi'}{\partial x} \right) - \varphi' \left(k \frac{\partial \varphi^*}{\partial x} \right) \right\} \Big|_{x=0}^{x=1} dt - \int\limits_0^T dt \int\limits_0^1 \delta k \frac{\partial \varphi'}{\partial x} \frac{\partial \varphi^*}{\partial x} dx$$

$$= \int\limits_0^T dt \int\limits_0^1 \delta k \frac{\partial \varphi'}{\partial x} \frac{\partial \varphi^*}{\partial x} dx \quad (2.3.24)$$

Use (2.3.23) and (2.3.24) to transform (2.3.22) into

$$-\frac{1}{c^2} \int\limits_0^1 h(x) \varphi'(x,T) dx - \frac{1}{c^2} \int\limits_0^1 g'(x) \varphi^*(x,0) dx$$

$$+ \int\limits_0^T dt \int\limits_0^1 \delta k \frac{\partial \varphi'}{\partial x} \frac{\partial \varphi^*}{\partial x} dx + \int\limits_0^T dt \int\limits_0^1 \delta q \varphi' \varphi^* dx$$

$$= \int\limits_0^T dt \int\limits_0^1 f' \varphi^* dx - \int\limits_0^T dt \int\limits_0^1 p \varphi' dx \quad (2.3.25)$$

or

$$\int\limits_0^T dt \int\limits_0^1 \delta k \frac{\partial \varphi'}{\partial x} \frac{\partial \varphi^*}{\partial x} dx + \int\limits_0^T dt \int\limits_0^1 \delta q \varphi' \varphi^* dx$$

$$= \left(\int\limits_0^T dt \int\limits_0^1 f' \varphi^* dx + \frac{1}{c^2} \int\limits_0^1 g'(x) \varphi^*(x,0) dx \right)$$

$$- \left(\int\limits_0^T dt \int\limits_0^1 p \varphi' dx + \frac{1}{c^2} \int\limits_0^1 h(x) \varphi'(x,T) dx \right) \quad (2.3.26)$$

By virtue of (2.3.3),

$$\int\limits_0^T dt \int\limits_0^1 f' \varphi^* + \frac{1}{c^2} \int\limits_0^1 g'(x) \varphi^*(x,0) dx$$

$$= J + \int\limits_0^T dt \int\limits_0^1 \delta f \varphi^* dx + \frac{1}{c^2} \int\limits_0^1 \delta g u^*(x,0) dx \quad (2.3.27)$$

Then

$$\int_0^T dt \int_0^1 p\varphi' dx - \int_0^1 h(x)\varphi'(x,T)dx = J' = J + \delta J \qquad (2.3.28)$$

where J is defined by formula (2.3.12) and δJ by (2.3.17).

Substituting (2.3.27) and (2.3.28) into expression (2.3.36) obtain the formula of the perturbation theory

$$\delta J = -\int_0^T dt \int_0^1 \left(\delta k \frac{\partial \varphi'}{\partial x} \frac{\partial \varphi^*}{\partial x} + \delta q \varphi' \varphi^* - \delta f \varphi^* \right) dx$$

$$+\frac{1}{c^2} \int_0^1 \delta g \varphi^*(x,0)dx \qquad (2.3.29)$$

If perturbations are assumed as small, then approximating $\varphi' = \varphi$ in (2.3.29), obtain the formula of the small perturbation theory

$$\delta J = -\int_0^T dt \int_0^1 \left(\delta k \frac{\partial \varphi}{\partial x} \frac{\partial \varphi^*}{\partial x} + \delta q \varphi \varphi^* - \delta f \varphi^* \right) dx$$

$$+\frac{1}{c^2} \int_0^1 \delta g \varphi^*(x,0)dx \qquad (2.3.30)$$

The formula above is used in estimating the sensitivity of the functional J to small variations in input data k, q, f, and g in problem (2.3.1)–(2.3.3).

Examine an example, assuming that $k = 1$, $q = 0$, $f(x,t) = 0$, and $g(x) = \sin \pi x$. Then problem (2.3.1)–(2.3.3) has the form

$$\frac{1}{c^2}\frac{\partial^2 \varphi}{\partial t^2} = \frac{\partial^2 \varphi}{\partial x^2}, \quad x \in (0,1), \quad t \in (0,T] \qquad (2.3.31)$$

$$\varphi(0,t) = \varphi(1,t) = 0 \qquad (2.3.32)$$

$$\varphi(x,0) = 0, \quad \frac{\partial \varphi}{\partial t}(x,0) = \sin \pi x \qquad (2.3.33)$$

It is easy to see that the solution of this problem is a standing sine wave:

$$\varphi(x,t) = \frac{1}{\pi c} \sin \pi c t \sin \pi x \qquad (2.3.34)$$

Introduce a functional of solution $\varphi(x,t)$ as

$$J = \frac{1}{c^2} \int_0^1 \sin \pi x \cdot \varphi(x,T) dx \qquad (2.3.35)$$

The functional coincides with (2.3.12) at $p \equiv 0$, $h(x) = -\sin \pi x$.
    From (2.3.34), (2.3.35)

$$J = \frac{1}{c^2} \int_0^1 \frac{1}{\pi c} \sin \pi x \sin \pi c T \sin \pi x dx = \frac{\sin \pi c T}{2\pi c^3} \qquad (2.3.36)$$

Compute this functional otherwise as was seen above: use the solution of adjoint problem (2.3.8)–(2.3.10) which is then

$$\frac{1}{c^2} \frac{\partial^2 \varphi^*}{\partial t^2} = \frac{\partial^2 \varphi^*}{\partial x^2}, \quad x \in (0,1), \quad t \in [0,T] \qquad (2.3.37)$$

$$\varphi^*(0,t) = \varphi^*(1,t) = 0 \qquad (2.3.38)$$

$$\varphi^*(x,T) = 0, \quad \frac{\partial \varphi^*}{\partial t}(x,T) = \sin \pi x \qquad (2.3.39)$$

The solution of this problem is the function

$$\varphi^*(x,t) = \frac{1}{\pi c} \sin \pi c(T-t) \sin \pi x \qquad (2.3.40)$$

Then, remembering that $f \equiv 0$, $g(x) = \sin \pi x$, obtain, with formula (2.3.13):

$$J = \frac{1}{c^2} \int_0^1 \sin \pi x \cdot \varphi^*(x,0) dx$$

$$= \frac{1}{c^2} \int_0^1 \frac{1}{\pi c} \sin \pi c T \sin^2 \pi x dx = \frac{\sin \pi c T}{2\pi c^3} \qquad (2.3.41)$$

which coincides with (2.3.36).
    Now instead of problem (2.3.31)–(2.3.33) examine a perturbed problem

$$\frac{1}{c^2} \frac{\partial^2 \varphi'}{\partial t^2} = \frac{\partial}{\partial x}(1 + \varepsilon \cos \pi x)\frac{\partial \varphi'}{\partial x} - \varepsilon \sin \pi x \cos \pi x \cdot \varphi'(x,t) \qquad (2.3.42)$$

$$\varphi'(0,t) = \varphi'(1,t) = 0 \qquad (2.3.43)$$

$$\varphi'(x,0) = 0, \quad \frac{\partial \varphi'}{\partial t}(x,0) = \sin \pi x \qquad (2.3.44)$$

where $\varepsilon \in [0,1]$, and assume that sought is the functional

$$J' = \int_0^1 \sin \pi x \cdot \varphi'(x,T)dx \qquad (2.3.45)$$

Failing to solving perturbed problem (2.3.42)–(2.3.44) in an explicit form, use the formulas of the perturbation theory derived above. Problem (2.3.42)–(2.3.44) coincides with (2.3.19)–(2.3.21) if

$$k' = 1 + \delta k = 1 + \varepsilon \cos \pi x, \quad q' = \delta q = \varepsilon \sin \pi x \cos \pi x$$

$$f' = f = 0, \quad g' = g = \sin \pi x$$

Since the input data f and g are not perturbed, i. e., $\delta f = \delta g = 0$, obtain through the perturbation theory formula (2.3.29) the functional

$$J' = J + \delta J \qquad (2.3.46)$$

where $\quad \delta J = -\int_0^1 dt \int_0^1 \left(\delta k \frac{\partial \varphi'}{\partial x} \frac{\partial \varphi^*}{\partial x} + \delta q \varphi' \varphi^* \right) dx \qquad (2.3.47)$

Assuming that ε is small, $\varphi' = \varphi$ approximately in (2.3.47); then obtain the formula of small perturbation theory for this case:

$$\delta J = -\int_0^T dt \int_0^1 \left(\delta k \frac{\partial \varphi}{\partial x} \frac{\partial \varphi^*}{\partial x} + \delta q \varphi \varphi^* \right) dx \qquad (2.3.48)$$

Substitute the values $\delta k = \varepsilon \cos \pi x$, $\delta q = \varepsilon \sin \pi x \cos \pi x$ into (2.3.48) with regard to the functions as $\varphi(x,t)$ and $\varphi^*(x,t)$ in (2.3.34) and (2.3.40). Then

$$\delta J = -\int_0^T dt \int_0^1 \left\{ \varepsilon \cos \pi x \left(\frac{1}{c} \sin \pi ct \cos \pi x \right) \left(\frac{1}{c} \sin \pi c(T - t) \cos \pi x \right) \right.$$

$$\left. + \varepsilon \sin \pi x \cos \pi x \left(\frac{1}{\pi c} \sin \pi ct \sin \pi x \right) \left(\frac{1}{\pi c} \sin \pi c(T - t) \sin \pi x \right) \right\} dx$$

$$= -\frac{\varepsilon}{c^2} \int_0^T \sin \pi ct \sin \pi c(T - t)dt \left(\int_0^1 \cos^3 \pi x dx + \frac{1}{\pi^2} \int_0^1 \sin^2 \pi x \cos \pi x dx \right)$$

then

$$\delta J = 0 \qquad (2.3.49)$$

And thus obtain, from (2.3.46) and (2.2.41) with an accuracy of up to the value of the second order in ε

$$J' \approx J = \frac{\sin \pi c T}{2\pi c^3}$$

Assume now that the solution φ of problem (2.3.42)–(2.3.44) may be represented as a series

$$\varphi' = \varphi + \varepsilon\varphi_1 + \varepsilon^2\varphi_2 + \dots \qquad (2.3.50)$$

and construct a perturbation formula of a higher order of accuracy for the functional J'.

Substitute (2.3.50) into (2.3.46) to obtain

$$J' = J + \varepsilon J_1 + \varepsilon^2 J_2 + \dots \qquad (2.3.51)$$

where

$$J_1 = 0, \quad J_k = \int\limits_0^T dt \int\limits_0^1 \left(\cos \pi x \frac{\partial \varphi_{k-1}}{\partial x} \frac{\partial \varphi^*}{\partial x} + \sin \pi x \cos \pi x \cdot \varphi_{k-1}\varphi^* \right) dx$$

$$k = 2, 3, \dots$$

To formulate equations for φ_k, substitute (2.3.50) into (2.3.42)–(2.3.44) and equate terms at the same powers of ε, obtaining a system

$$\frac{1}{c^2}\frac{\partial^2 \varphi_k}{\partial t^2} = \frac{\partial^2 \varphi_k}{\partial x^2} + \frac{\partial}{\partial x}\cos \pi x \frac{\partial \varphi_{k-1}}{\partial x} - \sin \pi x \cos \pi x \cdot \varphi_{k-1}(x,t) \quad (2.3.52)$$

$$\varphi_k(0,t) = \varphi_k(1,t) = 0 \qquad (2.3.53)$$

$$\varphi_k(x,0) = \frac{\partial \varphi_k}{\partial t}(x,0) = 0 \qquad (2.3.54)$$

where $k = 1, 2, \dots, \varphi_0 = \varphi$.

Solving consecutively problems (2.3.52)–(2.3.54) and substituting φ_k into (2.3.45), find the corrections J_1, J_2, J_3, $\dots$ and so determine J' with a higher order of accuracy.

It is easy to see that substituting (2.3.51) immediately into (2.3.45) will produce, by virtue of the adjointment relationship, the same formulas for J_1, J_2, J_3, $\dots$.

CHAPTER 3

Nonlinear Equations

Estimating model sensitivity to problem's functionals on the basis of the perturbation theory led researchers to study nonlinear equations now increasingly wider applied. The formal Lagrange identity cannot however cover directly nonlinear operators. This was crucial in the development of the theory which proposed to linearize initial problems prior to the application of the Lagrange identity. If the problem is correctly stated and its solutions continuously depend on its parameters and input data, linearization may be usable in analysis of nonlinear problems in many cases.

Many other cases, however, are more complex and linearization may produce gross errors, particularly in non-stationary nonlinear problems. These will require application of the theory of nonlinear processes with no simplifications whatsoever. We will therefore generalize the adjoint problems theory and the perturbation theory on the basis of a specifically generalized Lagrange identity.

This author introduced equations adjoint to quasi-linear equations studying hydrodynamic equations of atmospheric processes in 1974 [76] and recently examined, together with V. I. Agoshkov, general questions of the theory and methods of constructing the perturbation theory, discussed in detail in [95]. We will discuss here the methodological aspect.

3.1. Nonlinear Equations and Adjoint Problems

Examine the problem

$$-\nu\frac{d^2u}{dx^2} + u\frac{du}{dx} = f(x), \quad 0 < x < 1 \tag{3.1.1}$$

$$u(0) = u(1) = 0 \tag{3.1.2}$$

Assume that the solution u of this problem is continuous and twice differentiable in the domain $0 \leq x \leq 1$, the coefficient ν is constant, and the function $f(x)$ is integrable with square in the domain $0 \leq x \leq 1$, i. e.,

$$\int\limits_0^1 f^2(x)dx \leq c < \infty$$

and is therefore an element of the real Hilbert space $H = L_2(0,1)$. Introduce the inner product for the functions $v, \; w \in H$ conventionally:

$$(v, w) = \int\limits_0^1 vw\,dx$$

Introduce for the functions u a set $D(A) \subset H$. Every element of the set is continuous, twice differentiable, and equals zero at the boundaries of the domain $0 \leq x \leq 1$. Write problem (3.1.1) and (3.1.2) on functions from $D(A)$ formally as

$$A(u)u = f \tag{3.1.3}$$

where $A(u)$ is the operator

$$A(u) = -\nu\frac{d^2}{dx^2} + u\frac{d}{dx} \tag{3.1.4}$$

operating in H with the definition domain $D(A)$. Notice that the operator $A(u)$ distinctly differs from that in a linear case by its dependence on the solution of problem (3.1.1) and (3.1.2) or, which is the same, (3.1.3). Therefore the construction of an operator adjoint in the Lagrange sense to (3.1.3) must be different, since adjoint operators, strictly speaking, are defined only for linear operators independent of the solution u.

However, if we assume that the solution of initial problem is determined and the function u can be regarded as known, the operator becomes also known as we construct the adjoint problem, and we can determine the adjoint operator conventionally using the Lagrange identity. Assume two elements: $v \in D(A)$ and $w \in D(A^*)$, where $D(A^*)$ is the definition domain of the adjoint operator; properties of this domain may be found from requirements of the Lagrange identity to be valid. The relationship is then

$$(A(u)v, w) = (v, A^*(u)w), \quad v \in D(A), \quad w \in D(A^*) \qquad (3.1.5)$$

Let us see, which conditions identity (3.1.5) requires of operator (3.1.4) to hold. Examine the left-hand part of identity (3.1.5). It is

$$\int_0^1 \left(u\frac{dv}{dx} - \nu\frac{d^2v}{dx^2} \right) w\, dx = uvw \Big|_{x=0}^{x=1} - \nu\frac{dv}{dx}w \Big|_{x=0}^{x=1}$$

$$- \int_0^1 v\frac{duw}{dx}dx + \nu \int_0^1 \frac{dv}{dx}\frac{dw}{dx}dx \qquad (3.1.6)$$

This is the effect of integration by parts. The first term outside the integral in the right-hand part of (3.1.6) turns to zero, since $v \in D(A)$ and every element of the set $D(A)$ is continuous, twice differentiable and turns to zero at the boundary of the domain at $x = 0$ and $x = 1$. Turn the second term outside the integral to zero assuming that

$$w(0) = w(1) = 0 \qquad (3.1.7)$$

Then

$$\int_0^1 \left(u\frac{dv}{dx} - \nu\frac{d^2v}{dx^2} \right) w\, dx = \int_0^1 v\frac{duw}{dx}dx + \nu \int_0^1 \frac{dv}{dx}\frac{dw}{dx}dx$$

$$- \int_0^1 v\frac{duw}{dx}dx + \nu v\frac{dw}{dx}\Big|_{x=0}^{x=1} - \nu \int_0^1 v\frac{d^2w}{dx^2}dx$$

$$= \int_0^1 v\left(-\frac{duw}{dx} - \nu\frac{d^2w}{dx^2} \right) dx \qquad (3.1.8)$$

The terms outside the integrals in (3.1.8) turn to zero by integrating by parts, since $v(x) \in D(A)$ and therefore $v(0) = v(1) = 0$.

Transformations which deriving (3.1.8) involved, require not only quadratic summability of the functions $w(x)$ on $(0,1)$ but also that of its derivatives dw/dx, and d^2w/dx^2, i. e.,

$$\int_0^1 \left\{ \left(\frac{d^2w}{dx^2}\right)^2 + \left(\frac{dw}{dx}\right)^2 + w^2(x) \right\} dx < +\infty$$

Adding boundary condition (3.1.7) for the function $w(x)$ to these properties, obtain the set $D(A^*)$. Introduce[1]

$$A^*(u)\cdot = -\nu\frac{d^2\cdot}{dx^2} - \frac{d}{dx}u\cdot \qquad (3.1.9)$$

Then (3.1.8) becomes

$$\int_0^1 \left(u\frac{dv}{dx} - \nu\frac{d^2v}{dx^2} \right) wdx = \int_0^1 v \left(-\frac{duw}{dx} - \nu\frac{d^2w}{dx^2} \right) dx \qquad (3.1.10)$$

or, formally, the Lagrange identity (3.1.5). Notably, the adjoint operator $A^*(u)$ defined by equality (3.1.9) operates in H with the definition domain $D(A^*)$.

Assuming that $u(x)$ is a known function of solution of the main problem, we may thus construct an adjoint operator satisfying the Lagrange identity.

Since we may now regard an adjoint operator as existing, we may now proceed to a more important issue: construction of adjoint problems linked to assigned functionals of the solution $u(x)$. Assume therefore that sought is not actually the solution $u(x)$, but rather its certain functional,

$$J = \int_0^1 pudx = (p, u) \qquad (3.1.11)$$

where $p(x)$ is an assigned characteristic of the measurement of the field $u(x)$. Assume that $p \in H = L_2(0, 1)$.

Then introduce formally the following adjoint problem:

$$-\nu\frac{d^2u^*}{dx^2} - \frac{duu^*}{dx} = p(x), \quad x \in (0, 1) \qquad (3.1.12)$$

---

[1]The symbol $(\cdot)$ denotes hereafter a function affected by an operator.

with conditions

$$u^*(0) = 0, \quad u^*(1) = 0 \tag{3.1.13}$$

Assume that $u^* \in D(A^*)$. Now transform: multiply equation (3.1.1) by u^* and equation (3.1.12) by u and subtract the results from each other to obtain

$$\int_0^1 u^* \left(u\frac{du}{dx} - \nu\frac{d^2u}{dx^2} \right) dx - \int_0^1 u \left(-\frac{duu^*}{dx} - \nu\frac{d^2u^*}{dx^2} \right) dx$$

$$= \int_0^1 fu^* dx - \int_0^1 pu\, dx \tag{3.1.14}$$

The left-hand part of relationship (3.1.14) turns to zero according to the Lagrange identity. It is easy to establish using (3.1.10) where u is chosen as v and solution u^* as w. Then from (3.1.14)

$$\int_0^1 fu^* dx = \int_0^1 pu\, dx \tag{3.1.15}$$

Or, with regard to (3.1.11), obtain duality formulas

$$J = \int_0^1 pu\, dx \tag{3.1.16}$$

$$J = \int_0^1 fu^* dx \tag{3.1.17}$$

The sought value of the functional J may be thus obtained by solving either the main or the adjoint problem.

Examine an example, where $f(x) = \pi^2 \sin \pi x + \frac{\pi}{2} \sin \pi x$, $\nu = 1$. Then the main problem, (3.1.1) and (3.1.2), has the form

$$-\frac{d^2u}{dx^2} + u\frac{du}{dx} = \pi^2 \sin \pi x + \frac{\pi}{2} \sin \pi x, \quad x \in (0,1) \tag{3.1.18}$$

$$u(0) = u(1) = 0 \tag{3.1.19}$$

It is easy to see, that the solution of this problem is

$$u(x) = \sin \pi x \tag{3.1.20}$$

Examine functional (3.1.11) at

$$p(x) = \begin{cases} 1, & x \in [0, 1/3] \\ 0, & x \notin [0, 1/3] \end{cases} \qquad (3.1.21)$$

Then

$$J = \int_0^1 pu\,dx = \int_0^{1/3} u(x)\,dx \qquad (3.1.22)$$

By virtue of (3.1.20), obtain from (3.1.22)

$$J = -\frac{1}{\pi}\left(\cos\frac{\pi}{3} - \cos 0\right) = \frac{1}{\pi}\left(1 - \frac{1}{2}\right) = \frac{1}{2\pi} \approx 0.159155 \qquad (3.1.23)$$

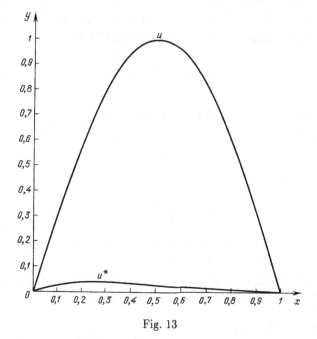

Fig. 13

This very functional, as has been shown, may be computed otherwise, using formula (3.1.17):

$$J = \int_0^1 f(x)u^*(x)\,dx \qquad (3.1.24)$$

where $u^*(x)$ is the solution of adjoint problem (3.1.12) and (3.1.13), which in this case has the form:

$$-\frac{d^2u^*}{dx^2} - \sin\pi x\frac{du^*}{dx} - \pi\cos\pi x u^* = p(x), \quad x \in (0,1) \qquad (3.1.25)$$

$$u^*(0) = u^*(1) = 0 \qquad (3.1.26)$$

The solution of this problem was found numerically by the finite element method involving piecewise linear basis functions on a uniform grid $x_i = ih$, where $i = 0, \ldots, N$, $h = 10^{-2}$. Fig. 13 shows the diagram of the function $u^*(x)$. Computing J from (3.1.24) using the Simpson rule, obtain

$$J \approx 0.159150 \qquad (3.1.27)$$

which coincides with (3.1.23) with an accuracy up to the fifth digit after the decimal point.

3.2. General Formulation of the Adjoint Problem

Examine now a general main nonlinear problem

$$A(u)u = f \qquad (3.2.1)$$

where for found solution u, $A(u)$ is a linear operator operating in a Hilbert space H with the definition domain $D(A)$, $u \in D(A)$, and $f \in H$. Notice, that the functions within $D(A)$ satisfy homogeneous boundary conditions typical of this problem.[2]

Examine next a functional of the solution u:

$$J = (u, p) \qquad (3.2.2)$$

where the function $p \in H$ is a characteristic of measurement variation of the field u, and $(\cdot, \cdot)$ is an inner product defined in H.

Remembering that we seek functional (3.2.2), formulate a problem adjoint to the main, (3.2.1)

$$A^*(u)u^* = p \qquad (3.2.3)$$

---

[2]If the boundary conditions are not homogeneous in the initial problem, they can be reduced to the homogeneous ones, as was noted earlier, by appropriate replacement of the function, the solution of the problem.

where the adjoint operator $A^*(u)$ satisfies the Lagrange identity (3.1.5):

$$(A(u)v, w) = (v, A^*(u)w), \quad v \in D(A), \quad w \in D(A^*) \qquad (3.2.4)$$

Notice, that the adjoint operator in (3.2.3) depends on the solution of the main problem. Transforming the left-hand part of (3.2.4) into the right-hand part of the Lagrange identity, (3.2.4), we also define the set of functions $w \in D(A^*)$ and their smoothness and verify if they meet the boundary conditions. This set will be the definition domain of the adjoint operator $A^*(u)$.

Now do the transformation: multiply the main equation (3.2.1) by u^* in H, the adjoint equation (3.2.3) by u and subtract the results from each other to obtain

$$(A(u)u, u^*) - (u, A^*(u)u^*) = (f, u^*) - (u, p) \qquad (3.2.5)$$

The left-hand part of this relationship turns to zero by virtue of the Lagrange identity (3.2.4); then

$$(f, u^*) = (u, p) \qquad (3.2.6)$$

Since the right-hand part of (3.2.6) has the functional J, obtain duality formulas

$$J = (u, p) \qquad (3.2.7)$$
$$J = (f, u^*) \qquad (3.2.8)$$

3.3. The Small Perturbation Theory

Examine now the small perturbation theory for the functionals of nonlinear problems.

Assume that at $f = f_0 \in H$ the initial problem, (3.2.1), has a solution $u_0 \in D(A)$:

$$A(u_0)u_0 = f_0 \qquad (3.3.1)$$

Refer to this problem as the unperturbed. Assume next that problem (3.3.1) is replaced by a perturbed problem

$$A(u)u = f \qquad (3.3.2)$$

where

$$f = f_0 + \varepsilon f_1, \quad f_1 \in H \tag{3.3.3}$$

ε is a small parameter.

Assume also that εf_1 is small in comparison to f_0:

$$\varepsilon \|f_1\| \ll \|f_0\| \tag{3.3.4}$$

where $\|f_i\| = (f_i, f_i)^{1/2}$, i. e., perturbations in the sources are small.

Assume that solution of perturbed problem (3.3.2) may be represented as a series

$$u = u_0 + \varepsilon u_1 + \dots \tag{3.3.5}$$

and substitute (3.3.3) and (3.3.5) into (3.3.2) to obtain

$$A(u_0 + \varepsilon u_1 + \dots)(u_0 + \varepsilon u_1 + \dots) = f_0 + \varepsilon f_1 \tag{3.3.6}$$

Expand the operator $A(u)$ assuming that it is sufficiently smooth, in the neighborhood of the element $u = u_0$, restricting consideration to members of the first order of smallness. Obtain then

$$A(u_0 + \varepsilon u_1 + \dots) = A(u_0) + \varepsilon \frac{dA(u_0)}{du_0} u_1 + O(\varepsilon^2) \tag{3.3.7}$$

where $\dfrac{dA(u_0)}{du_0} u_1 \equiv F(u_0, u_1)$ is an operator operating from H into H and linear in u_1. Substituting (3.3.7) into (3.3.6) and equating the terms at the same powers of parameter ε, obtain

$$A(u_0)u_0 = f_0, \quad A(u_0)u_1 + F(u_0, u_1)u_0 = f_1 \tag{3.3.8}$$

or

$$A(u_0)u_0 = f_0, \quad A(u_0)u_1 + \left(\frac{dA(u_0)}{du_0} u_1 \right) u_0 = f_1 \tag{3.3.9}$$

We have thus obtained, along with the main unperturbed equation, (3.3.1), an equation for the first variation. Solving this equation using (3.3.5) we obtain an approximate solution of the perturbed problem.

Now let us formulate the perturbation theory for the functional J in our problem. To do so, introduce the notations

$$J_0 = (u_0, p), \quad J_1 = (u_1, p) \tag{3.3.10}$$

where

$$J \approx J_0 + \varepsilon J_1 \qquad (3.3.11)$$

Introduce the linear operator

$$A_1(u_0) \cdot = A(u_0) \cdot + F(u_0, \cdot)u_0 \equiv A(u_0) \cdot + \left(\frac{dA(u_0)}{du_0} \cdot\right) u_0 \qquad (3.3.12)$$

operating from H into H. Then rewrite the equation for u_1 from (3.3.8) as

$$A_1(u_0)u_1 = f_1 \qquad (3.3.13)$$

Examine the adjoint equation

$$A_1^*(u_0)u_1^* = p \qquad (3.3.14)$$

where $A_1^*(u_0)$ is the operator adjoint to $A_1(u_0)$, p is the element defining the functional J_0. Multiply equation (3.3.13) by u_1^* in H and equation (3.3.14) by u_1 and subtract the results from each other to obtain

$$(A_1(u_0)u_1, u_1^*) - (u_1, A_1^*(u_0)u_1^*) = (f_1, u_1^*) - (u_1, p) \qquad (3.3.15)$$

Since the operators $A_1(u_0)$ and $A_1^*(u_0)$ are self-adjoint, they satisfy the Lagrange identity and the left-hand part of (3.3.15) turns therefore to zero. Then

$$(f_1, u_1^*) - (u_1, p) = 0$$

Whence, with regard to the second expression in (3.3.10), obtain two equivalent formulas:

$$J_1 = (u_1, p) \qquad (3.3.16)$$

$$J_1 = (f_1, u_1^*) \qquad (3.3.17)$$

The formulas obtained in (3.3.16), (3.3.17) are formulas of the small perturbation theory. Formula (3.3.17) is especially important in applications, since it links immediately functional variations with variation in the function f i. e., $\delta f = \varepsilon f_1$ and may be used many times for assessments, provided the solution u_1^* remains unchanged. All the research is certainly carried out in the conditions of the small perturbation theory.

Solving nonlinear problems to calculate variations of the functionals with small perturbations in f in the right-hand parts involves consequent solution of two problems. The first to be solved is the unperturbed nonlinear problem

$$A(u_0)u_0 = f_0 \tag{3.3.18}$$

and the operator

$$A_1(u_0)\cdot = A(u_0)\cdot + \left(\frac{dA(u_0)}{du_0}\cdot\right)u_0 \tag{3.3.19}$$

is found; then comes the solution of the adjoint problem

$$A_1^*(u_0)u_1^* = p \tag{3.3.20}$$

This done, compute the variation in the functional using the formula

$$\delta J = \varepsilon(f_1, u_1^*) \tag{3.3.21}$$

If the parameter ε is not separated explicitly in f, but the function f_1 is small in a certain sense as compared with f, we may assume $\varepsilon = 1$ everywhere.

Examine as an example the simple stationary problem (3.1.1) and (3.1.2). Assume that at $f = f_0$ this problem have the solution $u_0 \in D(A)$:

$$-\nu\frac{d^2u_0}{dx^2} + u_0\frac{du_0}{dx} = f_0(x), \quad x \in (0,1) \tag{3.3.22}$$

$$u_0(0) = u_0(1) = 0 \tag{3.3.23}$$

Assume now that problem (3.3.22) and (3.3.23) is substituted for a perturbed problem:

$$-\nu\frac{d^2u}{dx^2} + u\frac{du}{dx} = f(x), \quad x \in (0,1) \tag{3.3.24}$$

$$u(0) = u(1) = 0 \tag{3.3.25}$$

where $f(x) = f_0(x) + \varepsilon f_1(x)$, $f_1(x)$ is an assigned function, $\varepsilon \in [0,1]$. The operator $A(u)$ of (3.3.1) and (3.3.2) has then the form

$$A(u)\cdot = -\nu\frac{d^2\cdot}{dx^2} + u\frac{d\cdot}{dx} \tag{3.3.26}$$

It operates in $H = L_2(0,1)$ with the definition domain $D(A)$ introduced in § 3.1.

Assuming that the function εf_1 is small as compared with f_0, find the solution u of perturbed problem (3.3.24) and (3.3.25):

$$u = u_0 + \varepsilon u_1 + \varepsilon^2 u_2 + \ldots \qquad (3.3.27)$$

Then equality (3.3.6) holds and has, in this case, the form

$$-\nu \frac{d^2}{dx^2}(u_0 + \varepsilon u_1 + \ldots)$$

$$+ (u_0 + \varepsilon u_1 + \ldots)\frac{d}{dx}(u_0 + \varepsilon u_1 + \ldots) = f_0 + \varepsilon f_1 \qquad (3.3.28)$$

$$u_i(0) = u_i(1) = 0, \quad i = 0, 1, 2, \cdots \qquad (3.3.29)$$

Exclude the parentheses in the left-hand part of (3.3.28) and group together the terms at the same power of ε. Then, leaving out small terms beyond the first order in ε, obtain problems

$$-\nu \frac{d^2 u_0}{dx^2} + u_0 \frac{du_0}{dx} = f_0, \quad u_0(0) = u_0(1) = 0$$

$$\qquad (3.3.30)$$

$$-\nu \frac{d^2 u_1}{dx^2} + u_0 \frac{du_1}{dx} + \frac{du_0}{dx}u_1 = f_1, \quad u_1(0) = u_1(1) = 0$$

We have thus obtained, along with the main unperturbed problem (3.3.22) and (3.3.23) an equation for the first variation u_1. System (3.3.30) happens to be system (3.3.9) where the operator $F(u_0, u_1) = \dfrac{dA(u_0)}{du_0}u_1$ has the form

$$F(u_0, u_1) = \frac{dA(u_0)}{du_0}u_1 = u_1 \frac{d}{dx} \qquad (3.3.31)$$

Knowing the solutions u_0, u_1 of equations (3.3.30), we find an approximate solution of perturbed problem (3.3.24) and (3.3.25) using the formula

$$u \approx u_0 + \varepsilon u_1 \qquad (3.3.32)$$

Examine now the perturbation theory for the functional J in our problem. Assume this functional, like in § 3.1 as (see (3.1.11))

$$J = (u, p) = \int_0^1 u p\, dx \qquad (3.3.33)$$

where $p(x)$ is a given function.

Having obtained the solutions u_0, u_1 to problems (3.3.30), we may now determine an approximate value of the functional J using the formula

$$J \approx J_0 + \varepsilon J_1 \qquad (3.3.34)$$

where

$$J_0 = \int_0^1 u_0 p\, dx, \quad J_1 = \int_0^1 u_1 p\, dx$$

As seen earlier, linking a functional variation with variation in the source f requires application of another formula for J_1, namely (3.3.17):

$$J_1 = \int_0^1 f_1 u_1^* dx \qquad (3.3.35)$$

where u_1^* is the solution of adjoint problem (3.3.14).

The operator $A_1^*(u_0)$ in equation (3.3.14) is adjoint to the operator $A_1(u_0)$. From (3.3.12), (3.3.26) and (3.3.31) obtain

$$A_1(u_0) = -\nu \frac{d^2 \cdot}{dx^2} + u_0 \frac{d\cdot}{dx} + \frac{du_0}{dx} \cdot \equiv -\nu \frac{d^2 \cdot}{dx^2} + \frac{d(u_0 \cdot)}{dx} \qquad (3.3.36)$$

This operator operates in H with the definition domain $D(A)$. Verify easily that the technique as discussed in § 3.1 produces

$$A_1^*(u_0) \cdot = -\nu \frac{d^2 \cdot}{dx^2} - u_0 \frac{d\cdot}{dx} \qquad (3.3.37)$$

The adjoint operator $A_1^*(u_0)$ operates in H with the definition domain $D(A^*)$ as introduced in § 3.1. Then problem (3.3.14) takes the form

$$-\nu \frac{d^2 u_1^*}{dx^2} - u_0 \frac{du_1^*}{dx} = p \qquad (3.3.38)$$

$$u_1^*(0) = u_1^*(1) = 0 \qquad (3.3.39)$$

With the solution u_1^* to adjoint problem (3.3.38) and (3.3.39) known, determine the variation J_1 using formula (3.3.35) which links immediately variations in the functional with variation in the function f.

Examine a numerical example. Assume that $f_0(x) = \pi^2 \sin \pi x + \frac{\pi}{2} \sin 2\pi x$, $\nu = 1$, then the unperturbed problem, (3.3.22) and (3.3.23) has the form

$$-\frac{d^2 u_0}{dx^2} + u_0 \frac{du_0}{dx} = \pi^2 \sin \pi x + \frac{\pi}{2} \sin 2\pi x, \quad x \in (0,1) \qquad (3.3.40)$$

$$u_0(0) = u_0(1) = 0 \qquad (3.3.41)$$

As discussed in § 3.1 the solution to this problem is

$$u_0(x) = \sin \pi x$$

Examine the functional

$$J_0 = \int_0^1 p u_0 dx \qquad (3.3.42)$$

where

$$p(x) = \begin{cases} 1, & x \in [2/3, 1] \\ 0, & x \notin [2/3, 1] \end{cases} \qquad (3.3.43)$$

Obtain from (3.1.20) easily that

$$J_0 = \int_{2/3}^1 u_0(x) dx = -\frac{1}{\pi} \cos \pi x \Big|_{2/3}^1$$

$$= -\frac{1}{\pi}\left(-1 + \frac{1}{2}\right) = \frac{1}{\pi} \approx 0.159155 \qquad (3.3.44)$$

As an example of a perturbed problem consider problem (3.3.24) and (3.3.25) at $f = f_0 + \varepsilon f_1$, and $f_1 \equiv 1$ i. e., the problem

$$-\frac{d^2 u}{dx^2} + u \frac{du}{dx} = \pi^2 \sin \pi x + \frac{\pi}{2} \sin 2\pi x + \varepsilon \qquad (3.3.45)$$

$$u(0) = u(1) = 0 \qquad (3.3.46)$$

If we seek to calculate functional (3.3.33):

$$J = \int_0^1 p u \, dx \qquad (3.3.47)$$

employ formula (3.3.34):

$$J \approx J_0 + \varepsilon J_1 \qquad (3.3.48)$$

where J_1 is defined by formula (3.3.35):

$$J_1 = \int_0^1 f_1 u_1^* dx \qquad (3.3.49)$$

Since $f_1 \equiv 1$,

$$J_1 = \int_0^1 u_1^*(x) dx \qquad (3.3.50)$$

i. e., just compute the integral of the function u_1^* which happens to be the solution to adjoint problem (3.3.38) and (3.3.39). This problem has the form

$$-\frac{d^2 u_1^*}{dx^2} - \sin \pi x \frac{du_1^*}{dx} = p \qquad (3.3.51)$$

$$u_1^*(0) = u_1^*(1) = 0 \qquad (3.3.52)$$

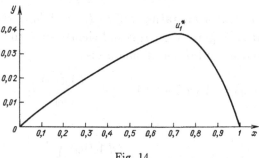

Fig. 14

Problem (3.3.51) and (3.3.52) was solved numerically by the finite element method using piecewise linear basic functions on a uniform grid $x_i = ih$ where $i = 0, \ldots, N$, and $h = 10^{-2}$. Fig. 14 shows the diagram of the function $u_1^*(x)$. Knowing $u_1^*(x)$, J_1 may be obtained using formula (3.3.50). Applying the Simpson rule, obtain

$$J_1 \approx 0.022875 \qquad (3.3.53)$$

Then from (3.3.44) and (3.3.48) finally obtain

$$J \approx 0.159155 + \varepsilon 0.022875$$

3.4. Case of Problem with Perturbed Operator

Let us now examine a more common case where not only the function f varies, but also the operator A does as input parameters or other input data alter. The initial unperturbed problem may be

$$A_0(u_0)u_0 = f_0 \tag{3.4.1}$$

where $A_0(u_0)$ is a linear operator operating within H with the definition domain $D(A_0)$, $u_0 \in D(A_0)$, and $f_0 \in H$.

Examine now the perturbed problem

$$A(u)u = f \tag{3.4.2}$$

where
$$\begin{aligned} f &= f_0 + \varepsilon f_1 \\ u &= u_0 + \varepsilon u_1 + \ldots \\ A(u) &= A_0(u) + \varepsilon \delta A(u) \end{aligned} \tag{3.4.3}$$

$\delta A(u)$ is a given operator operating within H, $D(\delta A) = D(A_0)$.

Assuming that the operator $A(u)$ is sufficiently smooth, obtain with accuracy up to the second order of smallness

$$A(u) \approx A_0(u_0) + \varepsilon \delta A(u_0) + \varepsilon \frac{dA_0(u_0)}{du_0} u_1 \tag{3.4.4}$$

Assuming that

$$A_1(u_0) \cdot = A_0(u_0) \cdot + \left(\frac{dA_0(u_0)}{du_0} \cdot \right) u_0 \tag{3.4.5}$$

substitute (3.4.4) and (3.4.3) into (3.4.2) and equate the terms at the same powers of ε to obtain for u_0 and u_1:

$$A_0(u_0)u_0 = f_0, \quad A_1(u_0)u_1 + \delta A(u_0)u_0 = f_1$$

This is thus an equation for the first variation u_1, obtained together with the main unperturbed equation, (3.4.1). By solving this equation, an approximate solution of perturbed problem (3.4.2) may be found with the formula

$$u \approx u_0 + \varepsilon u_1$$

Now formulate the perturbations theory for the functional J of the problem. Assume still, that

$$J_0 = (u_0, p), \quad J_1 = (u_1, p)$$

Then

$$J \approx J_0 + \delta J = J_0 + \varepsilon J_1$$

Examine now two problems:

$$A_1(u_0)u_1 + \delta A(u_0)u_0 = f_1 \tag{3.4.6}$$

$$A_1^*(u_0)u_1^* = p \tag{3.4.7}$$

Multiply equation (3.4.6) by u_1^* in H and equation (3.4.7) by u_1, subtract the results from each other and apply the Lagrange identity to obtain

$$(\delta A(u_0)u_0, u_1^*) = (f_1, u_1^*) - (u_1, p) \tag{3.4.8}$$

Then, given that $(u_1, p) = J_1$, obtain the formula of the perturbation theory

$$J_1 = (f_1, u_1^*) - (\delta A(u_0)u_0, u_1^*) \tag{3.4.9}$$

or

$$\delta J = \varepsilon(f_1 - \delta A(u_0)u_0, u_1^*) \tag{3.4.10}$$

Remember that

$$\delta J = \varepsilon J_1 \tag{3.4.11}$$

The small perturbation theory formula, (3.4.10), allows not only for variation of the function f, but also for that of the operator A.

Examine as an example another stationary problem (3.1.1) and (3.1.2). The unperturbed problem may have the form

$$-\nu_0 \frac{d^2 u_0}{dx^2} + u_0 \frac{du_0}{dx} = f_0(x), \quad x \in (0, 1) \tag{3.4.12}$$

$$u_0(0) = u_0(1) = 0 \tag{3.4.13}$$

where the operator $A_0(u_0)$ of (3.4.1) is assigned as

$$A_0(u_0) = -\nu_0 \frac{d^2}{dx^2} + u_0 \frac{d}{dx}, \quad \nu_0 = \text{const} > 0 \tag{3.4.14}$$

Examine perturbed problem (3.4.2) in the form

$$-\nu \frac{d^2 u}{dx^2} + u \frac{du}{dx} = f(x), \quad x \in (0,1) \tag{3.4.15}$$

$$u(0) = u(1) = 0 \tag{3.4.16}$$

where $f(x) = f_0(x) + \varepsilon f_1(x)$, $\nu = \nu_0 + \varepsilon \nu_1$, $\nu = const$, $f_1(x)$ is an assigned function. The operator $\delta A(u)$ of (3.4.3) is then simple:

$$\delta A(u) = -\nu_1 \frac{d^2}{dx^2} \tag{3.4.17}$$

Assume that the operators $A_0(u)$, $\delta A(u)$, $A(u)$ operate in a real Hilbert space $H = L_2(0,1)$ and have the definition domain $D(A)$ as described in § 3.1.

Assume the solution u of perturbed problem (3.4.15) and (3.4.16) as

$$u = u_0 + \varepsilon u_1 + \varepsilon^2 u_2 + \ldots$$

Substituting this expansion into (3.4.15) and (3.4.16), using the form of $f(x)$ and ν, and accepting only the first order small terms as concerns ε, formulate problems for u_0 and u_1:

$$-\nu_0 \frac{d^2 u_0}{dx^2} + u_0 \frac{du_0}{dx} = f_0, \quad u_0(0) = u_0(1) = 0 \tag{3.4.18}$$

$$-\nu_0 \frac{d^2 u_1}{dx^2} + \frac{d(u_0 u_1)}{dx} - \nu \frac{d^2 u_0}{dx^2} = f_1, \quad u_1(0) = u_1(1) = 0 \tag{3.4.19}$$

Rewrite the latter equation as

$$A_1(u_0)u_1 + \delta A(u_0)u_0 = f_1 \tag{3.4.20}$$

where the operator $A_1(u_0)$ is determined by formula (3.4.5) and may then be written as

$$A_1(u_0) \cdot = -\nu_0 \frac{d^2 \cdot}{dx^2} + u_0 \frac{d \cdot}{dx} + \frac{du_0}{dx} \cdot = -\nu_0 \frac{d^2 \cdot}{dx^2} + \frac{d(u_0 \cdot)}{dx} \tag{3.4.21}$$

This operator operates within $H = L_2(0,1)$ with the definition domain $D(A)$. Knowing the solution of problems (3.4.18) and (3.4.19), find an approximate solution of perturbed problem (3.4.15) and (3.4.16) with the formula

$$u \approx u_0 + \varepsilon u_1 \qquad (3.4.22)$$

Assume now as sought the value of the functional J of the solution of the perturbed problem

$$J = (u, p) = \int_0^1 up\,dx \qquad (3.4.23)$$

where $p(x)$ is a known function.

On the one hand, it may be written that

$$J \approx J_0 + \varepsilon J_1 \qquad (3.4.24)$$

where

$$J_0 = (u_0, p) = \int_0^1 u_0 p\,dx, \quad J_1 = (u_1, p) = \int_0^1 u_1 p\,dx \qquad (3.4.25)$$

where u_0, u_1 are the solutions of problems (3.4.18), (3.4.19).

On the other hand, the correction J_1 may be found as shown above, by solving an adjoint problem. To do so, make use of formula (3.4.9), where u_1^* is the solution of problem (3.4.7). It is easy to see that the problem in this case has the form

$$-\nu_0 \frac{d^2 u_1^*}{dx^2} - u_0 \frac{du_1^*}{dx} = p, \quad x \in (0,1) \qquad (3.4.26)$$

$$u_1^*(0) = u_1^*(1) = 0 \qquad (3.4.27)$$

Knowing u_1^*, find J_1 using formula (3.4.9) with regard to (3.4.17):

$$J_1 = (f_1, u_1^*) - (\delta A(u_0)u_0, u_1^*) = \int_0^1 \left(f_1 + \nu_1 \frac{d^2 u_0}{dx^2} \right) u_1^*\,dx \qquad (3.4.28)$$

In another example, examine the unperturbed problem again in the form of (3.4.12), (3.4.13) at $f_0(x) = \pi^2 \sin \pi x + \frac{\pi}{2} \sin 2\pi x$, and $\nu_0 = 1$. Problem (3.4.1) will then have the form

$$-\frac{d^2 u_0}{dx^2} + u_0 \frac{du_0}{dx} = \pi^2 \sin \pi x + \frac{\pi}{2} \sin 2\pi x, \quad x \in (0,1) \qquad (3.4.29)$$

$$u_0(0) = u_0(1) = 0 \qquad (3.4.30)$$

Use formula (3.4.14) to determine the operator $A_0(u_0)$:

$$A_0(u_0) = -\frac{d^2}{dx^2} + u_0\frac{d}{dx} \qquad (3.4.31)$$

The solution of problem (3.4.29) and (3.4.30) has the form

$$u_0(x) = \sin \pi x \qquad (3.4.32)$$

Examine the perturbed problem in the form of (3.4.2) at

$$f = f_0 + \varepsilon f_1, \quad A(u) = A_0(u) + \varepsilon \delta A(u)$$

where

$$f_1(x) = 2 + (1 - 2x)\sin \pi x + 200\pi \cos \pi x + 200\varepsilon(1 - 2x)$$
$$\delta A(u) = (x^2 - x + 200)\frac{d}{dx} \qquad (3.4.33)$$

The perturbed problem in this case is thus written as

$$-\frac{d^2u}{dx^2} + u\frac{du}{dx} + \varepsilon(x^2 - x + 200)\frac{du}{dx} = \pi^2 \sin \pi x + \frac{\pi}{2}\sin 2\pi x$$

$$+ \varepsilon[2 + (1 - 2x)\sin \pi x + 200\pi \cos \pi x + 200\varepsilon(1 - 2x)] \qquad (3.4.34)$$

$$u(0) = u(1) = 0 \qquad (3.4.35)$$

Assuming that the solution u of this problem is represented as a series in powers of ε:

$$u = u_0 + \varepsilon u_1 + \varepsilon^2 u_2 + \dots \qquad (3.4.36)$$

substituting (3.4.36) into (3.4.34) and accepting only first-order small terms concerning ε, obtain problems for u_0 and u_1. Unperturbed problem (3.4.29) and (3.4.30) will be the problem for u_0 and (3.4.20) the problem for u_1:

$$A_1(u_0)u_1 + \delta A(u_0)u_0 = f_1 \qquad (3.4.37)$$

Use formula (3.4.21) to determine the operator $A_1(u_0)$:

$$A_1(u_0)\cdot = -\frac{d^2\cdot}{dx^2} + \frac{d(u_0\cdot)}{dx}. \qquad (3.4.38)$$

The operator $\delta A(u_0)$ is assigned in (3.4.33). Then rewrite equation (3.4.37) as

$$-\frac{d^2 u_1}{dx^2} + \frac{d(u_0 u_1)}{dx} = f_1 - \delta A(u_0) u_0, \quad u_1(0) = u_1(1) = 0 \qquad (3.4.39)$$

or

$$-\frac{d^2 u_1}{dx^2} + \sin \pi x \frac{du_1}{dx} + \pi \cos \pi x u_1 = 2 + (1 - 2x) \sin \pi x$$

$$+ x(1 - x)\pi \cos \pi x + 200\varepsilon(1 - 2x) \qquad (3.4.40)$$

$$u_1(0) = u_1(1) = 0 \qquad (3.4.41)$$

Knowing solution u_1 to this one, find the solution to perturbed problem (3.4.34) and (3.4.35) using the formula

$$u \approx u^{(1)} = u_0 + \varepsilon u_1 \qquad (3.4.42)$$

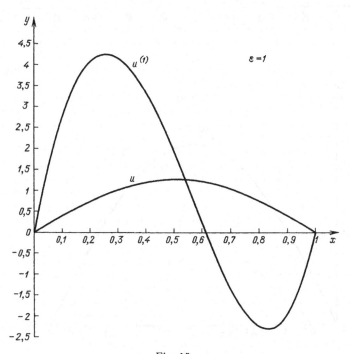

Fig. 15

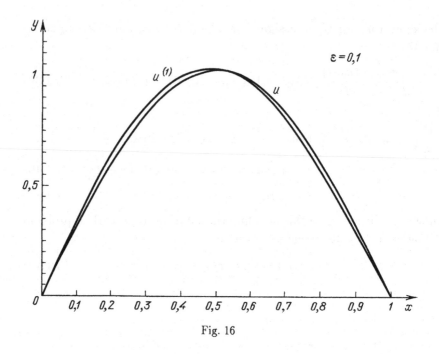

Fig. 16

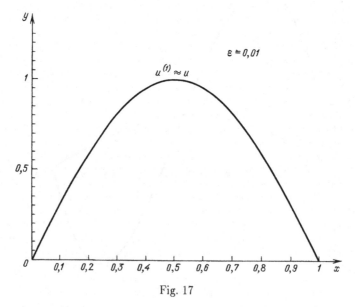

Fig. 17

Problem (3.4.40) and (3.4.41) was solved by the finite element method ısing piecewise linear basic functions on the uniform grid $x_i = ih$ where $= 0, \ldots, N$, $h = 10^{-2}$. Values of the function $u^{(1)} = u_0 + \varepsilon u_1$ were ›uilt at different ε ($\varepsilon = 1$, $\varepsilon = 0.1$, $\varepsilon = 0.01$). The function $u^{(1)}$ was :ompared with the accurate solution u of perturbed problem (3.4.34) ınd (3.4.35) known in this case in the explicit form:

$$u(x) = \sin \pi x + \varepsilon x(1 - x)$$

Fig. 15–17 show the diagrams of the functions $u^{(1)}$ and u for different As seen in the diagrams, $u^{(1)}$ converges to u at $\varepsilon \to 0$. The diagrams or $u^{(1)}$ and u virtually coincide at $\varepsilon = 0.001$ (see Fig. 17).

3.5. On The Higher-Order Perturbation Theory

Let us now briefly discuss the theory of an order higher than the first ɔrder. This one employs an algorithm as presented, for example, in [95]. Essentially, the algorithm is as follows.

Examine the main nonlinear equation

$$A(u)u = f \tag{3.5.1}$$

and assume that

$$A(u) = A_0(u) + \varepsilon \delta A(u), \quad f = f_0 + \varepsilon f_1 + \varepsilon^2 f_2 + \ldots \tag{3.5.2}$$

where $f_i \in H$ ($i = 1, 2, \ldots$) and the operators $A_0(u)$ and $\delta A(u)$ are as defined in § 3.4.

Represent the solution u as

$$u = u_0 + \varepsilon u_1 + \varepsilon^2 u_2 + \ldots \tag{3.5.3}$$

where ε is a formal parameter. Use (3.5.2) and (3.5.3) to write the main equation, (3.5.1), as

$$A(u_0 + \varepsilon u_1 + \varepsilon^2 u_2 + \ldots)(u_0 + \varepsilon u_1 + \varepsilon^2 u_2 + \ldots) = f_0 + \varepsilon f_1 + \varepsilon^2 f_2 + \ldots \tag{3.5.4}$$

The expression in the left-hand part of this relationship may be represented as a series in power of ε after specific transformations connected with the form of the operator A. Equate the terms at the same powers of ε in the left- and right-hand parts to obtain

$$
\begin{aligned}
A_0(u_0)u_0 &= f_0 \\
A_1(u_0)u_1 &= f_1 + B_1(u_0) \\
A_1(u_0)u_2 &= f_2 + B_2(u_0, u_1) \\
A_1(u_0)u_3 &= f_3 + B_3(u_0, u_1, u_2)
\end{aligned}
\qquad (3.5.5)
$$

. . . .

where

$$
A_1(u_0) \cdot = A_0(u_0) \cdot + \left(\frac{dA_0(u_0)}{du_0} \cdot \right) u_0
$$

and $B_1(u_0)$, $B_2(u_0, u_1)$, $B_3(u_0, u_1, u_2)$, etc. are expressions which depend not on relevant solutions, but rather only on the solutions already obtained.

Consequently solving equations (3.5.5), the solution of perturbed problem (3.5.1) may be built in the form of (3.5.3) (provided that series (3.5.3) converges in H).

Examine now the perturbation theory for the functional

$$
J = (u, p)
$$

where u is the solution of problem (3.5.1).

Using representation (3.5.3), write J as

$$
J = J_0 + \varepsilon J_1 + \varepsilon^2 J_2 + \dots
$$

where

$$
J_i = (u_i, p), \quad i = 0, 1, 2, \dots
$$

Knowing u_0, u_1, $u_2, \dots$ (which are determined in system (3.5.5)) the value of the functional J may be also determined with the formula

$$
J = \sum_{i=0}^{\infty} \varepsilon^i (u_i, p)
$$

Another representation of the corrections J_i may be sometimes usable through solutions of corresponding adjoint problems. It will be sufficient then to examine just one adjoint equation

$$
A_1^*(u_0)u_1^* = p
\qquad (3.5.6)
$$

where $A_1^*(u_0)$ is the operator adjoint to $A_1(u_0)$ in (3.5.5). Assume u_1^* as the solution of adjoint problem (3.5.6). Every equation in (3.5.5), starting with the second, multiply by u_1^* in H and equation (3.5.6) multiply consequently by $u_1, u_2, \ldots$; subtract the results from each other to obtain

$$0 = (f_1 + B_1(u_0), u_1^*) - (u_1, p)$$
$$0 = (f_2 + B_2(u_0, u_1), u_1^*) - (u_2, p)$$
$$0 = (f_3 + B_3(u_0, u_1, u_2), u_1^*) - (u_3, p)$$
$$\cdots\cdots\cdots\cdots\cdots\cdots\cdots\cdots\cdots\cdots$$

Whence, remembering that $J_i = (u_i, p)$, obtain formulas

$$J_0 = (u_0, p)$$
$$J_1 = (f_1 + B_1(u_0), u_1^*)$$
$$J_2 = (f_2 + B_2(u_0, u_1), u_1^*)$$
$$J_3 = (f_3 + B_3(u_0, u_1, u_2), u_1^*)$$
$$\cdots\cdots\cdots\cdots\cdots\cdots\cdots\cdots\cdots$$

These representations for J_i may be usable as every correction J_i is immediately linked here to the corresponding correction f_i in given right-hand part of f.

Examine, as a simple example, the problem

$$\frac{du}{dx} + au^2 = f(x), \quad x \in (0,1)$$
$$u(0) = 0 \tag{3.5.7}$$

where $a = const$, $f(x)$ is an assigned function. Assume that $f = f_0 + \varepsilon f_1$. Assume, that the solution u of problem (3.5.7) is bounded and representable in the form

$$u = u_0 + \varepsilon u_1 + \varepsilon^2 u_2 + \ldots \tag{3.5.8}$$

Substitute the expressions for f and u into the initial problem to obtain

$$\frac{d}{dx}(u_0 + \varepsilon u_1 + \varepsilon^2 u_2 + \ldots) + a(u_0 + \varepsilon u_1 + \varepsilon^2 u_2 + \ldots)^2 = f_0 + \varepsilon f_1$$

Make necessary computing and represent the left-hand part of the relationship obtained as a series in powers of ε to obtain

$$\left(\frac{du_0}{dx} + au_0^2\right) + \varepsilon\left(\frac{du_1}{dx} + 2au_0u_1\right) + \varepsilon^2\left(\frac{du_2}{dx} + 2au_0u_2 + au_1^2\right) + \dots$$

$$= f_0 + \varepsilon f_1 \quad (3.5.9)$$

Equate terms at the same powers of ε and obtain a system of problems

$$\frac{du_0}{dx} + au_0^2 = f_0; \quad u_0 = 0 \quad \text{at} \quad x = 0$$

$$\frac{du_1}{dx} + 2au_0u_1 = f_1; \quad u_1 = 0 \quad \text{at} \quad x = 0 \qquad (3.5.10)$$

$$\frac{du_2}{dx} + 2au_0u_2 = -au_1^2; \quad u_2 = 0 \quad \text{at} \quad x = 0$$

. . . .

Solve these equations consequently and obtain u in the form of series (3.5.8). If this series converges at ε assigned in the right-hand part of f, then it will be the sought solution to this problem.

Notice, that the operator $A(u)$ in problem (3.5.1) has then the form

$$A(u)\cdot = A_0(u)\cdot = \frac{d\cdot}{dx} + au\cdot \qquad (3.5.11)$$

and the operator $\delta A(u)$ in (3.5.2) is identically equal to zero operator:

$$\delta A(u) \equiv 0$$

The operator $A_1(u_0)$ in (3.5.5) is formally written as

$$A_1(u_0)\cdot = \frac{d\cdot}{dx} + 2au_0\cdot$$

As yet another example, examine problem (3.1.1) and (3.1.2), i. e.,

$$-\nu\frac{d^2u}{dx^2} + u\frac{du}{dx} = f(x), \quad 0 < x < 1$$

$$(3.5.12)$$

$$u(0) = u(1) = 0$$

Assume that the unperturbed problem has the form

$$-\nu_0\frac{d^2u_0}{dx^2} + u_0\frac{du_0}{dx} = f_0, \quad 0 < x < 1$$

$$(3.5.13)$$

$$u_0(0) = u_0(1) = 0$$

Assume also that the function f and the parameter ν in the perurbed problem are represented as a series in powers of ε:

$$f = f_0 + \varepsilon f_1 + \varepsilon^2 f_2 + \ldots, \quad \nu = \nu_0 + \varepsilon \nu_1 \qquad (3.5.14)$$

Then seek the solution of problem (3.5.12) as

$$u = u_0 + \varepsilon u_1 + \varepsilon^2 u_2 + \ldots \qquad (3.5.15)$$

Substitute expansions (3.5.14) and (3.5.15) into (3.5.12) to obtain

$$-(\nu_0 + \varepsilon \nu_1) \frac{d^2}{dx^2} (u_0 + \varepsilon u_1 + \varepsilon^2 u_2 + \ldots)$$

$$+ (u_0 + \varepsilon u_1 + \varepsilon^2 u_2 + \ldots) \frac{d}{dx} (u_0 + \varepsilon u_1 + \varepsilon^2 u_2 + \ldots)$$

$$= f_0 + \varepsilon f_1 + \varepsilon^2 f_2 + \ldots \qquad (3.5.16)$$

Eliminate the brackets and gather the terms at the same power of ε in the left-hand part of (3.5.16) to obtain

$$u_0 \frac{du_0}{dx} - \nu_0 \frac{d^2 u}{dx^2} + \varepsilon \left(u_0 \frac{du_1}{dx} + u_1 \frac{du_0}{dx} - \nu_0 \frac{d^2 u_1}{dx^2} - \nu_1 \frac{d^2 u_0}{dx^2} \right)$$

$$+ \varepsilon^2 \left(u_0 \frac{du_2}{dx} + u_1 \frac{du_1}{dx} + u_2 \frac{du_0}{dx} - \nu_0 \frac{d^2 u_2}{dx^2} - \nu_1 \frac{d^2 u_1}{dx} \right)$$

$$= f_0 + \varepsilon f_1 + \varepsilon^2 f_2 + \ldots \qquad (3.5.17)$$

Equate the terms of the same order in ε in the left- and right-hand parts of (3.5.17), obtain a system of equations

$$-\nu_0 \frac{d^2 u_0}{dx^2} + u_0 \frac{du_0}{dx} = f_0, \quad u_0(0) = u_0(1) = 0$$

$$-\nu_0 \frac{d^2 u_1}{dx^2} + \frac{du_0 u_1}{dx} = f_1 + \nu_1 \frac{d^2 u_0}{dx^2}, \quad u_1(0) = u_1(1) = 0$$

$$-\nu_0 \frac{d^2 u_2}{dx^2} + \frac{du_0 u_2}{dx} = f_2 + \nu_1 \frac{d^2 u_1}{dx^2} - u_1 \frac{du_1}{dx}, \quad u_2(0) = u_2(1) = 0 \quad (3.5.18)$$

. . . .

Solving these equations, find consequently u_1, u_1, u_2, ... Substitute these into (3.5.15) to determine the solution u of perturbed problem (3.5.12). Notice here, that, according to (3.5.16), the operators $A_0(u)$, $\delta A(u)$ in (3.5.2) have the form

$$A_0(u) \cdot = -\nu_0 \frac{d^2 \cdot}{dx^2} + u \frac{d \cdot}{dx}, \quad \delta A(u) \cdot = -\nu_1 \frac{d^2 \cdot}{dx^2}$$

and the operator $A_1(u_0)$ in (3.5.5) is written as

$$A_1(u_0) \cdot = -\nu_0 \frac{d^2 \cdot}{dx^2} + u_0 \frac{d \cdot}{dx} + \frac{du_0}{dx} \cdot = -\nu_0 \frac{d^2 \cdot}{dx^2} + \frac{d(u_0 \cdot)}{dx}$$

Examine now the perturbation theory for the functional

$$J = (u, p) = (u_0, p) + \varepsilon(u_1, p) + \varepsilon^2(u_2, p) + \ldots$$

$$\equiv J_0 + \varepsilon J_1 + \varepsilon^2 J_2 + \ldots \quad (3.5.19)$$

Introduce equations adjoint to (3.5.18). It is evident, that using the technique discussed above, we obtain

$$-\nu_0 \frac{d^2 u_0^*}{dx^2} - \frac{du_0 u_0^*}{dx} = p, \quad u_0^*(0) = u_0^*(1) = 0$$

$$-\nu_0 \frac{d^2 u_1^*}{dx^2} - u_0 \frac{du_1^*}{dx} = p, \quad u_1^*(0) = u_1^*(1) = 0 \quad (3.5.20)$$

$$-\nu_0 \frac{d^2 u_2^*}{dx^2} - u_0 \frac{du_2^*}{dx} = p, \quad u_2^*(0) = u_2^*(1) = 0$$

. . . .

Whence we see that $u_1^* = u_2^* = u_3^* = \ldots$.

Find now variations of the functionals through solutions of adjoint equations. Multiply every equation in (3.5.18) by $u_0^*, u_1^*, u_2^*, \ldots$ and every equation in (3.5.20) by $u_0, u_1, u_2, \ldots$ and subtract resulting pairs from each other. Then use the Lagrange identity to obtain

$$0 = (f_0, u_0^*) - (u_0, p)$$

$$0 = \left(f_1 + \nu_1 \frac{d^2 u_0}{dx^2}, u_1^*\right) - (u_1, p)$$

$$0 = \left(f_2 + \nu_1 \frac{d^2 u_1}{dx^2} - u_1 \frac{du_1}{dx}, u_2^*\right) - (u_2, p) \qquad (3.5.21)$$

$$\cdots\cdots\cdots\cdots\cdots\cdots\cdots\cdots\cdots\cdots\cdots\cdots\cdots$$

Whence, remembering that

$$(u_0, p) = J_0, \quad (u_1, p) = J_1, \quad (u_2, p) = J_2, \quad \ldots$$

taking into account that $u_2^* = u_3^* = \ldots = u_1^*$, obtain formulas

$$J_0 = (f_0, u_0^*)$$

$$J_1 = \left(f_1 + \nu_1 \frac{d^2 u_0}{dx^2}, u_1^*\right) = (f_1, u_1^*) - \nu_1 \left(\frac{du_0}{dx}, \frac{du_1^*}{dx}\right) \qquad (3.5.22)$$

$$J_2 = \left(f_2 + \nu_1 \frac{d^2 u_1}{dx^2} - u_1 \frac{du_1}{dx}, u_1^*\right) = (f_2, u_1^*) - \nu_1 \left(\frac{du_1}{dx}, \frac{du_1^*}{dx}\right) - \left(u_1 \frac{du_1}{dx}, u_1^*\right)$$

$$\cdots\cdots\cdots\cdots\cdots\cdots\cdots\cdots\cdots\cdots\cdots\cdots\cdots\cdots\cdots\cdots\cdots$$

If all the J_i are found, then

$$J = J_0 + \varepsilon J_1 + \varepsilon^2 J_2 + \ldots \qquad (3.5.23)$$

It may appear at first sight that solving an adjoint problem, corresponding to the main nonlinear one, has no advantage, since, prior to solving the adjoint problem, the solution must be sought to the main nonlinear problem. Once the solution to the main problem is found, then all the functionals we may be interested in, could be found. That is right, formally, but only formally.

As a matter of fact, a researcher in very complex problems (and these are that we are dealing with) must determine the sensitivity of the functional to various subdomains of the solution definition, in which

certain inputs are assigned. This dependence may not be obtained in
any other way; while it is just research in sensitivity that opens, in
many cases, the way to understanding most complex processes which
occur in examined systems, and therefore claims our special attention
to most important processes which do deserve it. We thus come in the
long run to new fuller and justified statements of problems. This is the
true value of adjoint equations technique as used in solution of linear
and nonlinear problems.

3.6. Other Approaches to Construction of Adjoint Operators in Nonlinear Problems

We discussed in § 3.1–3.5 elements of the theory of adjoint operators
and their applications in perturbation algorithms usable for solution of
nonlinear problems. To the knowledge of this author, there is rather
poor mathematical presentation of these subjects in scientific papers.
Even the definition of adjoint operator in nonlinear equations is not
commonly accepted. We shall dwell therefore on other approaches to
construction of adjoint operators in nonlinear problems as proposed
by V. S. Vladimirov and I. V. Volovich [33, 34], B. P. Maslov [118],
M. M. Weinberg [20], G. I. Marchuk and V. I. Agoshkov [225].

V. S. Vladimirov and I. V. Volovich introduced the adjoint linear
differential operator for a rather common system of nonlinear equations
with many independent variables in [33, 34], where they constructed
laws of conservation using a technique similar to that described in § 3.2.
In [33], they examined a nonlinear system of equations of the type

$$\mathcal{P}(u)u = 0 \qquad (3.6.1)$$

where $\mathcal{P}(u)$ is a linear differential operator depending on u (nonlinearly,
generally speaking).

The operator $\hat{\mathcal{P}}(u) = (\mathcal{P}(u))^*$ was constructed at a fixed u in a
conventional way (as is done for linear differential operators), formally
adjoint to $\mathcal{P}(u)$. Notice, that $\mathcal{P}(u)$ is also a linear differential operator
depending on u.

For construction of conservation laws, these authors examined in 33] a linear system of differential equations in the form

$$(\mathcal{P}(u))^* v = 0 \qquad (3.6.2)$$

which they referred to as *associated* to equation (3.6.1).

In other words, equation (3.6.2) is in fact adjoint to the main equation (3.6.1).

Examine, as an illustration, the Korteweg–de Vries equation:

$$\frac{\partial u}{\partial t} - 6u\frac{\partial u}{\partial x} + \frac{\partial^3 u}{\partial x^3} = 0, \quad t \in [0,T], \quad x \in (a,b) \qquad (3.6.3)$$

Rewrite it as

$$\left(\frac{\partial}{\partial t} - 6u\frac{\partial}{\partial x} + \frac{\partial^3}{\partial x^3}\right) u = 0$$

or as in (3.6.1)

$$\mathcal{P}(u)u = 0$$

where

$$\mathcal{P}(u) = \left(\frac{\partial}{\partial t} - 6u\frac{\partial}{\partial x} + \frac{\partial^3}{\partial x^3}\right)$$

Formally, the adjoint operator $\mathcal{P}(u) = (\mathcal{P}(u))^*$ will evidently have the form:

$$(\mathcal{P}(u))^* \cdot = -\frac{\partial \cdot}{\partial t} - 6\frac{\partial(u\cdot)}{\partial x} + \frac{\partial^3 \cdot}{\partial x^3}$$

Then associated system (3.6.2) is written in the form

$$\frac{\partial v}{\partial t} - 6\frac{\partial uv}{\partial x} + \frac{\partial^3 v}{\partial x^3} = 0 \qquad (3.6.4)$$

The solution v of associated problem (3.6.4) depends on u which is a solution of the Korteweg–de Vries equation, (3.6.3). If, say, u is a solution of equation (3.6.3), then $v = \partial u/\partial x$ is a solution of adjoint problem (3.6.4). This fact, in particular, was used in [33] to construct laws of conservation using the Korteweg–de Vries equation.

V. P. Maslov suggested another approach to construction of adjoint operators in nonlinear problems (see [1]). This is basically reduction of

nonlinear equation to linear one. As a matter of fact, certain nonlinear equations in mathematical physics may be reduced to linear equations by appropriate substitution. If substitution is impossible, new
operations of addition and multiplication by a number may be introduced in examined linear functional spaces, so that nonlinear equations
convert into linear in relation to these operations. The linear theory
is applicable to such transformed equations. In particular, an adjoint
equation may be introduced, all the reasoning done, necessary formulas
obtained, and then one may go back to solution of the initial problem
through substitution. V. P. Maslov used this approach to build adjoint
operators in dealing with nonlinear mathematical physics problems.

To understand this approach, examine, as a simple example, a nonlinear problem

$$\frac{\partial w}{\partial t} = \frac{h}{2}\frac{\partial^2 w}{\partial x^2} - \frac{1}{2}\left(\frac{\partial w}{\partial x}\right)^2, \quad x \in (0,1), \quad t \in (0,T] \qquad (3.6.5)$$

$$w(t,0) = w(t,1) = 0 \qquad (3.6.6)$$

$$w|_{t=0} = w_0(x) \qquad (3.6.7)$$

where $h = \text{const} > 0$, and $w_0(x)$ is an assigned function.

Assume that problem (3.6.5)–(3.6.7) does have a solution and this
is a smooth function in $H = L_2(\Omega)$ where $\Omega = (0,1) \times (0,T)$.

Assume now that the computed value of the functional $J(w)$ is
sought of the solution w of the form

$$J(w) = \int_0^T \int_0^1 p(t,x) \exp\left\{-\frac{w(t,x)}{h}\right\} dx dt \qquad (3.6.8)$$

where $p \in H$ is given function.

Now seek a way to compute this functional without solving the
nonlinear problem (3.6.5)–(3.6.7). It happens to be possible if made
through solution of a problem adjoint to (3.6.5)–(3.6.7). Make a substitution for the unknown function in (3.6.5). Let it be

$$u = e^{-w/h} \qquad (3.6.9)$$

Since

$$\frac{\partial u}{\partial t} = -\frac{1}{h}e^{-w/h}\frac{\partial w}{\partial t}, \quad \frac{\partial u}{\partial x} = -\frac{1}{h}e^{-w/h}\frac{\partial w}{\partial x}$$

$$\frac{\partial^2 u}{\partial x^2} = -\frac{1}{h}e^{-w/h}\frac{\partial^2 w}{\partial x^2} + \frac{1}{h^2}e^{-w/h}\left(\frac{\partial w}{\partial x}\right)^2$$

then problem (3.6.5)–(3.6.7) is reduced by substitution (3.6.9) into a linear problem

$$\frac{\partial u}{\partial t} = \frac{h}{2}\frac{\partial^2 u}{\partial x^2}, \quad x \in (0,1), \quad t \in (0,T] \qquad (3.6.10)$$

$$u(t,0) = u(t,1) = 1 \qquad (3.6.11)$$

$$u|_{t=0} = u_0(x) \qquad (3.6.12)$$

where $u_0(x) = \exp\{-w_0(x)/h\}$.

The functional $J(w)$ takes then the form

$$J(w) = \int_0^T \int_0^1 p(t,x)u(t,x)\,dx\,dt \qquad (3.6.13)$$

Now construct an equation adjoint to (3.6.10) in the usual way and use it to calculate functional (3.6.13) applying the method described in Chapter 1.

Examine a problem adjoint to (3.6.10)–(3.6.12)

$$-\frac{\partial u^*}{\partial t} = \frac{h}{2}\frac{\partial u^*}{\partial x^2} + p(t,x), \quad x \in (0,1), \quad t \in [0,T) \qquad (3.6.14)$$

$$u^*(t,0) = u^*(t,1) = 0 \qquad (3.6.15)$$

$$u^*|_{t=T} = 0 \qquad (3.6.16)$$

where the function $p(x,t)$ determines the functional $J(w)$ in (3.6.8).

Subtract the inner product of (3.6.10) and u^* within H and that of (3.6.14) and u from each other. Integrate by parts and use the initial and boundary conditions (3.6.11), (3.6.12), (3.6.15), and (3.6.16) to obtain the following formula to calculate the functional $J(w)$:

$$J(w) = \int_0^1 u^*(0,x)\exp\{-w_0(x)h\}\,dx$$

$$-\frac{h}{2}\int_0^T \left[\frac{\partial u^*}{\partial x}(t,1) - \frac{\partial u^*}{\partial x}(t,0)\right]dt \qquad (3.6.17)$$

The value of $J(w)$ is thus expressed through the solution u^* of adjoint problem (3.6.14)–(3.6.16), which happens to be linear.

It is not always possible, though, to use this method to construct an adjoint operator. As a matter of fact, firstly, a nonlinear problem may not always be reduced to linear. Secondly, it often happens that the initial equation is nonlinear and the functional examined is linear, and, after corresponding substitution, the equation becomes linear and the functional nonlinear. The latter will not allow to use adjoint equations to compute the functional value.

Nevertheless, this method of construction of the adjoint operator found its application in research of a number of nonlinear problems in mathematical physics [1]. This work pays special attention to problems where it is possible to introduce new operations of addition and multiplication by a number in the functional spaces and convert nonlinear equations into linear in relation to these operations.

M. M. Weinberg studied operators in nonlinear problems in [20] and introduced a definition of an adjoint operator for a class of nonlinear operators having certain properties. We will cite the definition and illustrate it with an example below.

To simplify the task, we will operate within the Hilbert space X and examine nonlinear operators F operating in X and satisfying the condition $F(0) = 0$. Assume that every operator F is differentiable in Gateâux's sense. This means that, at all $h \in X$, the limit

$$\lim_{t \to 0} \frac{F(u + th) - F(u)}{t} = Th \qquad (3.6.18)$$

exists where $x \in X$, T is a bounded linear operator operating in X with the definition domain $D(T) = X$. The operator T is referred to as the *Gateâux derivative* of the operator F at the point $u \in X$ and is denoted as $F'(u)$:

$$T = F'(u)$$

Separate a class $\mathcal{U}$ from a set of operators F so that there is an operator $G \in \mathcal{U}$ at every $F \in \mathcal{U}$ for which the following equality holds:

$$(F'(u)v, w) = (v, G'(u)w) \qquad (3.6.19)$$

at all u, v, $w \in X$. Here $F'(u)$, $G'(u)$ are the Gateâux derivatives of F, G relatively, and $(\cdot, \cdot)$ is an inner product in the Hilbert space X.

Definition 1 (M. M. Weinberg). An operator $G \in \mathcal{U}$ satisfying (3.6.19) for any u, v, $w \in X$ is referred to as *the adjoint* to F.

If (3.6.19) holds, write

$$G = F^* \qquad (3.6.20)$$

The adjoint operator, in the sense of this definition, is unique. Properties of such adjoint operators are discussed in [20].

Examine an example. Assume that $X = L_2(a, b)$ is the space of real functions summable with square on the interval $[a, b]$. Let $F(u)$ be a simple nonlinear operator

$$F(u) = \sin u \qquad (3.6.21)$$

operating in $X = L_2(a, b)$ with the definition domain $D(F) = X$. Evidently, this operator is differentiable in Gateâux's sense; according to (3.6.18)

$$\lim_{t \to 0} \frac{F(u + th) - F(u)}{t} = \lim_{t \to 0} \frac{\sin(u + th) - \sin u}{t} = (\cos u)h$$

i. e., the Gateâux derivative is

$$F'(u) = \cos u \qquad (3.6.22)$$

Now seek an operator $G(u)$ for which (3.6.19) holds. In this case, this may be written as

$$(F'(u)v, w) = (v, G'(u)w) \qquad (3.6.23)$$

where $(\cdot, \cdot)$ is an inner product in $X = L_2(a, b)$, namely

$$(u, v) = \int_a^b u(x)v(x)dx, \quad u, v \in X$$

Evidently, if

$$G(u) = \sin u$$

then $G'(u) = \cos u$ and equality (3.6.23) holds:

$$(v \cos u, w) = (v, w \cos u) = \int_a^b vw \cos u \, dx \qquad (3.6.24)$$

Then, according to Definition 1, the operator $G(u) = \sin u$ is adjoint to the operator $F(u)$, and this one is unique. We found in this simple example, that $F(u) = G(u)$, i. e., the operator F is self-adjoint.

From Definition 1, self-adjoint operators are defined only for nonlinear operators within the $\mathcal{U}$ class. If a F operator does not belong to the $\mathcal{U}$ class, F has no adjoint operator in the sense of Definition 1. This makes it difficult to use the Definition 1, since such nonlinear operators (which do not belong to the $\mathcal{U}$ class) arise in a number of practical problems [95].

G. I. Marchuk and V. I. Agoshkov in [225] give a review of methods of construction of adjoint operators in nonlinear problems and one more definition of an adjoint operator which holds for a broad range of nonlinear operators and in fact generalizes the notion of adjoint operator in the sense of Definition 1. We shall present here this definition and exemplify it.

Assume, for the sake of simplicity, that nonlinear operator F operates in the Hilbert space X with the definition domain $D(F)$ dense in X, and that $F(0) = 0$. Assume also that the operator F has the Gateâux derivative $F'(u)$ at any point $u \in D(F)$, i. e., limit (3.6.18) holds at the point $u \in D(F)$ for every $h \in D(F)$ and $F'(u) = T$ where T is a bounded linear operator operating in X with the definition domain $D(A) = X$.

With limitations accepted for the operator F, the simplest Taylor formula holds with a residual term in the integral form (see [151]):

$$F(u) = F(u_0) + \int_0^1 F'(u_0 + t(u - u_0))(u - u_0)dt, \quad u, u_0 \in D(F) \quad (3.6.25)$$

If $u_0 = 0$ is set in (3.6.25), then, in view of $F(0) = 0$, obtain

$$F(u) = \int_0^1 F'(tu)u \, dt, \quad u \in D(F) \qquad (3.6.26)$$

Rewrite it as

$$F(u) = A(u)u \qquad (3.6.27)$$

where $A(u) = \int_0^1 F'(tu)dt$ is bounded linear operator operating in X with the definition domain $D(A) = D(F')$.

Formula (3.6.26) was accepted as a basis in [225] for the definition of an adjoint operator. Fixing an element $u \in D(F)$ in the conventional way (as done in the linear case in Chapter 1), introduce an adjoint operator $A^*(u) = (A(u))^*$ which would satisfy the Lagrange identity

$$(A(u)v, w) = (v, A^*(u)w) \qquad (3.6.28)$$

for all $v,\ w \in X$.

Definition 2 (G. I. Marchuk, V. I. Agoshkov). The operator $F^*(u) = A^*(u)u = \left(\int_0^1 F'(tu)dt\right)^* u$ is referred to as *adjoint* to F.

The adjoint operator built according to Definition 2 is unique [225]. Properties of such adjoint operators and solvability of adjoint equations are discussed in [225], as well as interrelation between Definitions 1 and 2. It was proven in [225], that if an operator belongs to $\mathcal{U}$, operators adjoint to F and constructed according to Definitions 1 and 2, coincide.

Examine an example. Assume that $X = L_2(a, b)$ and take for $F(u)$ operator

$$F(u) = \sin u$$

as examined in the previous example. In this case, $D(F) \equiv X$.

Build an adjoint operator $F^*(u)$ according to Definition 2. Since

$$F'(u) = \cos u$$

then

$$A(u) = \int_0^1 F'(tu)dt = \int_0^1 \cos tu\, dt = \frac{\sin u}{u}$$

Then representation (3.6.27) for the operator $F(u)$ may be written as

$$F(u) = A(u)u$$

where
$$A(u) = \frac{\sin u}{u}$$

Introduce the adjoint operator $A^*(u)$ satisfying identity (3.6.28). Evidently $A^*(u) = (\sin u)/u$, i. e., the operator $A(u)$ is self-adjoint. Then, according to Definition 2, the adjoint operator $F^*(u)$ will have the form
$$F^*(u) = A^*(u)u = \frac{\sin u}{u}u = \sin u$$
i. e., $F^*(u)$ also coincides with $F(u)$, predictably. As a matter of fact, we saw in the preceding example that the operator $F(u) = \sin u$ belongs to the $\mathcal{U}$ class, which means that adjoint operators constructed with Definitions 1 and 2 coincide, as has been checked.

Though, as noted above, nonlinear operators F often do not belong to the $\mathcal{U}$ class, and then Definition 2 is only applicable.

Besides, notice that formulating Definition 2 we assumed that the Gateâux derivative $F'(u)$ of the operator F is a bounded linear operator operating in X with the definition domain $D(F') = X$. This limitation may in effect be made less strict. In [225] Definition 2 of an operator adjoint to F is discussed while $F'(u)$ may be, generally speaking, an unbounded operator in X with the definition domain $D(F') \supset D(F)$. In this case formula (3.6.25) will be also valid for the assumption that the derivative $F'(u)$ is continuous (as a function of u) at all $u \in D(F)$. It makes Definition 2 applicable for a broader range of nonlinear problems.

Examine a related example. Assume that $(t, x) \in \Omega = (0,1) \times (0,1)$, $X = L_2(\Omega)$ is a space of real periodic in t and x functions with a period equal to one along each variable. The inner product in $X = L_2(a, b)$ has the conventional form:
$$(u, v) = \int\limits_{\Omega} uvdtdx$$

Examine the operator F in the form
$$F(u) = \frac{\partial u}{\partial t} + u\frac{\partial u}{\partial x} + au, \quad a = const > 0 \tag{3.6.29}$$

Assume that F operates in X with the definition domain $D(F) = C^1(\Omega) \subset X$. Here $C^1(\Omega)$ is a set of functions continuously differentiable on Ω. It is not difficult to check (see [94]) that the operator F has the Gateâux derivative at every point:

$$F'(u)v = \frac{\partial v}{\partial t} + u\frac{\partial v}{\partial x} + \left(a + \frac{\partial u}{\partial x}\right)v \qquad (3.6.30)$$

with $D(F') = D(F) = C^1(\Omega)$. Notice, that in this case the operator $F'(u)$ is not a bounded operator from X into X.

Construct the adjoint operator $F^*(u)$ using Definition 2. Represent $F(u)$ in the form of (3.6.27). For $A(u)$

$$A(u)v = \int_0^1 F'(\tau u)v d\tau = \int_0^1 \left[\frac{\partial v}{\partial t} + \tau u\frac{\partial v}{\partial x} + \left(a + \frac{\partial \tau u}{\partial x}\right)v\right]d\tau$$

$$= \frac{\partial v}{\partial t} + \frac{u}{2}\frac{\partial v}{\partial x} + \left(a + \frac{1}{2}\frac{\partial u}{\partial x}\right)v$$

Now construct the operator $A^*(u)$ satisfying the Lagrange identity (3.6.28). Integrate by parts for $w \in C^1(\Omega)$ (in view of the periodic character of the functions from X) and obtain

$$(A(u)v, w) = \left(\frac{\partial v}{\partial t} + \frac{u}{2}\frac{\partial v}{\partial x} + \left(a + \frac{1}{2}\frac{\partial u}{\partial x}\right)v, w\right)$$

$$= \left(v, -\frac{\partial w}{\partial t} - \frac{u}{2}\frac{\partial w}{\partial x} + aw\right)$$

The operator $A^*(u)$ on the functions $w \in C^1(\Omega)$ has thus the form

$$A^*(u)w = -\frac{\partial w}{\partial t} - \frac{u}{2}\frac{\partial w}{\partial x} + aw \qquad (3.6.31)$$

Then, according to Definition 2 the adjoint operator $F^*(u)$ is written as

$$F^*(u) = A^*(u)u = -\frac{\partial u}{\partial t} - \frac{u}{2}\frac{\partial u}{\partial x} + au, \quad u \in C^1(\Omega) \qquad (3.6.32)$$

Definition 2 leads us to the adjoint operator $F^*(u)$ in the form of (3.6.32). Notice, that in this case Definition 1 is not to be used since the operator F defined by equality (3.6.29), does not belong to the $\mathcal{U}$ class (see [94]).

We discussed in this section different approaches to determining adjoint operators in nonlinear problems. Each approach may be used, if applicable, in research in mathematical physics nonlinear problems. The variety of approaches provides a choice of methods for construction of adjoint operators depending on the goal of research (construction of conservation laws, calculation of functionals, formulation of perturbation algorithms, investigation in solvability of nonlinear problems, etc.).

The author hopes that this concise review of approaches to construction of adjoint operators in nonlinear problems will help the reader navigate in scientific writings on the subject.

CHAPTER 4

Inverse Problems and Adjoint Equations

Mathematical modelling complex problems provided both a wealth of techniques and new theoretical challenges for researchers, particularly identification of processes and solving inverse problems, i.e., finding coefficients and other model parameters, knowing problem functionals. Techniques for solving these problems are important in complex models where causative/sequential relationships are difficult to trace.

This author's research results have led him to conclude that statements and solution algorithms based on the perturbation theory formulas for functionals, using adjoint equations, happened to be the most productive in a certain sense. As a matter of fact, adjoint problems involving one or other functional are more informative once it comes to a process as a whole. Solutions of adjoint problems (adjoint functions), which are weight multipliers at unknown parameters, reflect contributions of processes within a system to investigated problem functionals.

The term "inverse problems" is applied to different types of mathematical physics problems in contemporary literature.

We will discuss two types of inverse problems. The first is problems of determination of the state of a process in previous moments. Initial distribution of heat in a body with a known heat field at a certain moment is an example. The other type are problems requiring to determine operators with known structures and unknown coefficients to be found by information about functionals of solutions. The inverse problem for the Sturm–Liouville equation, requiring to find a coefficient in differential equation of the second order from the properties of a spectral function of a boundary value problem, is an example of a problem of this type.

Inverse problems in mathematical physics often prove to be incorrectly stated, in the classical sense. Great variations in solutions may correspond to small variations in registered functionals. Adamar proposed the notion of correctness and supplied an example of an incorrect problem of mathematical physics, the Cauchy problem for the Laplace equation, at the beginning of the century.

The so-called incorrect problems had long been considered uninteresting and deserving little study until interpreting geophysical data brought about the interest and intensive study. Russian mathematicians S. G. Krein [56], M. M. Lavrentyev [57], A. N. Tikhonov [147] and others have greatly contributed to the development of the theory and methods of solution of problems of mathematical physics, not correct in the classical sense according to Adamar.

Solution of classically incorrect problems was proven to be stable in relation to data variations, if extra restrictions are imposed on sets of permissible solutions. Problems of this type were referred to as "conditionally correct."

Construction of approximate solutions of conditionally correct problems by approximate data brought about the notion of controlling families by Tikhonov. Basically this notion means that a conditionally correct problem is correlated with a family of classically correct problems (the controlling family) which depends on a parameter. If the parameter tends to a limit, the sequence of solutions of classically correct problems must tend to the solution of the conditionally correct problem. It was proven that, with a proper controlling parameter depending on the accuracy of data, the solution of a problem of the controlling family by approximate data will be an approximate solution of the conditionally correct problem.

Many authors examined a broad range of conditionally correct problems. We will discuss only a few. In § 4.1, we will introduce basic notions of the theory of conditionally correct problems. In § 4.2, 4.3 we will discuss regularization of problem of determination of input (initial) data of evolutionary equations. The rest of the chapter examines the problem of the restoration of structures of linear and nonlinear operators with perturbation theory methods. These studies are adjacent to the works by A. V. Balakrishnan and J.-L. Lions on identification theory [171–173, 214].

4.1. Basic Definitions and Examples

Examine solution of the problem

$$A\varphi = f \tag{4.1.1}$$

where A is a linear operator in the Banach space F with an unbounded inverse operator. Problem (4.1.1) may then be stated incorrectly, for, on the one hand, the solution of equation (4.1.1) may not exist for the arbitrary element $f \in F$, and, on the other hand, any large variations of the solution φ may correspond to small variations in the right-hand part f. Problem (4.1.1), however, often happens to be correct on a certain subspace $\Phi \subset F$. This means that the operator A has inverse operator bounded on Φ, i. e., the estimate

$$\|\varphi\| \le c\|A\varphi\|, \quad \varphi \in \Phi, \quad c = \text{const} > 0 \tag{4.1.2}$$

holds at all $\varphi \in \Phi$. Problem (4.1.1) is then referred to as *conditionally correct*. Let us analyze, applicably for this particular case, a very simple example of building an iteration algorithm of computations which will not remove the sequence of approximate solutions from the given subspace Φ.

Let F be a Hilbert space every element f of which is representable as a Fourier series in a certain complete biorthogonal system of functions $\{u_n\}$, $\{u_n^*\}$. Then

$$f = \sum_{n=1}^{\infty} f_n u_n \tag{4.1.3}$$

where

$$f_n = (f, u_n^*)$$

Assign the subspace Φ so that it includes only elements f of the Hilbert space which have no more than N (though not equal to zero) harmonics in expansion (4.1.3), corresponding to the most greatest perturbations:

$$f = \sum_{n=1}^{N} f_n u_n$$

Solve equation (4.1.1) with the iteration process

$$\varphi^{j+1} = \varphi^j - \tau(A\varphi^j - f), \quad \varphi^0 = 0 \qquad (4.1.4)$$

or (which is shorter)

$$\varphi^{j+1} = T\varphi^j + \tau f, \quad \varphi^0 = 0$$

where $T = E - \tau A$ is the step operator.

Assume that the solution of problem (4.1.1) does exist, is unique, and belongs to the subspace Φ. Assume also that the norm $\|T\|_F$ is defined over all of the Hilbert space by the equality

$$\|T\|_F^2 = \sup_{\varphi \in F} \frac{(T\varphi, T\varphi)}{(\varphi, \varphi)} > 1$$

and on the subspace Φ by the equality

$$\|T\|_\Phi^2 = \sup_{\varphi \in \Phi} \frac{(T\varphi, T\varphi)}{(\varphi, \varphi)} < 1$$

Given these assumptions, iterative-process (4.1.4) will converge in the functions φ^j from Φ and diverge in the functions of the whole of the space F. Therefore, if we want to realize the converging iterative process, the approximate solution φ^j should belong to Φ at every step of iterative process. It is very simple technically.

Assume $\varphi^{j-1} \in \Phi$ at a step to obtain a new element φ^j using recursion relationship (4.1.4). The function $T\varphi^{j-1}$ may then not belong to the subspace Φ . To ensure that $\varphi^j \in \Phi$, expand these functions into Fourier series

$$\varphi^j = \sum_{n=1}^{\infty} \varphi_n^j u_n$$

and omit all the members in this series with numbers $n > N$. To do so, it will be algorithmically very simple to determine just N first Fourier coefficients

$$\varphi_k^j = (\varphi^j, u_k^*), \quad k = 1, 2, \ldots, N$$

and build the final sum

$$\varphi^j = \sum_{k=1}^{N} \varphi_k^j u_k$$

If this process continues as iterative, φ^j functions which are approximate solutions of the problem will belong to the subspace Φ. With additional assumptions (e. g., that of orthogonality of the basis $\{u_k\}$) the sequence $\{\varphi^j\}$ converges to some function φ^∞ which is taken as an approximate solution of equation (4.1.1).

Examine now another aspect of solving conditionally correct problems, i. e., the input data assignment precision. In solving problems of mathematical physics, there are at least errors arising from approximation of problem (4.1.1) by a difference problem or inaccurate data on the operator A and function f.

Let $\bar{A}$ be an operator corresponding to a standard model, $\bar{f}$ an assigned vector, and $\bar{\varphi}$ a solution of the problem

$$\bar{A}\bar{\varphi} = \bar{f}$$

with

$$A = \bar{A} + \delta A, \quad f = \bar{f} + \delta f$$

and the problem $A\varphi = f$ is correct on the subspace Φ of the Hilbert space F, $A\Phi \subseteq \Phi$, A is symmetric positive operator. As for δA and δf, their a priori error is known on the subspace Φ:

$$\|\delta A\|_\Phi \le \varepsilon_1, \quad \|\delta f\|_\Phi \le \varepsilon_2 \qquad (4.1.5)$$

Then the solution of the equation $A\varphi = f$ may be reduced to an iterative process, which would generate new approximations belonging to the subspace Φ. If the operator A is a positive definite matrix, there exists quite a set of iterative processes converging to the solution of the problem $A\varphi = f$, as this author proved in [73]. The subspace of approximate solutions Φ coincides here with all of the Hilbert (Euclidean) space F. Iterative process (4.1.4) will converge, if the parameter τ is properly chosen. The process may be optimized, with regard to information (4.1.5) known a priory, choosing an appropriate number of steps j_0 of the iterative process.

Assume now that the operator A in the equation $A\varphi = f$ is symmetric and its spectrum has both positive and negative components.

Analysis reveals that iterative process (4.1.4) will diverge with the assumptions allowed. Indeed, let

$$\varphi = \sum_n \varphi_n u_n, \quad f = \sum_n f_n u_n \tag{4.1.6}$$

where $\{u_n\}$ be complete orthonormal system of eigenfunctions of the operator A. Substituting (4.1.6) into (4.1.4) and taking inner product of the result and u_n obtain recursion relationships for the Fourier coefficients

$$\varphi_n^{j+1} = \varphi_n^j - \tau(\lambda_n \varphi_n^j - f_n), \quad \varphi_n^0 = 0$$

or for the residual $\xi^j = A\varphi^j - f$

$$\xi_n^{j+1} = (1 - \tau\lambda_n)\xi_n^j, \quad \xi_n^0 = -f_n \tag{4.1.7}$$

Solve equation (4.1.7) to obtain

$$\xi_n^j = -(1 - \tau\lambda_n)^j f_n$$

Therefore,

$$\xi^j = -\sum_n (1 - \tau\lambda_n)^j f_n u_n$$

Iterative process (4.1.4) will only converge if

$$\lim_{j\to\infty} \xi^j = 0$$

If the eigenvalues of the operator A are only positive numbers from the interval

$$\alpha(A) \leq \lambda_n(A) \leq \beta(A)$$

the selection of τ from the interval

$$0 < \tau < 2/\beta \tag{4.1.8}$$

makes process (4.1.4) convergent.

The symmetric operator A has in this case both positive and negative eigenvalues. If τ is chosen within interval (4.1.8), all the residual harmonics corresponding to the positive λ will be damped, from iteration to iteration, at a rate of T_n^j, where $T_n = (1 - \tau\lambda_n)^j < 1$ and j is the power index. As concerns harmonics corresponding to the negative eigenvalues, since the components of the residual will grow

$$T_n^j = (1 - \tau\lambda_n)^j > 1 \tag{4.1.9}$$

which leads to the divergence of the process.

Iterative process (4.1.4) with a sequence of pilot functions φ^j belonging to the whole of the Hilbert space thus diverges.

Now let us formulate a few remarks concerning the practical approach to numerical solution of conditionally correct problems. These problems usually are reduced to systems of linear equations with ill-conditioned general structure matrices. They are solved as a rule by a multistep method of minimal residuals which provides a rapid convergence of the iterative process. The method of adjoint gradients is also applicable after the equations are symmetrized by the Gauss transformation. We will discuss this method later in connection with solution of inverse evolutionary problems (see § 4.3). In solving a problem by the iterative method, the process should be ceased at the step where the residual norm is approximately equal to the a priori input data error, i. e., $\|\xi^j\| \approx \varepsilon_1 + \varepsilon_2$. The maximum possible accuracy of solution is obtained for a priori errors in this case (see [73]).

Restricting the solutions $\{\varphi\}$ may thus provide stability of equation (4.1.1). Let us now formulate a general definition of conditional correctness of problem (4.1.1). Examine equation (4.1.1) in the Banach space F. Let M be a set from F. Let us refer to problem (4.1.1) as *conditionally correct* (on M) if there is an a priori estimate (at all $\varphi \in M$)

$$\|\varphi\| \leq w(\|A\varphi\|), \quad \varphi \in M \tag{4.1.10}$$

where $w(\varepsilon)$ is a continuous function and $w(0) = 0$. The set M is referred to as a *correctness set*. The choice of the set M is affected by physical considerations related to the statement of problem, efficiency of the computer and accuracy required of the results.

This definition prompts a method of stable solution of equation (4.1.1) which is minimization of the functional $\|A\varphi - f\|$ on the correctness set M. The element φ_0 which provides the minimum to this functional is referred to as a *quasi-solution*. It may be proven (see [47]) that certain additional assumptions (as, e. g. M is a convex compact while the space F, Hilbert in particular, is strictly convex) make the quasi-solution uniquely possible and continuously depending on the right-hand part $f \in F$. The idea of quasi-solution thus returns correctness to problem (4.1.1). Fig. 18 shows the situation where the right-hand part $f \notin AM$. The quasi-solution φ_0 is found then from the equation $A\varphi_0 = g$, where g is a projection of the element f on the set AM.

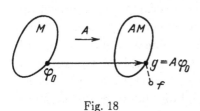

Fig. 18

The iterative and variational methods of the stable solution of equation (4.1.1) we discussed are special cases of the so-called controlling algorithms. Let us formulate a general definition thereof.

A family of linear operators R_α in the space F depending on numerical parameter α, $0 < \alpha \leq \alpha_0$ is a *controlling family* (or algorithm) for equation (4.1.1) on the set M_R, if the following conditions are satisfied:

$$\|R_\alpha\| < \infty \qquad (4.1.11)$$

at any $\alpha \in (0, \alpha_0]$ and

$$\|R_\alpha A\varphi - \varphi\| \to 0 \qquad (4.1.12)$$

for $\alpha \to 0$ and all $\varphi \in M_R$. The set M_R where relationship (4.1.12) is valid may be wider than the correctness set M of problem (4.1.1). If (4.1.12) tends uniformly to zero all over $\varphi \in M_R$, these sets usually coincide.

Let us see how an approximate solution of equation (4.1.1) may be found using the controlling family R_α, provided the exact solution $\varphi \in M_R$, and the εth approximation f_ε is known, $\|f - f_\varepsilon\| \leq \varepsilon$, instead of the exact right-hand part f.

Assign $\varphi_{\alpha\varepsilon} = R_\alpha f_\varepsilon$ and estimate $\|\varphi - \varphi_{\alpha\varepsilon}\|$. Then, by the triangle inequality,

$$\|\varphi - \varphi_{\alpha\varepsilon}\| = \|\varphi - R_\alpha f + R_\alpha(f - f_\varepsilon)\| \leq \|\varphi - R_\alpha A\varphi\|$$

$$+ \|R_\alpha\|\|f - f_\varepsilon\| \leq \|\varphi - R_\alpha A\varphi\| + \|R_\alpha\|\varepsilon \qquad (4.1.13)$$

Choose $\alpha(\varepsilon)$ so that $\|R_\alpha\|\varepsilon \to 0$ and $\alpha(\varepsilon) \to 0$ for $\varepsilon \to 0$.

Then the first term in (4.1.13) tends to zero if $\varepsilon \to 0$ because of (4.1.12), and the second by construction of the function $\alpha(\varepsilon)$. As a result, $\varphi_{\alpha(\varepsilon)\varepsilon} \to 0$ for $\varepsilon \to 0$.

The controlling family R_α thus provides a principal possibility of a stable solution of conditionally correct problem (4.1.1) with an approximate right-hand part. The parameter α is referred to as the regularization parameter. The number of iteration j plays the role of a regularization parameter in the iterative method as represented by (4.1.6).

An infinite number of regularizing algorithms may be usually found for problem (4.1.1). It will be natural to take into account the simplicity of the computing process, peculiarities of the initial problem, efficiency, etc. in practical computations.

Let us illustrate the notions we have just discussed with a simple example of an integral equation of the first type, the Volterra equation

$$A\varphi(t) \equiv \int_0^t A(t,\tau)\varphi(\tau)d\tau = f(t), \quad t \in [0,T] \qquad (4.1.14)$$

Interpretation of readings of physical instruments are frequently reduced to solving equation (4.1.14). Examine the operator A, as defined by (4.1.14), operating in the space $C[0,T]$ of continuous functions φ with the norm

$$\|\varphi\| \equiv \|\varphi\|_T = \max_{t\in[0,T]} |\varphi(t)|$$

The kernel $A(t,\tau)$ is assumed to be continuously differentiable in t, continuous in τ and differring from zero on the diagonal $t = \tau$. To simplify, assume that

$$A(t,t) \equiv 1 \qquad (4.1.15)$$

Equation (4.1.14) may then be reduced by differentiation in t to the Volterra equation of the second type

$$\varphi(t) + \int_0^t \frac{d}{dt}A(t,\tau)\varphi(\tau)d\tau = \frac{d}{dt}f(t) \qquad (4.1.16)$$

which may be solved by the successive approximation method. However, if, instead of f, only its ε-th approximation f_ε is known in the norm of $C[0,T]$: $\|f - f_\varepsilon\| \leq \varepsilon$, the differentiation problem becomes incorrect. Let us see that the family of operators R_α which make the function $f(t)$ correspond to the solution φ_α of the equation

$$\varphi_\alpha(t) + \int_0^t \Delta_\alpha A(t,\tau)\varphi_\alpha(\tau)d\tau = \Delta_\alpha f(t), \quad t \in [0,T_0] \tag{4.1.17}$$
$$T_0 = T - \alpha_0, \quad \alpha \in (0,\alpha_0], \quad \alpha_0 < T$$

where

$$\Delta_\alpha f(t) = \frac{f(t+\alpha) - f(t)}{\alpha}$$

$$\Delta_\alpha A(t,\tau)\varphi_\alpha(\tau) = \frac{A(t+\alpha,\tau)\varphi_\alpha(\tau) - A(t,\tau)\varphi_\alpha(\tau)}{\alpha}$$

will be regularizing in the interval $[0,T_0]$ for all continuously differentiable solutions φ of equation (4.1.14). The solution φ_α of the Volterra equation of the second type, (4.1.17) exists, is unique, and satisfies the estimate

$$\|\varphi_\alpha\|_{T_0} \leq e^{KT_0}\|\Delta_\alpha f\|_{T_0}$$

where

$$K = \max_{0 \leq \tau \leq t \leq T_0} |\Delta_\alpha A(t,\tau)|$$

and since

$$\|\Delta_\alpha f\|_{T_0} \leq \frac{2}{\alpha}\|f\|_T$$

then

$$\|\varphi_\alpha\|_{T_0} \equiv \|R_\alpha f\|_{T_0} \leq \frac{2e^{KT_0}}{\alpha}\|f\|_T < \infty \tag{4.1.18}$$

for $\alpha > 0$ and, consequently, the condition (4.1.11) is satisfied. In order to check the condition (4.1.12) it is enough to show, that

$$\|\varphi_\alpha - \varphi\|_{T_0} \to 0, \quad \text{for} \quad \alpha \to 0$$

Applying the operator Δ_α in equation (4.1.14) with regard to condition (4.1.15), obtain

$$\varphi(t) + \int_0^t \Delta_\alpha A(t,\tau)\varphi_\alpha(\tau)d\tau = \Delta_\alpha f(t) - g_\alpha(t) \tag{4.1.19}$$

where

$$g_\alpha(t) = \frac{1}{\alpha} \int\limits_t^{t+\alpha} (A(t+\alpha,\tau) - A(\tau,\tau))\varphi(\tau)d\tau$$

$$+ \frac{1}{\alpha} \int\limits_t^{t+\alpha} (\varphi(\tau) - \varphi(t))d\tau \quad (4.1.20)$$

Subtract equality (4.1.19) from (4.1.17) to obtain an equation for $u_\alpha = \varphi_\alpha - \varphi$:

$$u_\alpha + \int\limits_0^t \Delta_\alpha A(t,\tau)u_\alpha(\tau)d\tau = g_\alpha(t), \quad t \in [0,T_0] \quad (4.1.21)$$

By (4.1.21) and (4.1.20)

$$\|\varphi - \varphi_\alpha\|_{T_0} \equiv \|u_\alpha\|_{T_0} \leq e^{KT_0}\|g_\alpha\|_{T_0} \quad (4.1.22)$$

$$\|g_\alpha\|_{T_0} \leq \frac{1}{\alpha}K\|\varphi\|_T\alpha^2 + \frac{1}{\alpha}\left\|\frac{d\varphi}{dt}\right\|_T\alpha^2$$

$$= \alpha\left(K\|\varphi\|_T + \left\|\frac{d\varphi}{dt}\right\|_T\right) \to 0 \quad (4.1.23)$$

at $\alpha \to 0$. Therefore the operators $\{R_\alpha\}$ make a regularizing family for equation (4.1.14) on a set of continuously differentiable functions φ. Unlike common variational algorithms, this regularizing algorithm retains the Volterra property of the initial equation.

4.2. Solution of Inverse Evolutionary Problems with Constant Operators

This section discusses two stable techniques for solving an inverse evolutionary problem:

$$\frac{d\varphi}{dt} - A\varphi = 0, \quad \varphi(0) = g, \quad A \geq 0$$

The first technique is basically the Fourier's method reduced to solution of a spectral problem. In the second, the initial incorrect problem is reduced to solution of a sequence of correct (direct) evolutionary problems:

$$\frac{d\varphi}{dt} + A\varphi = 0, \quad \varphi(0) = g, \quad A \geq 0$$

Examine first the Fourier's method. Let A be a positive definite matrix which does not depend on time and has a real spectrum in the interval $\alpha(A) \leq \lambda \leq \beta(A)$, and vector-function φ is a solution of the Cauchy problem:

$$\frac{d\varphi}{dt} - A\varphi = 0, \quad 0 \leq t \leq t_0 \tag{4.2.1}$$
$$\varphi = g \quad \text{at} \quad t = 0$$

where g is an assigned value of a vector in the initial moment of time. Examine two spectral problems:

$$Au = \lambda u, \quad A^* u^* = \lambda u^* \tag{4.2.2}$$

Assume that they define two biorthogonal bases of the eigenfunctions $\{u_n\}$ and $\{u_n^*\}$. Represent functions φ and g as Fourier sums:

$$\varphi = \sum_n \varphi_n u_n, \quad g = \sum_n g_n u_n \tag{4.2.3}$$

Substitute these sums into (4.2.1) and take the inner product of the result and u_n^* to obtain a system of ordinary differential equations for Fourier coefficients:

$$\frac{d\varphi_n}{dt} - \lambda_n \varphi_n = 0 \tag{4.2.4}$$
$$\varphi_n = g_n \quad \text{at} \quad t = 0, \quad n = 1, 2, \ldots, N$$

All the solutions of equations (4.2.4) have the form of

$$\varphi_n = g_n e^{\lambda_n t}, \quad n = 1, 2, \ldots, N \tag{4.2.5}$$

whence the solution of problem (4.2.1) may be represented by the sum

$$\varphi(t) = \sum_{n=1}^{N} g_n e^{\lambda_n t} u_n \tag{4.2.6}$$

We have thus established that the solution of problem (4.2.1) is represented as a Fourier sum where each term grows exponentially in time, depending on the value of n-th eigenvalue λ_n.

If we are interested in a physically definite solution of this problem in the interval of time $0 \leq t \leq t_0$, examine a correctly stated problem similar to (4.2.1):

$$\frac{d\varphi}{dt} - A\varphi = 0, \quad 0 \leq t \leq t_0$$
$$\varphi = h \quad \text{at} \quad t = t_0 \tag{4.2.7}$$

Just as above,

$$\varphi = \sum_{n=1}^{N} h_n e^{-\lambda_n(t_0-t)} u_n \tag{4.2.8}$$

where $h_n = (h, u_n^*)$. Require that solution (4.2.8) for $t = 0$ coincides with the vector g from problem (4.2.1), whence obtain a certain link between the Fourier coefficients of the function h and those of the function g:

$$g_n = h_n e^{-\lambda_n t_0} \tag{4.2.9}$$

The function g is simply restored with the function h:

$$g = \sum_{n=1}^{N} h_n e^{-\lambda_n t_0} u_n \tag{4.2.10}$$

Small errors in the function g correspond to small errors in h (or h_n). But our problem is inverse to the one we are examining. Information on the function g is available, and we need to restore the function h with the formula

$$h = \sum_{n=1}^{N} h_n e^{\lambda_n t_0} u_n \tag{4.2.11}$$

If we had accurate information about function g and could calculate with an infinite number of digits, the restoration of the function h by formula (4.2.11) would not be difficult. In this given case, we know the function g with a certain error known a priory, and use a computer which has a limited number of digits; hence errors due to rounding off. Given these two facts, the task of computation of h with (4.2.11) is not so simple.

As you try to process experimental data on the basis of inverse evolutionary problem (4.2.1), assume knowing in advance the system of eigenfunctions u_n. You may single out useful information and estimate the error for each Fourier component g_n with a sufficient accuracy, expanding the input data (functions g) with this system.

If the problem is statistical and allows repeats, the accuracy of initial data may be enhanced in g_n (even if the error of single measurement surpasses the usable information) on the basis of developed correlation analysis methods. Preliminary processing of measurement data allows conclusions on a scale of a systematic error (or random error, if the measurement is single) in g_n. Therefore

$$g_n = \bar{g}_n(1 + \delta_n)$$

at any n, where $\bar{g}_n$ is the exact value (unknown a priory), and δ_n is a relative error which we will consider as known.

The error δ_n usually happens to be minimum for the longest perturbation waves and increases rapidly toward high harmonics which describe as a rule small features of the solution. Therefore, the coefficients g_n describe, starting from a certain number, mostly an error in the input data. From formula (4.2.11) it follows that just the highest-frequency components have the biggest exponential weight. Hence we may most probably get an erroneous result, unless we exclude the parasite harmonics in advance, since g_n contains virtually no useful information for such harmonics. However, multiplied by big coefficients $e^{\lambda_n t_0}$, they may greatly contribute to h and distort (and sometimes irreparably) the solution of the problem. The first and main problem is thus to determine informativeness of the coefficients g_n.

Assume that we have established, on the basis of a priory information, that n_0 first coefficients g_n have a relative error smaller than η, i. e., $\delta_n < \eta$, where η is the maximum admissible error. Then an algorithm for restoration of function (4.2.11) proves to be similar to that we examined in problem (4.2.1) as we constructed the elements of subspace Φ. We just have to exclude those harmonics from series (4.2.11), which are parasite. We shall have as a result

$$h = \sum_{n=1}^{n_0} g_n e^{\lambda_n t_0} u_0 \qquad (4.2.12)$$

Use the iterative method as a basis for an algorithm for the solution of particular spectral problem. If a set of first (the largest-scale) eigenfunctions u_n or u_n^* and corresponding eigenvalues is necessary, the orthogonalization algorithm is usable as described in [73].

Re-examine the Cauchy problem (see [165])

$$\frac{d\varphi}{dt} - A\varphi = 0, \quad \varphi(0) = g, \quad 0 \le t \le t_0 \qquad (4.2.13)$$

where A is a self-adjoint unlimited positive operator in the Hilbert space F. In other words, we do not make the finite-difference approxmation in spatial variables in advance. Problem (4.2.13) is classically .ncorrect in this case. Let us assume that problem (4.2.13) does have a solution φ which belongs to the set $M = \{\varphi(t): \|\varphi(t)\| \le m, \ t \in [0, t_0]\}$, though instead of the exact initial condition g we have but the approximation to g:

$$\|g - g_\varepsilon\| \le \varepsilon \qquad (4.2.14)$$

There is a well-know inequality (see S. G. Krein [56]):

$$\|\varphi(t)\| \le \|\varphi(0)\|^{1-t/t_0}\|\varphi(t_0)\|^{t/t_0} \qquad (4.2.15)$$

therefore the problem of finding $\varphi(t)$ is conditionally correct on the set M for any fixed $t \in (0, t_0)$, since by virtue of (4.2.15)

$$\|\varphi(t)\| \le m^{t/t_0}\|g\|^{1-t/t_0}, \quad \varphi \in M \qquad (4.2.16)$$

Prove that the operators R_α defined by the formula

$$R_\alpha = \left(e^{-At_0} + \alpha E\right)^{-t/t_0}, \quad \alpha > 0 \qquad (4.2.17)$$

constitute a regularizing family on the correctness set M. Since the operator R_α is a function of self-adjoint positive operator A according to (4.2.17), use the spectral expansion to obtain

$$\|R_\alpha\| \le \max_{\lambda \ge 0} \left(e^{-\lambda t_0} + \alpha\right)^{-t/t_0} = \alpha^{-t/t_0} < \infty \qquad (4.2.18)$$

Check conditions (4.1.12) (see § 4.1; the operator e^{-At} plays the role of operator A and $f \equiv g$ since $e^{-At}\varphi(t) = g$): suffice it to say that

$$\|R_\alpha e^{-At}\varphi(t) - \varphi(t)\| \to 0 \quad \text{for} \quad \alpha \to 0, \quad \varphi \in M \qquad (4.2.19)$$

Use again the spectral expansion of the operator A to obtain

$$\|\varphi(t) - R_\alpha e^{-At}\varphi(t)\|$$

$$= \|e^{At}g - R_\alpha g\| \le \max_{\lambda \ge 0} e^{\lambda(t-t_0)} \left[1 - \left(1 + \alpha e^{\lambda t_0}\right)^{-t/t_0}\right] \|e^{At_0}g\|$$

To estimate the value of the right-hand part of this inequality, use

$$(1+x)^\tau - 1 \le \tau x(1+x)^\tau (1+\tau x)^{-1}$$

which holds at $x \ge 0$ and $\tau \in [0,1]$. Since $\varphi \in M$ i. e., $\|\varphi(t_0)\| = \|e^{At_0}g\| \le m$, obtain

$$\max_{\lambda \ge 0} e^{\lambda(t-t_0)} \left[1 - \left(1 + \alpha e^{\lambda t_0}\right)^{-t/t_0}\right] \|e^{At_0}g\|$$

$$\le \alpha^{1-t/t_0} \max_{\alpha \le x < \infty} x^{t/t_0-1} \left[1 - (1+x)^{-t/t_0}\right] m$$

$$\le \alpha^{1-t/t_0} \max_{\alpha \le x < \infty} \frac{t}{t_0} x^{t/t_0} \left(1 + \frac{t}{t_0}x\right)^{-1} m = \frac{t}{t_0}\left(1 - \frac{t}{t_0}\right)^{1-t/t_0} \alpha^{1-t/t_0} m$$

Therefore

$$\|\varphi(t) - R_\alpha g\| \le \frac{t}{t_0}\left(1 - \frac{t}{t_0}\right)^{1-t/t_0} \alpha^{1-t/t_0} m \to 0 \qquad (4.2.20)$$

at $\alpha \to 0$ and $t < t_0$. This proves that $\{R_\alpha\}$ is a regularizing family. Use estimates (4.2.14), (4.2.18), (4.2.20), and the triangle inequality to obtain

$$\|\varphi(t) - R_\alpha g_\varepsilon\| \le \|\varphi(t) - R_\alpha g\| + \|R_\alpha\| \cdot \|g - g_\varepsilon\|$$

$$\le \frac{t}{t_0}\left(1 - \frac{t}{t_0}\right)^{1-t/t_0} \alpha^{1-t/t_0} m + \alpha^{-t/t_0}\varepsilon$$

Simple computations show that the smallest value of the right-hand part of this inequality is present if

$$\alpha_0 = \left(1 - \frac{t}{t_0}\right)^{-2+t/t_0} \frac{\varepsilon}{m}$$

Since at $t \in (0, t_0)$

$$\left(1 - \frac{t}{t_0}\right)^{-(1-t/t_0)^2} \le e^{1/(2e)}$$

then

$$\|\varphi(t) - R_{\alpha_0}g\| \leq e^{1/(2e)}m^{t/t_0}\varepsilon^{1-t/t_0} \qquad (4.2.21)$$

Estimate (4.2.21) for deviation of approximate solution of problem (4.2.13), constructed with the use of operator R_{α_0}, from the exact solution $\varphi(t)$ differs from estimate (4.2.16) of stability on M just by the factor $e^{1/(2e)} \approx 1.21$. In this sense, the method to solve problem (4.2.13) proposed here is optimal.

For practical computation of the component

$$R_{\alpha_0}g_\varepsilon \equiv \left(e^{-At_0} + \alpha_0 E\right)^{-t/t_0} g_\varepsilon$$

assume, appropriately, that

$$R_{\alpha_0}g_\varepsilon \approx Q_n\left(e^{-At_0}\right) g_\varepsilon$$

where $Q_n(x)$ is a polynomial of the best approximation of power n to the function $(x + \alpha_0)^{-t/t_0}$ in the interval $0 \leq x \leq 1$, that is

$$\max_{x \in [0,1]} \left|(x + \alpha_0)^{-t/t_0} - Q_n(x)\right| = \inf \max_{x \in [0,1]} \left|(x + \alpha_0)^{-t/t_0} - P_n(x)\right|$$

while the exact lower bound is assumed all over the algebraic polynomials P_n of power n. The operator $Q_n\left(e^{-At_0}\right)$ is a polynomial in e^{-At_0}, i. e., it is enough to know how to compute the components e^{-kAt_0} at $k = 1, 2, \ldots, n$ in order to compute the component $Q_n\left(e^{-At_0}\right) g_\varepsilon$. In the meantime, $e^{-kAt_0}g_\varepsilon$ is a solution of the correct Cauchy problem

$$\frac{d\psi}{dt} + A\psi = 0, \quad \psi(0) = g_\varepsilon, \quad t \geq 0 \qquad (4.2.22)$$

at $t = kt_0$. We thus reduced incorrect problem (4.2.13) to a sequence of correct problems (4.2.22) effective methods of solution of are discussed in [73]. It may be proven that the final error $\left\|\varphi(t) - Q_n\left(e^{-At_0}\right) g_\varepsilon\right\|$ of the method is estimated from above by the value (see [73])

$$\left(1 - \frac{t}{t_0}\right)^{-(1-t/t_0)^2} \varepsilon^{1-t/t_0}m^{t/t_0}$$

$$+ \frac{2^{2t/t_0+1}}{\Gamma(t/t_0)}(n+1)^{t/t_0-1}\frac{\beta^{n+1+t/t_0}}{(1-\beta^2)^{1+t/t_0}}\|g_\varepsilon\| \qquad (4.2.23)$$

where $\beta = 1 + 2\alpha_0 - 2\sqrt{\alpha_0 + \alpha_0^2}$, and Γ is gamma-function. Notice that the power of polynomials Q_n is not large in practical interesting cases. It follows from (4.2.23) that if, for example,

$$\varepsilon/\|g_\varepsilon\| = 0.1, \quad \|g_\varepsilon\|/m = 0.1$$

it is enough to use the polynomial Q_2 to obtain the accuracy $2\varepsilon^{1-t/t_0}m^{t/t_0}$; at

$$\varepsilon/\|g_\varepsilon\| = 0.05, \quad \|g_\varepsilon\|/m = 0.1$$

it is enough to use the polynomial Q_4. There are two explicit formulas for computation of the coefficients of the polynomial Q_n, which depend only on the ratios ε/m, t/t_0 and are independent of the operator A or the input data g_ε.

4.3. Inverse Evolutionary Problem with Operator Depending on Time

Examine the evolutionary problem

$$\frac{d\varphi}{dt} - A(t)\varphi = 0, \quad 0 \le t \le t_0 \tag{4.3.1}$$
$$\varphi = g \quad \text{at} \quad t = 0.$$

with the operator $A > 0$ depending on time. Assume again that problem (4.3.1) is a result of reduction of a mathematical physics problem in spatial variables to a system of ordinary differential equations. The Fourier method is not applicable in this case, so the numerical methods should be applied.

Let us discuss a possible algorithm of the numerical solution. Put a model problem in correspondence to problem (4.3.1), close to the latter in a certain sense:

$$\frac{d\bar{\varphi}}{dt} - \bar{A}\bar{\varphi} = 0, \quad 0 \le t \le t_0 \tag{4.3.2}$$
$$\bar{\varphi} = g \quad \text{at} \quad t = 0$$

where $\bar{A} > 0$ is an operator which does not depend on time and has a positive spectrum

$$\alpha(\bar{A}) \le \lambda(\bar{A}) \le \beta(\bar{A})$$

and close to the operator $A(t)$ in a certain sense. Assume for the sake of definiteness that

$$A(t) = \bar{A} + \delta A(t) \tag{4.3.3}$$

where

$$\|\delta A(t)\| \ll \|\bar{A}\| \tag{4.3.4}$$

at any t from the interval $0 \le t \le t_0$.

Problem (4.3.2) will provide later necessary information for computation process in solving main problem (4.3.1).

Use the methods discussed in § 4.2 to determine m informative (from the point of view of errors in input data) eigenelements u_n, u_n^* and eigenvalues λ_n $(n = 1, 2, \ldots, m)$. All the remaining Fourier series harmonics for g_n $(n = m + 1,\ m + 2, \ldots, N)$ must be disregarded since errors in their coefficients exceed (sometimes considerably) the useful information. Then assume the function

$$\bar{g} = \sum_{n=1}^{m} g_n u_n \tag{4.3.5}$$

as g, where

$$g_n = (g, u_n^*)$$

As a result, the solution $\bar{\varphi}$ of model problem in the interval $0 \le t \le t_0$ may be represented as

$$\bar{\varphi}(t) = \sum_{n=1}^{m} g_n e^{\lambda_n t} u_n \tag{4.3.6}$$

Now try to solve model problem (4.3.2) numerically. Examine, for example, the difference scheme of the second order of accuracy as related to $\Delta t = \tau$:

$$\frac{\bar{\varphi}^{j+1} - \bar{\varphi}^j}{\tau} - \bar{A}\frac{\bar{\varphi}^{j+1} + \bar{\varphi}^j}{2} = 0, \quad j = 1, 2, \ldots, j_0, \quad \bar{\varphi}^0 = g \tag{4.3.7}$$

Seek a solution to problem (4.3.7) using the Fourier method, assuming to have the full set of eigenelements u_n and u_n^*. The assumption is necessary only for theoretical analysis and a priori information on the solution's behavior. Then

$$\bar{\varphi}^j = \sum_{n=1}^{N} \bar{\varphi}_n^j u_n \tag{4.3.8}$$

As a result, obtain recursion relationships for the Fourier coefficients using (4.3.7)

$$\bar{\varphi}^{j+1} = \frac{1 + \tau\lambda_{n/2}}{1 - \tau\lambda_{n/2}}\bar{\varphi}_n^j, \quad j = 1, 2, \ldots, j_0$$

(4.3.9)

$$\bar{\varphi}_n^0 = g_n$$

Therefore,

$$\bar{\varphi}_n^j = \left[\frac{1 + \tau\lambda_n/2}{1 - \tau\lambda_n/2}\right]^j g_n$$

(4.3.10)

Thus,

$$\bar{\varphi}^j = \sum_{n=1}^{N} T_n^j g_n u_n$$

(4.3.11)

where

$$T_n = \frac{1 + \tau\lambda_n/2}{1 - \tau\lambda_n/2}$$

Assume the step τ as chosen with regard to a condition which requires that the denominator in the expression for T_n does not turn to zero at none of the values of n, e. g.,

$$\tau < 2/\beta(\bar{A})$$

(4.3.12)

Notice that this condition fits the condition of approximation of large perturbations.

A formal analysis of the solution of the model problem as represented in (4.3.11) shows that all the values of $T_n > 1$ and the amplitudes of high-frequency harmonics corresponding to large numbers of n increase rapidly with the number. Therefore $T_n \gg 1$, and especially $T_n^j \gg 1$ for them. Since we have disregarded all the Fourier series harmonics beginning with $n = m + 1$ in processing the input data g, it appears at first sight as sufficient for the Fourier sum

$$\bar{g} = \sum_{n=1}^{m} g_n u_n$$

to generate the solution with the same number of terms

$$\bar{\varphi}^j = \sum_{n=1}^{m} T_n^j g_n u_n$$

(4.3.13)

This might be true in reality, if our computers could compute using an infinite number of digits. Since the length of the computer word is, however, limited, components g_n will appear immediately at $n > m$ in computations because of errors of rounding off. Though small, their large weight in the solution is in proportion to $T_n^j \gg 1$. These errors may eventually distort considerably the main solution of a problem. To avoid the damaging increase of the errors in high-frequency components of the Fourier series, a construction is necessary for automatic conversion of any element of the vector space F into an element of some subspace Φ.

Define Φ, assuming that an element belongs to the subset Φ if the amplitudes of the last $N - m$ harmonics of the Fourier sum of this element on a system of the functions u_n increase in the process of numerical solution of a problem from step to step and no faster than several amplitudes of the last informative harmonics numbered m. In construction of such a subspace, errors of rounding off on its elements will increase no faster than the amplitude of mth harmonics. This will provide a correct computation scheme. Several authors proposed to examine the operator $\bar{A}_\varepsilon = \bar{A} - \varepsilon \bar{A}^2$ instead of $\bar{A}$ in model problem (4.3.2). Instead of problem (4.3.2) the problem is in this case

$$\frac{d\bar{\varphi}_\varepsilon}{dt} - \bar{A}\bar{\varphi}_\varepsilon = -\varepsilon\bar{A}^2\bar{\varphi}_\varepsilon, \quad 0 \le t \le t_0$$
$$\bar{\varphi}_\varepsilon = g \quad \text{at} \quad t = 0 \tag{4.3.14}$$

where ε is still an arbitrarily set parameter. Select it so that the solution of the problem stays within the set Φ. To simplify the analysis, assume that $\bar{A} = \bar{A}^*$. Examine the difference scheme

$$\frac{\bar{\varphi}_\varepsilon^{j+1} - \bar{\varphi}_\varepsilon^j}{\tau} - (\bar{A} - \varepsilon\bar{A}^2)\frac{\bar{\varphi}_\varepsilon^{j+1} + \bar{\varphi}_\varepsilon^j}{2} = 0, \quad \bar{\varphi}_\varepsilon^0 = g \tag{4.3.15}$$

Find the solution of problem (4.3.15) using the Fourier series in eigenfunctions of the operator $\bar{A}$. Thus

$$\bar{\varphi}_\varepsilon^j = \sum_{n=1}^{N} \left[\frac{1 + \tau\lambda_n/2 - \varepsilon\tau\lambda_n^2/2}{1 - \tau\lambda_n/2 - \varepsilon\tau\lambda_n^2/2}\right]^j g_n u_n \tag{4.3.16}$$

Select the parameter ε so that the relative error in harmonic number m did not exceed η due to introduction of the operator $\varepsilon\bar{A}^2$ (η may be usually assumed as $\eta < 1$ depending on relation in harmonic $n \equiv m$ between the relevant information and errors we do not account for (the "noise")). This condition set, obtain

$$\eta \frac{\tau \lambda_m}{2} = \varepsilon \frac{\tau \lambda_m^2}{2} \tag{4.3.17}$$

whence

$$\varepsilon = \eta / \lambda_m \tag{4.3.18}$$

We arrive thus at a definition of a value most important a priory and necessary for further computation. It is easy to see, that the amplitudes of all the harmonics $n > m$ will increase with time no faster than T_m^j at the given parameter ε in (4.3.18).

Finally we need one more a value. Examine

$$\bar{\varphi}^j = \sum_{n=1}^{m} g_n e^{\lambda_n t} u_n \tag{4.3.19}$$

$$\bar{\varphi}_\varepsilon^j = \sum_{n=1}^{N} g_n T_n^j(\varepsilon) u_n \tag{4.3.20}$$

where

$$T_n(\varepsilon) = \frac{1 + \tau \lambda_n / 2 - \varepsilon \tau \lambda_n^2 / 2}{1 - \tau \lambda_n / 2 - \varepsilon \tau \lambda_n^2 / 2}$$

Since the solution $\bar{\varphi}_\varepsilon^j$ belongs to Φ, it may be substituted with no great error for

$$\bar{\varphi}_\varepsilon^j = \sum_{n=1}^{m} g_n T_n^j(\varepsilon) u_n \tag{4.3.21}$$

where we restrict our consideration to the first m members only. The solution may be found as (4.3.21) in a constructive way using the obtained system of functions u_n and u_n^* $(1, 2, \ldots, m)$. Find the values $\bar{\varphi}^j$ and $\bar{\varphi}_\varepsilon^j$ at $j = 1, 2, \ldots, j_0$ from expressions (4.3.19) and (4.3.21). Then introduce vectors

$$\bar{\varphi} = \begin{bmatrix} \bar{\varphi}^1 \\ \bar{\varphi}^2 \\ \ldots \\ \bar{\varphi}^{j_0} \end{bmatrix}, \quad \bar{\varphi}_\varepsilon = \begin{bmatrix} \bar{\varphi}_\varepsilon^1 \\ \bar{\varphi}_\varepsilon^2 \\ \ldots \\ \bar{\varphi}_\varepsilon^{j_0} \end{bmatrix}$$

and calculate the norm

$$\|\bar{\varphi} - \bar{\varphi}_\varepsilon\| = \delta \qquad (4.3.22)$$

This is the last of the sought a priori values. Two other values, τ and ε, are defined by formulas (4.3.12) and (4.3.18).

Now formulate the numerical algorithm for solving the original problem, (4.3.1). Construct the following approximation of the problem, taking into account the analysis above:

$$\frac{\varphi^{j+1} - \varphi^j}{\tau} - (A_j - \varepsilon A_j^2)\frac{\varphi^{j+1} + \varphi^j}{2} = 0, \quad \varphi^0 = g \qquad (4.3.23)$$

where τ and ε are selected on the basis of simple model studied here:

$$\tau < 2/\beta(\bar{A}), \quad \varepsilon = \eta/\lambda_m(\bar{A}). \qquad (4.3.24)$$

Introduce vectors

$$\varphi = \begin{bmatrix} \varphi^1 \\ \varphi^2 \\ \cdots \\ \varphi^{j_0} \end{bmatrix}, \quad f = \begin{bmatrix} -R_0 g \\ 0 \\ \cdots \\ 0 \end{bmatrix}$$

and a matrix

$$\Lambda = \begin{bmatrix} -S_0 & 0 & 0 & 0 & \cdots & 0 & 0 \\ R_1 & -S_1 & 0 & 0 & \cdots & 0 & 0 \\ 0 & R_2 & -S_2 & 0 & \cdots & 0 & 0 \\ 0 & 0 & R_3 & -S_3 & \cdots & 0 & 0 \\ \cdots & \cdots & \cdots & \cdots & \cdots & \cdots & \cdots \\ 0 & 0 & 0 & 0 & \cdots & R_{j_0-1} & -S_{j_0-1} \end{bmatrix}$$

where

$$S_j = E - \frac{\tau}{2}(A_j - \varepsilon A_j^2), \quad R_j = E + \frac{\tau}{2}(A_j - \varepsilon A_j^2), \quad A_j = A(t_{j+1/2})$$

and obtain a problem

$$\Lambda\varphi = f \qquad (4.3.25)$$

Symmetrize problem (4.3.25) by multiplying both parts by Λ^*:

$$\Lambda^*\Lambda\varphi = \Lambda^*f \qquad (4.3.26)$$

and formulate an iterative process. The method of adjoint gradients may be used for it; it does not require a priori knowledge of the boundaries of the spectrum $\Lambda^*\Lambda$.

Formulation of successive approximations method does not, however, make the description of the algorithm complete. It is still necessary to determine the optimum number of iterations k_0 which provide the maximum possible accuracy for conditions given a priori. Since so great a number cannot be found at a fairly high degree of precision, assume that the a priori estimate of approximation (4.3.22) obtained for the model problem fits problem (4.3.1), too. Assume therefore that

$$\|\varphi - \varphi_\varepsilon\| = \delta \qquad (4.3.27)$$

where φ is the exact solution of problem (4.3.1) at the nodes of the grid, and φ_ε is the solution of the difference problem with a regularizing operator. Then it would be logical to continue this iteration process until the iterative process error exceeds the error of approximation (4.3.27). The iteration should be stopped as the errors become equal. The simplest algorithmic procedure for it is the following. Introduce the residual vector ξ^k with the formula

$$\xi^k = \Lambda^*(\Lambda\varphi^k - f) = \Lambda^*\Lambda(\varphi^k - \varphi) \qquad (4.3.28)$$

Then the estimate is

$$\|\xi^k\| \le \|\Lambda^*\Lambda\| \|\varphi^k - \varphi\| \qquad (4.3.29)$$

The value $\|\varphi^k - \varphi\|$ should apparently be equivalent to δ which leads to the requirement

$$\|\xi^k\| \le \delta\|\Lambda^*\Lambda\| \qquad (4.3.30)$$

This means, that computations should be continued until the residual norm $\|\xi^k\|$ is comparable with the value in the right-hand part of inequality (4.3.30) and thus a parametric estimation for k_0 is obtained:

$$\|\xi^{k_0}\| \le \beta(\Lambda^*\Lambda)\delta \qquad (4.3.31)$$

Solving inverse evolutionary problems clearly requires a great deal of preparatory work involving study of various simple models which provide a priori information necessary for building a good computation algorithm. More complicated situations emerge in some particular cases. However, the discussion above gives an idea of certain principles of formation of numerical methods on the basis of study of arising errors and analysis of the algorithm using simple models. We have so far discussed just one point of view on regularizing process, but this one represents possible approaches to numerical solution of inverse problems.

To conclude, it should be said that the methods and ideas discussed above are applicable to numerical solution of the Cauchy problem in equations of the elliptic type. These problems are stated incorrectly in classical sense and require for their solution methods developed in the theory of conditionally correct problems.

4.4. Statement of Inverse Problems on the Basis of Adjoint Equations Methods and Perturbation Theory

The statements of certain inverse problems on the basis of adjoint functions and perturbation theory methods play an increasing role in the development of computing algorithms, especially for solving complex problems in mathematical physics, where it is difficult to estimate a priori the influence of individual factors on problem solutions. This becomes very important in planning the experiments where obtaining the most informative sets of functionals is essential.

Let us first discuss a few issues of the linear theory of measurements.

The linear theory of measurements is currently becoming very important for organization of information systems. Measurement techniques allow to produce sets of data about a process (functionals), analyze the process and control it. The functionals in fact help in interpreting physical processes.

We will not discuss simple measurements, like those of voltage and amperage in elements of electric circuits, etc. We will focus only on complex physical phenomena and processes to be understood and evaluated quantitatively with a required degree of accuracy. Such tasks always arise, especially in new fields of technology. For instance, methods for measurement of neutron multiplication in a reactor cannot be developed, unless the physics of the chain reaction and diffusion of neutrons is clear in full detail and its equations describing the operation of a nuclear reactor when different conditions vary, are known.

Measurement techniques and instruments improve of course with the development of theory of physical process. New or improved measurement techniques follow as a rule progress in the theory and experimentation.

Could more or less general approach to measurement techniques be developed and applied to different processes which allows formally to describe the algorithm mathematically? Such approach may be formulated at least for problems with linear operators. We will be discussing below just that type of problems.

The perturbation theory may be a likely basis for the main theory of measurements of variations of physical values. Essentially, if we study a complex physical process using instruments with certain physical characteristics, the readings will be linked with the field of the studied physical value and be thus functionals of the field. In most cases, though, the goal of the experiment is not the field of physical value, but deviations from this field caused by perturbations, usually small. This means, that measurements require a degree of accuracy sufficient to register such deviations of a field from a "standard." Suppose that this first prerequisite concerning the instrument is met, we have fairly accurate measurements of reading deviations from the standard. The question is, if this information suffices to interpret the experiment adequately, and if we can restore the information about perturbed state of the system accurately. Unfortunately it is very difficult to give a positive answer to this question in most cases. As a matter of fact, the problems of restoration of information about the field of physical values by measuring instruments are incorrectly stated problems of mathematical physics as a rule.

To overcome that principal difficulty arising in processing experimental data, reading deviations should be linked immediately from the very beginning with deviations in parameters of the studied process. The error in the characteristic in question will then be directly proportional to that in the reading deviation, i. e., to the variation in the functional. We will then be using the maximum information of the instrument as we interpret the process. This is the approach we will be using as we discuss below the theory which will repeat and widen to a certain extent the matters discussed earlier.

Examine a function $\varphi(x)$ satisfying the equation

$$A\varphi(x) = q(x) \qquad (4.4.1)$$

where A is some linear operator, and $q(x)$ is the distribution of sources in a medium. We will denote with x a totality of all variables of the problem (temporal and spatial coordinates, energy and velocity direction), assuming that the functions φ and q are real.

For the sake of definiteness assume, for instance, that the process we are studying is linked with diffusion or transport of substance, though conclusions of the theory go far beyond problems of this type.

Introduce a Hilbert space of real functions H with an inner product

$$(g, h) = \int g(x)h(x)dx \qquad (4.4.2)$$

where the integration is made all over the definition domain Ω of the functions g and h. Assume that $D(A) \in H$ is a given definition domain of the operator A and contains a set of functions among which we seek the solution of problem (4.4.1).

A physical value, which is a functional of $\varphi(x)$, is usually sought in physical problems. Any value happening to be a linear continuous functional of $\varphi(x)$ may be expressed as an inner product. For instance, if we are interested in a measurement of some process in a medium with the instrument characteristic $\Sigma(x)$, the value is

$$J_\Sigma[\varphi] = \int \varphi(x)\Sigma(x)dx = (\varphi, \Sigma) \qquad (4.4.3)$$

We will thus examine physical values which may be expressed as linear continuous functionals of $\varphi(x)$:

$$J_p[\varphi] = (\varphi, p)$$

where the value p characterizes the physical process we are interested in. As before, together with operator A introduce an adjoint operator

A^* determined from

$$(g, Ah) = (h, A^*g) \qquad (4.4.4)$$

for any functions $g \in D(A)$ and $h \in D(A^*)$, where $D(A)$ and $D(A^*)$ are the definition domains respectively for the operators A and A^*. Add to the main equation, (4.4.1), a formally nonhomogeneous adjoint equation

$$A^*\varphi_p^* = p(x) \qquad (4.4.5)$$

where $p(x)$ is a certain still arbitrary function, and $\varphi_p^* \in D(A^*)$. Substituting functions h and g in formula (4.4.4) with solutions of equations (4.4.1) and (4.4.5) φ and φ_p^*, obtain

$$(\varphi_p^*, A\varphi) = (\varphi, A^*\varphi_p^*) \qquad (4.4.6)$$

or, using equations (4.4.1)–(4.4.5),

$$(\varphi_p^*, q) = (\varphi, p) \qquad (4.4.7)$$

in other words,

$$J_q[\varphi_p^*] = J_p[\varphi]$$

Therefore we can obtain the functional $J_p[\varphi]$ either solving equation (4.4.1) and determining this value with the formula

$$J_p[\varphi] = (\varphi, p) \qquad (4.4.8)$$

or solving equation (4.4.5) and determining the same value with the formula

$$J_p[\varphi] = J_q[\varphi_p^*] = (\varphi_p^*, q) \qquad (4.4.9)$$

Hence every linear functional $J_p[\varphi] = (\varphi, p)$ can be provided with a corresponding function $\varphi_p^*(x)$ satisfying equation (4.4.5), the function $p(x)$, characterizing the process, being a free member of this equation.

Assume a "unit power source" located at the point x_0, i. e.,

$$q(x) = \delta(x - x_0) \qquad (4.4.10)$$

Since[1]

$$(\varphi(x), \delta(x - x_0)) = \varphi(x_0) \qquad (4.4.11)$$

---

[1] Here the Dirak function $\delta(x-x_0)$ is understood as the functional $(\delta(x-x_0), \varphi) = \int \delta(x-x_0)\varphi(x)dx = \varphi(x_0)$ where $\varphi(x)$ is continuous function [30] (see also the footnote at page 40).

then

$$J_p[\varphi] = J_{q=\delta(x-x_0)}[\varphi_p^*] = \varphi_p^*(x_0) \qquad (4.4.12)$$

Hence the adjoint function $\varphi_p^*(x)$ describes the dependence of the functional $J_p[\varphi] = (\varphi, p)$ on the unit power source location.

Assume a physical system (or instrument) measuring a value $J_p[\varphi]$ which is a linear functional of the solution linked, for example, with the density φ of particles in a substance. If a quantity of particles enter a point of the system (or, on the contrary, a quantity leaves) the measured value of $J_p[\varphi]$ will relatively increase (or decrease). The variation will depend on the point where the number of particles varies. As is clear from the matter discussed above, the dependence is described by an adjoint function $\varphi_p^*(x)$ satisfying equation (4.4.5). Therefore the adjoint function $\varphi_p^*(x)$ reflects the contribution of particles located at a certain point of a system to the functional $J_p[\varphi]$ we are seeking.

As pointed out in Chapter 1, understanding of the adjoint function $\varphi_p^*(x)$ as an importance of information provides a clear interpretation of the perturbation theory for any functional $J_p[\varphi]$, too. If we alter in effect the number of particles in an element of volume Δx near the point x by δN, the corresponding variation in the value J_p will be expressed as

$$\delta J_p = \delta N \varphi_p^*(x) \qquad (4.4.13)$$

If small parameter variations occur in the system, so that the operator A transforms into $A + \delta A$, it will correspond to the variation in the number of particles in every element Δx by a value $\delta N = -\Delta x \delta A \varphi$. Write the total variation in the functional J_p as

$$\delta J_p = - \int \varphi_p^*(x) \delta A \varphi(x) dx \qquad (4.4.14)$$

Equation (4.4.13) makes it possible to measure distribution of value function in a system by varying the number of particles at different points x of the system and measuring the corresponding variation in the value J_p. The notion of imporatnce we introduced may be useful in the theory of measuring instruments. An instrument is usually designed to measure an isolated variable J_p. Quite a definite importance function $\varphi_p^*(x)$ may therefore be introduced, once measured or calculated for every instrument. If the distribution of substance and that of its value are known, equation (4.4.14) may be used for the measurements in two ways. First, by measuring the values δJ_p at different variations in the medium parameters δA we may use (4.4.14) to determine the values δA, provided the structure of the operator A is known, but no parameters which belong to this operator, i. e., characteristics of interaction of particles and the matter are known. For example, this technique may be (and in fact is) used to measure cross-sections of interaction of neutrons and the matter in different samples, by placing the samples into the instrument and determining $\delta\Sigma = \delta A$ by variations in the variable J_p. Second, (4.4.14) provides for corrections for the measured variable J_p on the account of various perturbing factors in the instrument.

The formulas given above may also lead to a reciprocity theorem for the Green functions, respectively $G(x, x_0)$ and $G^*(x, x_1)$ for the main and adjoint equations. The function $G(x, x_0)$ satisfies equation (4.4.1) for $q(x) = \delta(x - x_0)$, and function $G^*(x, x_1)$ satisfies equation (4.4.5) for $p(x) = \delta(x - x_1)$.

Substituting into formula (4.4.7)

$$\varphi(x) = G(x, x_0), \quad \varphi_p^*(x) = G^*(x, x_1)$$

and above expressions for q and p, obtain

$$G(x_1, x_0) = G^*(x_0, x_1) \tag{4.4.15}$$

which is the formulation of the reciprocity theorem.

Let us now discuss briefly the perturbation theory for linear functionals. We will restate it.

If the properties of a medium a field interacts with vary, i. e., the operator of equation (4.4.1) transforms into

$$A' = A + \delta A$$

the field $\varphi(x)$ also varies as a value of the functional $J_p[\varphi]$:

$$\varphi(x) \to \varphi'(x), \quad J_p[\varphi] \to J_p = J_p + \delta J_p$$

Establish a link between variation in the operator δA and variation in the functional δJ_p. A perturbed system is described as

$$A'\varphi' = (A + \delta A)\varphi' = q \tag{4.4.16}$$

The adjoint function of a perturbed system corresponding to the functional J_p is described as

$$A^*\varphi_p^* = p \qquad (4.4.17)$$

Multiplying equation (4.4.16) by φ^*, equation (4.4.17) by φ', subtracting the results from each other, and using the definition of adjoint operator to equation (4.4.4), obtain in the left-hand part

$$(\varphi_p^*, A'\varphi') - (\varphi', A^*\varphi_p^*) = (\varphi_p^*, \delta A\varphi') \qquad (4.4.18)$$

and, in accordance with equation (4.4.7), in the right-hand part

$$(\varphi_p^*, q) - (\varphi', p) = J_p[\varphi] - J_p[\varphi'] = -\delta J_p \qquad (4.4.19)$$

Equate (4.4.18) and (4.4.19) to obtain the general equation for the functional increment:

$$\delta J_p = -(\varphi_p^*, \delta A\varphi') \qquad (4.4.20)$$

Examine instead of equations (4.4.16) and (4.4.17) the perturbed adjoint equation

$$(A^* + \delta A^*)\varphi_p^{*\prime} = p \qquad (4.4.21)$$

and unperturbed main equation (4.4.1). Obtain similarly

$$\delta J_p = -(\varphi, \delta A^*\varphi_p^{*\prime}) \qquad (4.4.22)$$

which is undoubtedly equivalent to (4.4.20).

Notice an important feature of application of perturbation theory formulas: since these formulas are written for functional variation, where the error is allowed to several percents, there is no need to know the exact solution of main and adjoint problems to compute the variation – it will be sufficient to use approximate solutions.

If the perturbation of the operator A (and, consequently, A^*) is so small that it does not distort much the functions φ and φ_p^*, we can substitute approximately φ and φ_p^* in formulas (4.4.20) and (4.4.22) for φ' and $\varphi_p^{*\prime}$ respectively and obtain two equivalent formulas of the small perturbation theory:

$$\delta J_p = -(\varphi_p^*, \delta A\varphi) \qquad (4.4.23)$$

$$\delta J_p = -(\varphi, \delta A^*\varphi_p^*) \qquad (4.4.24)$$

The perturbation theory formulas we obtained may have another very important application, besides that for evaluation of various effects and analysis of measurements.

Simplified models are quite often substituted for studied complex systems in theoretical examination and practical computations. The precondition for the substitution is that it must not make certain basic characteristics of the system vary. An example is substitution of constant coefficient for varying coefficients in differential equations. The effective boundary conditions technique, involving substitution of simplified conditions for original, which results in correct values of selected functionals, also belongs to these methods.

The perturbation theory formulas obtained above make it possible to formulate a general approach to many problems. Examine a system characterized by the operator A; the functional J_p remains the most essential quantity. If the simple model we are looking for is characterized by the operator $A' = A + \delta A$, the following conditions must be satisfied to make sure that J_p does not vary in the transition from the original system to simple model:

$$J_p = - \left(\varphi_p^*, [A' - A] \varphi' \right) = 0 \qquad (4.4.25)$$

i. e., $(\varphi_p^*, A'\varphi') = (\varphi_p^*, A\varphi')$. If we seek several quantities J_{p_1}, J_{p_2}, etc., we obtain several conditions of the type (4.4.25) with the solutions $\varphi_{p_1}^*$, $\varphi_{p_2}^*$, etc.

The requirement (4.4.25) does not determine uniquely the sought equivalent model, but is a prerequisite and may help find it, together with other conditions. In particular, if the model's operator A' the form of which may be found from physical considerations, contains at least one or several parameters, conditions (4.4.25) may be used to determine the parameters.

Let us now discuss numerical techniques for solving inverse problems and the problem of experiment planning.

Assume a set of functionals (measurements) J_{p_i} $(i = 1, 2, \ldots, n)$, and that the measurements are different in character; e. g., characteristics are measured by the same instrument at different "points" of the solution definition domain, or the instruments register different characteristics of the process. Assume, to make it simpler, that the statistical errors are excluded from the measurements, and we deal with a system of preprocessed data.

Supply every functional J_{p_i} with a corresponding importance function for the unperturbed problem, i. e., the model with the known operator A and its definition domain. Solve n different problems

$$A^*\varphi_{p_i}^* = p_i, \quad i = 1, 2, \ldots, n \tag{4.4.26}$$

Find n importance functions $\varphi_{p_i}^*$ in advance and solve one main problem with model "unperturbed" operator A adjoint to A^*:

$$A\varphi = q \tag{4.4.27}$$

Assume that $\varphi \in D(A)$ and $\varphi^* \in D(A^*)$. Make n formulas of the small perturbation theory:

$$\left(\varphi_{p_i}^*, \delta A\varphi\right) = -\delta J_{p_i}, \quad i = 1, 2, \ldots, n \tag{4.4.28}$$

where δA is the difference between the operator A' in question and the model operator A.

Assume now that the operator A is known:

$$A = \sum_{k=1}^{m} [\alpha_k A_k + B_k(\beta_k C_k)] \tag{4.4.29}$$

where A_k, B_k and C_k are elementary linear operators, e. g., either differentiation or integration operators, or combination of both, and $\alpha_k(x)$ and $\beta_k(x)$ are the sought coefficients known in the unperturbed, or model, problem.

Now our goal is to restore the coefficients α_k' and β_k' in

$$A' = \sum_{k=1}^{m} [\alpha_k' A_k + B_k(\beta_k' C_k)] \tag{4.4.30}$$

Use expressions (4.4.29) and (4.4.30) to obtain

$$\delta A = \sum_{k=1}^{m} [\delta\alpha_k A_k + B_k(\delta\beta_k C_k)] \tag{4.4.31}$$

where

$$\delta\alpha_k = \alpha_k' - \alpha_k, \quad \delta\beta_k = \beta_k' - \beta_k$$

Substitute (4.4.31) into (4.4.28). Then, given corresponding conditions, arrive at a system of equations

$$\sum_{k=1}^{m} \left[\left(\varphi_{p_i}^*, \delta\alpha_k A_k\varphi \right) + \left(B_k^*\varphi_{p_i}^*, \delta\beta_k C_k\varphi \right) \right] = -\delta J_{p_i}$$
$$i = 1, 2, \ldots, n \tag{4.4.32}$$

The next task is parametrizing the variations $\delta\alpha_k$ and $\delta\beta_k$. Examine at first a simple case where $\delta\beta_k = 0$ and $\delta\alpha_k$ are constants. Then (4.4.32) transforms into a problem of linear algebra

$$\sum_{k=1}^{m} \delta\alpha_k \left(\varphi_{p_i}^*, A_k\varphi \right) = -\delta J_{p_i}, \quad i = 1, 2, \ldots, n \tag{4.4.33}$$

where $(\varphi_{p_i}^*, A_k\varphi)$ are elements of a matrix computable at assigned $\varphi, \varphi_{p_i}^*$ and A_k.

Assume y as a vector with components $\delta\alpha_k$, F be a vector with components $-\delta J_{p_i}$, and $a_{ik} = (\varphi_{p_i}^*, A_k\varphi)$ as elements of a matrix Λ. Obtain equation

$$\Lambda y = F \tag{4.4.34}$$

If the number n of functionals is equal to the number m of sought variations in the coefficients α_k, system (4.4.34) provides in principle the value of $\delta\alpha_k$. If n is greater than m, system (4.4.34) is overdefined and its solution (if it exists) is found by the method of the least squares assuming that y provides a minimum to the quadratic functional:

$$\|\Lambda y - F\|^2 = \min \tag{4.4.35}$$

The vector y, which minimizes this functional, is sometimes referred to as the quasi-solution of equation (4.4.34). If $n = m$, the methods discussed in [73] (in connection with the analysis of iterative processes where input data are not accurate) solve system (4.4.34).

If $\delta\alpha_k$ and $\delta\beta_k$ are functions, the solution of the inverse problem is provided by a paramatrizing technique. Basically, it is this: assume that a priori analysis (usually statistical and correlational) of the behavior of physical parameters produced a complete orthogonal systems of functions $u_{k,l}(x)$ and $v_{k,l}(x)$ usable for fairly good approximation to functions α_k and β_k at a small $n(k)$, so that

$$\delta\alpha_k(x) = \sum_{l=1}^{n(k)} a_{k,l} u_{k,l}(x)$$

(4.4.36)

$$\delta\beta_k(x) = \sum_{l=1}^{n(k)} b_{k,l} v_{k,l}(x)$$

where $a_{k,l}$ and $b_{k,l}$ are sought coefficients.

Substitute expressions (4.4.36) into (4.4.32) to obtain

$$\sum_{k=1}^{m} \sum_{l=1}^{n(k)} \left[a_{k,l} \left(\varphi_{p_i}^*, u_{k,l} A_k \varphi \right) + b_{k,l} \left(B_k^* \varphi_{p_i}^*, v_{k,l} C_k \varphi \right) \right] = -\delta J_{p_i} \qquad (4.4.37)$$

$$i = 1, 2, \ldots, n$$

Regularize now the quantities $a_{k,l}$ and $b_{k,l}$ and redenote them as y_j $(j = 1, 2, \ldots)$. Introduce a matrix Λ which makes the equation

$$\Lambda y = F$$

equivalent to system (4.4.37) and obtain again linear-algebraic problem (4.4.37). Solve it to find $a_{k,l}$ and $b_{k,l}$ and, consequently, $\delta\alpha_k$ and $\delta\beta_k$.

We have examined only a case where a solution of the model problem is close to that of the real, i. e., φ might be substituted for φ' and thus the small perturbation theory applied. If the unperturbed (or model) process differs considerably from the real, the algorithm discussed above may be considered only as the first approximation to the solution of the inverse problem. After the variations $\delta\alpha_k$ and $\delta\beta_k$ have been found, it will be possible to correct the coefficients α_k and β_k and find

$$\alpha_k' = \alpha_k + \delta\alpha_k, \quad \beta_k' = \beta_k + \delta\beta_k$$

This done, it will be necessary to solve the "perturbed" problem

$$A'\varphi' = f \qquad (4.4.38)$$

with the operator

$$A' = \sum_{k=1}^{m} [\alpha_k' A_k + B_k(\beta_k' C_k)]$$

and to proceed to a new approximation in solving the inverse problem
with a more common perturbation formula instead of (4.4.32):

$$\sum_{k=1}^{m} \left[\left(\varphi_{p_i}^*, \delta\alpha_k A_k \varphi' \right) + \left(B_k^* \varphi_{p_i}^*, \delta\beta_k C_k \varphi' \right) \right] = -\delta J_p, \quad i = 1, 2, \ldots, n \quad (4.4.39)$$

and repeat the computation cycle to refine the variations $\delta\alpha_k$ and $\delta\beta_k$.
We will refer to this as the second approximation in the solution of the
an inverse problem. The procedure may be continued, of course. The
convergence of the successive approximations techniques may be proven
with regard to specific information about the elementary operators A_k
of the problem and definition domains $D(A)$ and $D(A^*)$.

A simple example will illustrate the algorithm. Examine a problem

$$-\frac{d}{dx}\beta(x)\frac{d\varphi'}{dx} + \alpha(x)\varphi' = f(x),$$
$$\varphi'(0) = \varphi'(1) = 0 \quad (4.4.40)$$

with unknown coefficients $\alpha(x)$ and $\beta(x)$ for which a priori assumption
is made. Assume, for instance, that they are continuous functions in
the definition domain $0 \le x \le 1$ of the solution and their approximate
values $\bar{\alpha}$ and $\bar{\beta}$ are known, i. e.,

$$\alpha(x) = \bar{\alpha} + \delta\alpha(x), \quad \beta(x) = \bar{\beta} + \delta\beta(x) \quad (4.4.41)$$

If the values of $\alpha(x)$ and $\beta(x)$ are selected in a model on the basis of
a priori information more precisely, there is no need to suppose that
they are equal to the constants $\bar{\alpha}$ and $\bar{\beta}$.

Besides, it may be concluded from the preliminary analysis that
$\delta\alpha(x)$ and $\delta\beta(x)$ may be represented as finite sums

$$\delta\alpha(x) = \sum_{l=1}^{n(1)} a_l u_l(x), \quad \delta\beta(x) = \sum_{l=1}^{n(1)} b_l v_l(x) \quad (4.4.42)$$

where $\{u_i(x)\}$ and $\{v_i(x)\}$ are complete orthonormal systems of functions (e. g., trigonometrical polynomials, Legendre polynomials, etc.).

Let $p_1(x)$, $p_2(x)$, $\ldots$, $p_n(x)$ be measurement characteristics and
every measurements registers the functional

$$J_{p_i}'[\varphi'] = \int_0^1 p_i(x)\varphi'(x)dx, \quad i = 1, 2, \ldots, n \quad (4.4.43)$$

The functions $p_i(x)$ may be referred to as instrument characteristics for a given measurement.

Introduce an unperturbed, or model, problem corresponding to problem (4.4.40):

$$-\frac{d}{dx}\bar{\beta}\frac{d\varphi}{dx} + \bar{\alpha}\varphi = f, \quad \varphi(0) = \varphi(1) = 0 \qquad (4.4.44)$$

Formulate, along with (4.4.44), n adjoint problems related to the chosen model

$$-\frac{d}{dx}\bar{\beta}\frac{d\varphi^*_{p_i}}{dx} + \bar{\alpha}\varphi^*_{p_i} = p_i(x),$$

$$\varphi^*_{p_i}(0) = \varphi^*_{p_i}(1) = 0, \quad i = 1, 2, \ldots, n \qquad (4.4.45)$$

According to the general theory

$$J_{p_i}[\varphi] = \int_0^1 p_i(x)\varphi(x)dx = \int_0^1 f(x)\varphi^*_{p_i}(x)dx \qquad (4.4.46)$$

Now assume that model problems (4.4.44) and (4.4.45) are solved. Then the functional variations δJ_{p_i} may be found with the formula

$$\delta J_{p_i} = J'_{p_i} - J_{p_i}, \quad i = 1, 2, \ldots, n \qquad (4.4.47)$$

where J'_{p_i} is the instrument measurement with characteristic p_i corresponding to (4.4.43) (where φ' is unknown) and J_{p_i} is a functional computed theoretically on the basis of any of relationships in (4.4.46). The accuracy of measurements here must guarantee computation of variations δJ_{p_i}.

Examine now small perturbations theory formulas (4.4.32):

$$A_k = E, \quad B_k = -\frac{d}{dx}, \quad C_k = \frac{d}{dx}, \quad m = 1$$

Given boundary conditions for $\varphi^*_{p_i}$ and φ,

$$\int_0^1 \left(\delta\alpha\varphi\varphi^*_{p_i} + \delta\beta\frac{d\varphi}{dx}\frac{d\varphi^*_{p_i}}{dx} \right) dx = -\delta J_{p_i} \qquad (4.4.48)$$

Substitute the expressions for $\delta\alpha(x)$ and $\delta\beta(x)$ in (4.4.42) into (4.4.48) to obtain

$$\sum_{l=1}^{n(1)} \left(a_l \int_0^1 u_l \varphi \varphi_{p_i}^* dx + b_l \int_0^1 v_l \frac{d\varphi}{dx} \frac{d\varphi_{p_i}^*}{dx} dx \right) = -\delta J_{p_i} \qquad (4.4.49)$$

$$i = 1, 2, \ldots, n.$$

If $n = 2n(1)$, system (4.4.49) is fully defined.

Solve this system to find the coefficients a_l and b_l and obtain the first approximation to the quantities α' and β' on the basis of representation (4.4.42). These quantities can be refined by the successive approximations method discussed above. More complex inverse problems may be stated and solved in the same way, including the problem of determination of source perturbations δf.

Let us now discuss planning of complex experiments. The problem may be formulated as follows: select among all possible (or practically realizable) sets of measurements the most informative from the point of view of solving a concrete inverse problem of restoration of required medium characteristics (coefficients of equations). This problem proves very complicated in terms of general optimization. Certain particular approaches to its solution may however be examined.

Let us suppose that a model of the unperturbed problem is built before the experiment to describe linear functionals of the solution. The information available a priori supplies a conclusion about required precision of measurements of the functionals. Assume that the measurements of the functionals δJ_{p_i} meet the precision requirements. Then aggregates of measurements are examined and chosen to provide the best conditioning for the matrix Λ. The obtained system of linear equations is then solved easily, so such a plan of the experiment is likely the best in a certain sense, within a given set of plans (we do not discuss costs here, though it may be decisive in experimentation planning). If the given set of informative functionals conditioning the matrix Λ the best does not meet the requirement of a high precision of measurement of the functionals J_{p_i}, a more complex still problem of experiment planning arises with assigned restrictions imposed on precision of measurements allowed by the resolution of the instruments. This will be a peculiar optimization problem with imposed restrictions.

4.5. Formulation of Perturbation Theory
for Complex Nonlinear Models

It is not usually simple to build a mathematical model describing a complex process or phenomena. Models must account as a rule for many and various effects by no means all described with a proper accuracy. This means the use in one or another moment of a simplified mathematical formulation which describes only a small number of process characteristics and disregards many, sometimes important, details. However, such examinations provide general descriptions of processes with the required accuracy as a rule. As concerns the estimation of the effect unaccounted by the mathematical model, it may be done using the perturbations theory defined in a special way as discussed in § 4.4.

This section discusses a more or less common approach to building and analysis of mathematical models.

Compared with problems discussed above the new feature in this section will be transition from linear processes to those described by nonlinear equations. We will prove that nonlinear processes allow approaches used in dealing with linear problem, provided certain approximations. Various methods of linearization may be applied, of course.

Examine a steady-state process described by an equation in the following operator form:

$$A\varphi = f \qquad (4.5.1)$$

where A is a matrix-differential operator depending on the solution vector function $\varphi \in D(A)$ and input data $\alpha_1, \alpha_2, \ldots, \alpha_n$, functions of coordinates; f is an assigned vector of sources, functions of coordinates and given parameters $\beta_1, \beta_2, \ldots, \beta_m$. Hence

$$A = A(\varphi, \alpha_1, \alpha_2, \ldots \alpha_n), \quad f = f(\beta_1, \beta_2, \ldots, \beta_m)$$

Let $\bar{\alpha}_1, \bar{\alpha}_2, \ldots, \bar{\alpha}_n$, and $\bar{\beta}_1, \bar{\beta}_2, \ldots, \bar{\beta}_m$ be data corresponding to a standard state of a system. The unperturbed state of the system will then be described as

$$A(\bar{\varphi}, \bar{\alpha}_1, \bar{\alpha}_2, \ldots \bar{\alpha}_n)\bar{\varphi} = f(\bar{\beta}_1, \ldots, \bar{\beta}_m). \qquad (4.5.2)$$

Assume that

$$\bar{A} = A(\bar{\varphi}, \bar{\alpha}_1, \bar{\alpha}_2, \ldots \bar{\alpha}_n), \quad \bar{f} = f(\bar{\beta}_1, \ldots, \bar{\beta}_m)$$

and rewrite (4.5.2) as

$$\bar{A}\bar{\varphi} = \bar{f} \qquad (4.5.3)$$

Assume now that the solution of equation (4.5.3) is known.

Assume also that the real state of a system we will refer to as the perturbed state hereinafter, is governed by equation (4.5.1) where input data differ slightly from the standard, i. e.,

$$\alpha_i = \bar{\alpha}_i + \delta\alpha_i, \quad \beta_i = \bar{\beta}_i + \delta\beta_i$$

where the deviations $\delta\alpha_i$ and $\delta\beta_i$ are assumed as small, in a certain sense, as compared with $\bar{\alpha}_i$ and $\bar{\beta}_i$ relatively.

Then instead of (4.5.2) we have

$$A(\bar{\varphi} + \delta\varphi, \bar{\alpha}_i + \delta\alpha_i)(\bar{\varphi} + \delta\varphi) = f(\bar{\beta}_j + \delta\beta_j) \qquad (4.5.4)$$

The following representation is used here:

$$\varphi = \bar{\varphi} + \delta\varphi$$

Assume a priori that $\delta\varphi$ is "far less" than $\bar{\varphi}$. If the operator A, the solution, and the input data are sufficiently smooth, examine the expansions

$$A(\bar{\varphi} + \delta\varphi, \bar{\alpha}_i + \delta\alpha_i) = \bar{A} + \frac{\overline{\partial A}}{\partial\varphi}\delta\varphi + \frac{\overline{\partial A}}{\partial\alpha_i}\delta\alpha_i + \dots$$

$$\qquad (4.5.5)$$

$$f(\bar{\beta}_j + \delta\beta_j) = \bar{f} + \frac{\overline{\partial f}}{\partial\beta_j}\delta\beta_j$$

Substitute (4.5.5) into (4.5.4) and retain the members of the first order only to obtain[2]

$$\bar{A}\bar{\varphi} + \left(\bar{A} + \frac{\overline{\partial A}}{\partial\varphi}\bar{\varphi}\right)\delta\varphi + \frac{\overline{\partial A}}{\partial\alpha_i}\bar{\varphi}\delta\alpha_i = \bar{f} + \frac{\overline{\partial f}}{\partial\beta_j}\delta\beta_j \qquad (4.5.6)$$

Use equation (4.5.3) to obtain

$$\left(\bar{A} + \frac{\overline{\partial A}}{\partial\varphi}\bar{\varphi}\right)\delta\varphi = \frac{\overline{\partial f}}{\partial\beta_j}\delta\beta_j - \frac{\overline{\partial A}}{\partial\alpha_i}\bar{\varphi}\delta\alpha_i \qquad (4.5.7)$$

---

[2] we suppose hereinafter that $\left(\frac{\overline{\partial A}}{\partial\varphi}\delta\varphi\right)\bar{\varphi} = \left(\frac{\overline{\partial A}}{\partial\varphi}\bar{\varphi}\right)\delta\varphi$, where $\frac{\overline{\partial A}}{\partial\varphi}$ is the Gateàux derivative of the operator A at the point $\bar{\varphi}$ (see Chapter 3).

This is the basic equation for determining small deviations in the solution φ from the unperturbed state.

Assume now that the operator $\bar{A}$ of the unperturbed state and the source $\bar{f}$ are known with a limited accuracy, i. e.,

$$\bar{A} = \bar{\Lambda} + \varepsilon, \quad \bar{f} = \bar{F} + \xi \tag{4.5.8}$$

where ε is the operator of model error

$$\varepsilon = \bar{A} - \Lambda$$

and ξ is the error of sources vector function

$$\xi = \bar{f} - \bar{F}$$

Assume that

$$\|\bar{A}\| \gg \|\varepsilon\|, \quad \|\bar{f}\| \gg \|\xi\| \tag{4.5.9}$$

The norms of the operator and vector-functions are defined in corresponding spaces here.

Substitute (4.5.8) into (4.5.7) to obtain

$$\left(\bar{\Lambda} + \frac{\overline{\partial\Lambda}}{\partial\varphi}\bar{\varphi}\right)\delta\varphi = \frac{\overline{\partial F}}{\partial\beta_j}\delta\beta_j - \frac{\overline{\partial\Lambda}}{\partial\alpha_i}\bar{\varphi}\delta\alpha_i + \eta \tag{4.5.10}$$

where

$$\eta = \frac{\overline{\partial\xi}}{\partial\beta_j}\delta\beta_j - \frac{\overline{\partial\varepsilon}}{\partial\alpha_i}\bar{\varphi}\delta\alpha_i - \left(\varepsilon + \frac{\overline{\partial\varepsilon}}{\partial\varphi}\bar{\varphi}\right)\delta\varphi \tag{4.5.11}$$

Assuming (4.5.9), rewrite (4.5.10) as

$$\left(\bar{\Lambda} + \frac{\overline{\partial\Lambda}}{\partial\varphi}\bar{\varphi}\right)\delta\varphi = \frac{\overline{\partial F}}{\partial\beta_j}\delta\beta_j - \frac{\overline{\partial\Lambda}}{\partial\alpha_i}\bar{\varphi}\delta\alpha_i + O(\|\varepsilon\| + \|\xi\|)$$

Hence

$$\left(\bar{\Lambda} + \frac{\overline{\partial\Lambda}}{\partial\varphi}\bar{\varphi}\right)\delta\varphi = \frac{\overline{\partial F}}{\partial\beta_j}\delta\beta_j - \frac{\overline{\partial\Lambda}}{\partial\alpha_i}\bar{\varphi}\delta\alpha_i \tag{4.5.12}$$

with an accuracy up to the small quantities.

Equation (4.5.12) is a model for computation of deviation from the unperturbed state of a system where the input data vary by quantities $\delta\alpha_i$ and $\delta\beta_j$.

Introduce the following notations:

$$L = \bar{\Lambda} + \frac{\overline{\partial \Lambda}}{\partial \varphi} \bar{\varphi}, \quad \delta F = \frac{\overline{\partial F}}{\partial \beta_j} \delta \beta_j - \frac{\overline{\partial \Lambda}}{\partial \alpha_i} \bar{\varphi} \delta \alpha_i \qquad (4.5.13)$$

Then obtain finally

$$L\delta\varphi = \delta F \qquad (4.5.14)$$

and rewrite the formal solution of the problem as

$$\delta\varphi = L^{-1}\delta F \qquad (4.5.15)$$

Formula (4.5.15) of the perturbation theory is very convenient when it is necessary to define deviations in the solution for only one input data set. When directed variations in the model state are planned, it is very important to make a series of test computations. We should take into account the fact that a great number of different solutions is necessary for estimation of the sensitivity of a model to variations in parameters or for obtaining optimum correlations between parameters. Broader applicable perturbation theories are thus necessary for functionals to study thoroughly the mathematical model with varying input data.

Just as we do in a linear case, add to the main equation

$$L\delta\varphi = \delta F \qquad (4.5.16)$$

the adjoint equation

$$L^*\varphi^* = p \qquad (4.5.17)$$

where L and L^* are the operators adjoint in the Lagrange sense:

$$(Lg, h) = (g, L^*h) \qquad (4.5.18)$$

Here g and h are elements of the Hilbert space in the definition domain of the operators L and L^* respectively. Assume the function p as still undefined.

Multiply (4.5.16) and (4.5.17) by φ^* and φ respectively and subtract the results from each other to obtain

$$(L\delta\varphi, \varphi^*) - (\delta\varphi, L^*\varphi^*) = (\delta F, \varphi^*) - (\delta\varphi, p) \qquad (4.5.19)$$

The expression in the left-hand part of (4.5.19) equals zero in view of (4.5.18), hence

$$(\delta\varphi, p) = (\delta F, \varphi^*) \qquad (4.5.20)$$

Examine now a set of linear functionals

$$\delta J_n = (\delta\varphi, p_n) \qquad (4.5.21)$$

If, in particular, $p_n = \delta(x - x_0)$, then

$$\delta J_n = \delta\varphi(x_n) \qquad (4.5.22)$$

Denote the adjoint function φ^* corresponding to p_n as φ_n^*.

Equality (4.5.20) thus produces a set of functionals

$$\delta J_n = (\delta F, \varphi_n^*) \qquad (4.5.23)$$

Assume to have chosen N functionals J_1, J_2, ... J_N in advance and solved respectively N adjoint problems

$$L^*\varphi_n^* = p_n, \quad n = 1, 2, \ldots, N \qquad (4.5.24)$$

By (4.5.23), there is no need to compute the variations $\delta\varphi$ corresponding to different sets of parameters $\delta\alpha_i$ and $\delta\beta_i$, since it will be possible to compute immediately values of the functionals δJ_n at any perturbations in the input data using φ_n^* $(n = 1, 2, \ldots, N)$.

PART II

Problems of Environment and Optimization
Methods on the Basis of Adjoint Equations

The number of complex problems which may be investigated with a help of adjoint equations technique is enormous indeed. These are the problems in quantum mechanics, nuclear power, nonlinear kinetic processes and many others.

We shall discuss in detail in this book just the problems of environment and climate. They are intermingled and become the central problems for the life on the Earth.

The first problem is the environment and its protection. In order to "protect" one should comprehend the processes of diffusion and propagation of a substance in real atmosphere with real discharges of contaminating substances which convert, migrating in the atmosphere, from harmless components into harmful ones and vice versa, polluting the oceans as well as the continents. It is necessary to construct the complete system of these processes' kinetics.

These problems were in the center of our attention for many years. The following conception was suggested: since the world will evolve, and the civilization will develop, we should know in advance, how to combine growing industrial power generated by the civilization, with the problems of protection of nature and environment. The conception implies, that considering any project, one should first of all precompute the changes in the environment, the implementation of the project will bring with. Even now one needs to make the necessary investments in order not only to compensate the damages to nature and losses of resources, but also to improve nature and environment. Of course, large expenditures will be required. But, the problem is so important, that this global efforts will pay back. Indeed, it is possible to come to the agreement on nuclear weapon ban. There is an urgent need as well to reach the agreement on preserving the ecosystem which supports the life on the Earth. Therefore, this question becomes common to all mankind. The questions of that kind are considered already in the world scientific society. The United Nations Organization appeals even to conclude the treaties on utilization of the nature of all countries in the interests of all mankind. A nature of each country and its ecological components are the constituent part of the whole mankind's property. When it will be understood by everybody, new effective methods to solve the above problems will appear. They will allow both to develop the civilization, and to preserve and increase natural wealth.

In order to solve these very problems, mathematical techniques and imitation models are developed. The adjoint equations considered in this monograph are of great help here, since they allow to estimate the degree of a damage induced by contaminating substances to the biosphere and the changes in planet's ecological regime as a whole. Actually, the scientists have already prepared a mathematical apparatus for imitation calculations both on local and global scale. Our task now is to demonstrate that these models are adequate to the system of measurements of planet's resources changes under the influence of humane activity.

The second problem is that of atmosphere and ocean interaction. Approaching gradually to the statement of problems on atmosphere and ocean interaction the researchers accumulated gradually the knowledge about atmosphere's dynamics and ocean's dynamics as of two more or less independent systems. Later, as the knowledge about these two systems was accumulated, new mathematical apparatus was employed which allowed more detailed description of complex processes. Powerful computers made it possible to pass from simple models to complex hydrothermodynamic models of these two systems which resulted in the development of coupled models of atmosphere and ocean. A key to the solution of problem of climate change on our planet lies namely in the interaction between these two systems. This author had proposed some time ago a model of interaction between atmosphere and ocean which takes into account the peculiarities of this most complicated system. The essence of this model is as follows.

The waters heated in equatorial zones of Atlantic Ocean and Caribbean basin are transported by the Gulf Stream to northern latitudes of the Atlantic Ocean, to the Iceland region. They neighbor there with cold waters and, mixing with them, compose vertically unstable system. The mechanism appears transferring powerful heat flux stored in equatorial waters to the atmosphere of northern regions of

Atlantic Ocean. The air masses heated by oceanic waters meet cold polar air streams in this region of the ocean and form a system which is very unstable hydrodynamically. Cyclones and anticyclones appear to resolve this instability and the planetary west-east transport direct these atmospheric formations carrying storage of heat and moisture to the Europe and the Asian continent.

The same process occurs in the Pacific Ocean. The same scheme acts there. But the main instrument transporting an energy is yet the Kuroshio current. And the region where the heat brought by this current from South China See is released is here the region of the Pacific Ocean close to the Aleutian Islands. The cyclonic activity there is directed to the North-American continent.

It turned out, that intensity of global cyclonic activity depends essentially on the amount of heat energy stored by the Gulf Stream and the Kuroshio in the zones where this energy is generated.

This idea made it possible to construct the conception of so called energetically active zones in the ocean. The national project was formed in USSR on the basis of this conception which is carried out beginning from 1981. It implies the study of oceanic zones where the interaction between the ocean and atmosphere is the most active. We hope that further development of the experiment with dynamical factors influential in the system atmosphere-ocean-continent will allow to turn to the solution of climate change problem. It seems that this problem will grow more and more important in the last decade of this century. There are all the necessary facilities – powerful computers, satellite observations, research ships, and great interest of the scientists of all countries to the investigation of this global problem. These studies become essentially of international character.

Ever increasing role of satellite measurements of characteristics of atmosphere, ocean and environment have made this author to find it appropriate to include into this monograph his studies on application of adjoint problems to data processing problems. These problems are of global character and, being connected with space research, they are developed vigorously. There is no doubts, the observation of the Earth as a planet will be needed, of its resources, state of atmosphere, oceans, and continents, to support the environment protection, geo-physical monitoring, and analysis of climate and its changes.

I would like to call for further comprehensive development of these problems. They are extremely difficult, expensive, but the humanity can not economize on them. Correction of our errors will be manifold more expensive and some times it will be just impossible to correct them. Therefore, we deal here with high-level science on global processes which grows more general both in problems and in consequences for the life on our planet. This problem should unite all the people on the Earth.

Note in conclusion, that we shall employ strict mathematical analysis of mathematical statements of global problems in minimal form, concentrating our attention on complex algorithmic aspects. A reader equipped with investigations from the Part I of this book can carry out himself the studies he is interested in. To this end one should introduce corresponding spaces of functions, describe them thoroughly, seeing that all the algorithms one uses are completely justified.

CHAPTER 5

Analysis of Mathematical Models

in Environmental Problems

Environment, its condition and pollution control are becoming pivotal problems of science, since life on this planet is globally and regionally a universal concern. The extent of industrial pollution of the atmosphere and the sea is so great that is harming living conditions in large industrialized areas and beginning to produce global effects changing the radiation balance of the Earth by larger concentrations of carbon dioxide and aerosols and thinning the ozone layer in the atmosphere. Polluting emissions and acid rains affect greater still environmental processes in industrialized areas.

Winds, with their small-scale fluctuations, carry pollutants in the atmosphere. The averaged flux of pollutants carried by air masses consists, as a rule, of an advective and a turbulent components, therefore the averaged fluctuations may be represented as diffusion on the background of the principal averaged motion.

We will discuss in this chapter models of transport and diffusion of substances, main and adjoint equations and approaches to the perturbation theory and to solving inverse problems arising in formulating the models and dealing with practical problems of environmental pollution control.

5.1. Atmospheric Pollutants Transport Equation.
Uniqueness of Solution

Let $\varphi(x, y, z, t)$ be concentration of an aerosol pollutant carried by an air flux in the atmosphere. Define the solution of the problem in the cylindrical domain Ω with a surface S composed by the side surface Σ of the cylinder, bottom base Σ_0 (at $z = 0$), and top base Σ_H (at $z = H = const$). If $\mathbf{u} = u\mathbf{i} + v\mathbf{j} + w\mathbf{k}$ (where $\mathbf{i}$, $\mathbf{j}$ and $\mathbf{k}$ are unit vectors in the coordinate axes x, y and z respectively) is the velocity vector of air particles as a function of x, y, z and t, formulate the transport of the pollutant down the trajectory of air particles at the same unchanged concentration simply as

$$\frac{d\varphi}{dt} = 0$$

or, if expanded,

$$\frac{\partial \varphi}{\partial t} + u\frac{\partial \varphi}{\partial x} + v\frac{\partial \varphi}{\partial y} + w\frac{\partial \varphi}{\partial z} = 0 \qquad (5.1.1)$$

Since the mass conservation law holds for the bottom part of atmosphere with a high degree of accuracy, as expressed by the continuity equation

$$\frac{\partial u}{\partial x} + \frac{\partial v}{\partial y} + \frac{\partial w}{\partial z} = 0 \qquad (5.1.2)$$

obtain

$$\frac{\partial \varphi}{\partial t} + \text{div } \mathbf{u}\varphi = 0 \qquad (5.1.3)$$

Assume from now on, if not considered otherwise, that $\text{div } \mathbf{u} = 0$. Assume also that

$$w = 0 \quad \text{at} \quad z = 0, \quad z = H \qquad (5.1.4)$$

As we formulated (5.1.3), we used an identity which holds if the functions φ and $\mathbf{u}$ are differentiable:

$$u\frac{\partial \varphi}{\partial x} + v\frac{\partial \varphi}{\partial y} + w\frac{\partial \varphi}{\partial z} = \text{div } \mathbf{u}\varphi - \varphi\text{div } \mathbf{u} \qquad (5.1.5)$$

The last term in equality (5.1.5) turns to zero because of (5.1.2), and

$$u\frac{\partial\varphi}{\partial x} + v\frac{\partial\varphi}{\partial y} + w\frac{\partial\varphi}{\partial z} = \text{div}\,\mathbf{u}\varphi \qquad (5.1.5')$$

We shall use this important relationship often below.

Add initial data to equation (5.1.3)

$$\varphi = \varphi_0 \quad \text{at} \quad t = 0 \qquad (5.1.6)$$

and conditions on the boundary S of the domain Ω

$$\varphi = \varphi_s \quad \text{on} \quad S \quad \text{at} \quad u_n < 0 \qquad (5.1.7)$$

where φ_0 and φ_s are given functions and u_n is a projection of the vector u on the external normal to the surface S. Relationship (5.1.7) prescribes a solution in the part of S where air masses import into the domain Ω the substance we are studying. The exact solution of problem (5.1.3) is possible only if values of the functions u, v, and w are known in space and at all moments of time. If information about velocity vector components is short, approximations should be used as discussed below.

Equation (5.1.3) may be generalized. If, for example, some of the substance reacts with the environment or decays as it propagates, the process may be interpreted as absorption of the substance proportional to the value of σ. Then equation (5.1.3) becomes

$$\frac{\partial\varphi}{\partial t} + \text{div}\,\mathbf{u}\varphi + \sigma\varphi = 0 \qquad (5.1.8)$$

where $\sigma \geq 0$ is a quantity inversely proportional to time. The meaning of this quantity will be clearer, if $u = v = w = 0$ is assumed in (5.1.8) which then becomes $\partial\varphi/\partial t + \sigma\varphi = 0$. The solution to this one will be the function $\varphi = \varphi_0 e^{-\sigma t}$. It is clear now that σ is a quantity inversely proportional to the length of time interval during which the concentration of the substance decreases by the factor of e from the initial.

If there are the sources of the pollutant in the definition domain of the solution φ, described by the function $f(x, y, z, t)$, equation (5.1.8) becomes

$$\frac{\partial\varphi}{\partial t} + \text{div}\,\mathbf{u}\varphi + \sigma\varphi = f \qquad (5.1.9)$$

Let us now examine statement of the problem and conditions linked vith equation (5.1.9). Multiply this equation by φ and integrate the esult in time over the interval $0 \leq t \leq T$ and in spacial variables over he domain Ω to obtain

$$\int \frac{\varphi^2}{2}\big|_{t=T} d\Omega - \int_\Omega \frac{\varphi^2}{2}\big|_{t=0} d\Omega + \int_0^T dt \int_\Omega \text{div } \frac{\mathbf{u}\varphi^2}{2} d\Omega + \sigma \int_0^T dt \int_\Omega \varphi^2 d\Omega$$

$$= \int_0^T dt \int_\Omega f\varphi d\Omega \quad (5.1.10)$$

By the Ostrogradski–Gauss formula,

$$\int_\Omega \text{div}\frac{u\varphi^2}{2} d\Omega = \int_S \frac{u_n\varphi^2}{2} dS \quad (5.1.11)$$

Notice, that by virtue of (5.1.4), u_n at $z = 0$, $z = H$ turns to zero, so integration over side cylindrical surface Σ may be substituted for integration over S in (5.1.11). But, to remain general, retain the notation S, since condition (5.1.4) holds. Assuming that

$$\begin{aligned} \varphi &= \varphi_0 \quad \text{at} \quad t = 0 \\ \varphi &= \varphi_s \quad \text{on} \quad S \text{ at } \ u_n < 0 \end{aligned} \quad (5.1.12)$$

where φ_0 and φ_s are prescribed functions, and substituting these functions into (5.1.10), obtain

$$\int_\Omega \frac{\varphi_T^2}{2} d\Omega + \int_0^T dt \int_S \frac{u_n^+\varphi^2}{2} dS + \sigma \int_0^T dt \int_\Omega \varphi^2 d\Omega$$

$$= \int_\Omega \frac{\varphi_0^2}{2} d\Omega - \int_0^T dt \int_S \frac{u_n^-\varphi_s^2}{2} dS + \int_0^T dt \int_\Omega f\varphi d\Omega \quad (5.1.13)$$

where

$$u_n^+ = \begin{cases} u_n, & u_n > 0 \\ 0, & u_n < 0 \end{cases}$$

$$u_n^- = u_n - u_n^+, \quad \varphi_0 = \varphi|_{t=0}, \quad \varphi_T = \varphi|_{t=T}$$

Identity (5.1.13) is basic in checking the uniqueness of solution to problem (5.1.9) and (5.1.12). If we assume that two different solutions, e. g., φ_1 and φ_2, satisfy equation (5.1.9) and boundary conditions (5.1.12), we will obtain the following problem for their difference $\omega = \varphi_1 - \varphi_2$

$$\frac{\partial \omega}{\partial t} + \operatorname{div} \mathbf{u}\omega + \sigma\omega = 0 \qquad (5.1.14)$$

$$
\begin{aligned}
\omega &= 0 \quad \text{at} \quad t = 0 \\
\omega &= 0 \quad \text{on} \quad S \quad \text{at} \quad u_n < 0
\end{aligned}
\qquad (5.1.15)
$$

Identity (5.1.13) for the function ω becomes

$$\int_\Omega \frac{\omega_T^2}{2} d\Omega + \int_0^T dt \int_S \frac{u_n^+ \omega^2}{2} dS + \sigma \int_0^T dt \int_\Omega \omega^2 d\Omega = 0 \qquad (5.1.16)$$

Since all the terms in (5.1.16) are positive at $\omega \neq 0$, this expression is equal to zero only at $\omega = 0$, i. e., at $\varphi_1 = \varphi_2$. The solution is therefore unique. This conclusion holds of course if all the operations and transformations used to prove it are justified. It is clearly so if we assume only that the solution φ of the problem, functions u, v and w are differentiable and integrals do exist in (5.1.13).

The problem

$$\frac{\partial \varphi}{\partial t} + \operatorname{div} \mathbf{u}\varphi + \sigma\varphi = 0 \qquad (5.1.17)$$

$$
\begin{aligned}
\varphi &= \varphi_0 \quad \text{at} \quad t = 0 \\
\varphi &= \varphi_s \quad \text{on} \quad S \quad \text{at} \quad u_n < 0
\end{aligned}
\qquad (5.1.18)
$$

has thus the unique solution in a class of functions $\varphi(x, y, z, t)$ continuously differentiable in all their variables with continuous initial data $\varphi_0(x, y, z)$, boundary conditions $\varphi_s(x, y, z, t)$, and continuous and differentiable coefficients $\mathbf{u}(x, y, z, t)$ satisfying the condition $\operatorname{div} \mathbf{u} = 0$, and piecewise continuous function σ. We will assume from now on that all these conditions are satisfied. Equation (5.1.17) may be written as

$$\frac{\partial \varphi}{\partial t} + A\varphi = f \qquad (5.1.19)$$

where $A\varphi = \operatorname{div} \mathbf{u}\varphi + \sigma\varphi$. The operator A may be assumed as operating in a real Hilbert space $L_2(\Omega)$ with a definition domain $D(A)$ which is a set of continuously differentiable functions in x, y and z.

5.2. Stationary Equation for Propagation of Substances

Let us now discuss propagation of pollutants. If the equation coefficients u, v and w and other inputs, f and φ_s do not depend on time, a stationary problem corresponding to (5.1.17) and (5.1.18) formulates as

$$\operatorname{div} \mathbf{u}\varphi + \sigma\varphi = f \qquad (5.2.1)$$

$$\varphi = \varphi_s \quad \text{on} \quad S \quad \text{for} \quad u_n < 0 \qquad (5.2.2)$$

It is easy to check that the identity corresponding to (5.1.13) is

$$\int_S \frac{u_n^+ \varphi^2}{2} dS + \sigma \int_\Omega \varphi^2 d\Omega = -\int_S \frac{u_n^- \varphi_s^2}{2} dS + \int_\Omega f\varphi d\Omega \qquad (5.2.3)$$

It is possible to use the method discussed in § 5.1, to verify that the solution of problem (5.2.1) and (5.2.2) is unique.

Problem (5.2.1) and (5.2.2) thus describes a particular case of substance transport with input data constant in time. A set of such particular solutions corresponding to various stationary input data $\mathbf{u}$, f, and φ_s of the problem may be used in description of more complex real physical processes. Assume motions of air masses of various types taking place in a given region at different periods of time and consider these motions as steady-state within a period of characteristic time of existence. Every such period is followed by transformation of motion of air masses and a new steady state settles. Since transformations in circulations take far less time than a certain type of motion exists, it may be assumed that types of motion change instantly. Assume n of such types and obtain a system of independent equations

$$\operatorname{div} \mathbf{u}_i\varphi_i + \sigma\varphi_i = f \qquad (5.2.4)$$

$$\varphi_i = \varphi_{is} \quad \text{on} \quad S \quad \text{for} \quad u_{in} < 0, \quad i = 1, \ldots, n \qquad (5.2.5)$$

Problem (5.2.4) and (5.2.5) where φ_{is} is a value of the function φ_i at the boundary S, u_{in} is a projection of velocity vector of ith type on the external normal to the boundary corresponding to each of intervals $t_i < t < t_{i+1}$ of length Δt_i.

Assume now that all the problems (5.2.4) and (5.2.5) are solved. Then the solution of the problem of the average distribution of pollutants in the period $T = \sum\limits_{i=1}^{n} \Delta t_i$ is a linear combination

$$\tilde{\varphi} = \frac{1}{T} \sum_{i=1}^{n} \varphi_i \Delta t_i \qquad (5.2.6)$$

Problem (5.2.4)–(5.2.6) may be referred to as a statistical model.

The solution of stationary problems of the (5.2.1), (5.2.2) and (5.2.4), (5.2.5) types has much in common with the solution of problem of the average substance distribution for a certain period of time T on the basis of a specifically stated nonstationary problems. Examine the problem

$$\frac{\partial \varphi}{\partial t} + \operatorname{div} \mathbf{u}\varphi + \sigma\varphi = f \qquad (5.2.7)$$

$$\varphi = \varphi_s \quad \text{on} \quad S \quad \text{at} \quad u_n < 0$$
$$\qquad (5.2.8)$$
$$\varphi(\mathbf{r}, T) = \varphi(\mathbf{r}, 0), \quad \mathbf{r} = (x, y, z)^T \in \Omega$$

Assume the functions $\mathbf{u}$ and φ_s to be independent of t as in (5.2.1) and (5.2.2).

The solution uniqueness of problems (5.2.7) and (5.2.8) (if the functions are assumed as sufficiently smooth) can be established as in § 5.1.

Integrate equation (5.2.7) within the interval $[0, T]$ to obtain

$$\operatorname{div} \mathbf{u}\bar{\varphi} + \sigma\bar{\varphi} = f, \quad \bar{\varphi} = \frac{1}{T} \int_0^T \varphi \, dt \qquad (5.2.9)$$

whence by virtue of the uniqueness of solution of problem (5.2.1) and (5.2.2) the solution of problem (5.2.7) and (5.2.8) averaged for the period T coincides with the solution of problem (5.2.1) and (5.2.2).

Examine a more complex case. Let the function $\mathbf{u}$ which is smooth enough on $[0, T]$, be independent of time over every interval $t_i + \tau \le t \le t_{i+1}$ ($i = 0, \ldots, n-1$) and coincide with $\mathbf{u}_i$ from (5.2.4). Assume the time τ of circulation transformation to be far shorter than Δt_i, i. e.,

$$\tau \ll \Delta t_i \qquad (5.2.10)$$

Solve non-steady-state problem (5.2.7) and (5.2.8) related to the vector **u** as defined above:

$$\frac{\partial \varphi}{\partial t} + \operatorname{div} \mathbf{u}\varphi + \sigma\varphi = f, \quad \operatorname{div} \mathbf{u} = 0 \qquad (5.2.11)$$

$$\varphi = \varphi_s \quad \text{on} \quad S \quad \text{at} \quad u_n < 0 \qquad (5.2.12)$$

where φ_s is linked to φ_{is} from (5.2.5) like **u** does to $\mathbf{u}_i$ from (5.2.4). Assume also that **u** and φ_s are functions periodic in time and their period T is one year. Examine problem (5.2.11), (5.2.12) with the condition

$$\varphi(\mathbf{r}, T) = \varphi(\mathbf{r}, 0), \quad \mathbf{r} = (x, y, z)^T \qquad (5.2.13)$$

Having solved problem (5.2.11)–(5.2.13), deduce the mean annual distribution of the substance as

$$\bar{\varphi} = \frac{1}{T} \int_0^T \varphi\, dt \qquad (5.2.14)$$

It is clear that the statements of nonstationary problem (5.2.11)–(5.2.13) and problem (5.2.4) and (5.2.5) are closely linked.

Examine equation (5.2.11) in the interval $[t_i + \tau, t_{i+1}]$:

$$\frac{\partial \varphi}{\partial t} + \operatorname{div} \mathbf{u}_i\varphi + \sigma\varphi = f \qquad (5.2.15)$$

with condition (5.2.12):

$$\varphi = \varphi_s \quad \text{on} \quad S \quad \text{at} \quad u_{in} < 0 \qquad (5.2.16)$$

Let the function φ take the following value in the moment of time $t = t_i + \tau$:

$$\varphi(t_i + \tau) = \varphi^0 \qquad (5.2.17)$$

Denote the following function as ω_i:

$$\omega_i = \varphi - \varphi_i \qquad (5.2.18)$$

where φ_i is the solution of problem (5.2.4) and (5.2.5). This function is a solution of the following problem in the interval $t \in [t_i + \tau, t_{i+1}]$:

$$\frac{\partial \omega_i}{\partial t} + \text{div }\mathbf{u}_i\omega_i + \sigma\omega_i = 0$$

$$\omega_i = 0 \quad \text{on} \quad S \quad \text{for} \quad u_{in} < 0 \qquad (5.2.19)$$

$$\omega_i(t_i + \tau) = \varphi^0 - \varphi_i$$

Multiply the first equation in (5.2.19) by ω_i and integrate the result over the domain Ω to obtain

$$\|\omega_i\| \left(\frac{d}{dt}\|\omega_i\| + \sigma\|\omega_i\|\right) + \int\limits_S \frac{\omega_i^2 u_{in}^+}{2} dS = 0 \qquad (5.2.20)$$

where $\|\omega\| = \left(\int\limits_\Omega \omega^2 d\Omega\right)^{1/2}$. The following equality follows from (5.2.20):

$$\|\omega_i\| \le \exp\{-\sigma(t - t_i - \tau)\}\|\varphi_0 - \varphi_i\| \qquad (5.2.21)$$

The solution of problem (5.2.11)–(5.2.13) may be fairly close to a solution of related problem (5.2.4) and (5.2.5) on every interval in $t_i \le t \le t_{i+1}$, starting at a certain moment of time. Generally, those functions may differ essentially depending on Δt_i. The quantity Δt_i $(i = 1, \dots, n)$ is natural of the class of problems we are examining, if the interval, where φ and φ_i differ little, is much larger than its supplement to Δt_i. Given this, introduce the quantity ε and assume the density distributions, φ^1 and φ^2, of the pollutant as coinciding, provided

$$\|\varphi^1 - \varphi^2\| \le \varepsilon \qquad (5.2.22)$$

Denote the time required for propagation of the substance φ to become steady in the interval $t_i + \tau \le t \le t_{i+1}$ (in the sense of definition (5.2.22)) as τ_i. Then

$$\tau_i = \frac{1}{\sigma} \ln \frac{\|\varphi^0 - \varphi_i\|}{\varepsilon} \ll \Delta t_i \qquad (5.2.23)$$

Similarly, and taking into account (5.2.21), obtain an estimate for the function φ in the interval $t_i \le t \le t_i + \tau$:

$$\|\varphi\| \le e^{-\sigma(t-t_i)}(\|\varphi_{i-1}\| + \varepsilon) + \frac{\|f\|}{\sigma}(1 - e^{-\sigma(t-t_i)}) \qquad (5.2.24)$$

Examine the mean annual value of the substance φ:

$$\bar{\varphi} = \frac{1}{T}\int_0^T \varphi dt = \frac{1}{T}\sum_{i=1}^n \varphi_i \Delta t_i + \frac{1}{T}\sum_{i=1}^{n-1}\int_{t_i+\tau+\tau_i}^{t_i+1} w_i dt$$

$$-\frac{1}{T}\sum_{i=1}^n \varphi_i(\tau+\tau_i) + \frac{1}{T}\sum_{i=0}^{n-1}\int_{t_i}^{t_i+\tau+\tau_i}\varphi dt \quad (5.2.25)$$

Hence by virtue of (5.2.21)–(5.2.24), the difference between the solutions of problems (5.2.4)–(5.2.5) averaged over the interval $[0,T]$, and those of problem (5.2.11)–(5.2.13) satisfies the inequality $\|\bar{\varphi} - \tilde{\varphi}\| \leq \varepsilon + (\varepsilon + 3\|f\|/\sigma)r/\Delta t$ where $r = \max(\tau + \tau_i)$, $\Delta t = T/n$. Given condition (5.2.10), select Δt and τ_i so that $r/\Delta t \leq \varepsilon$, to obtain

$$\|\bar{\varphi} - \tilde{\varphi}\| \leq (1 + 3\|f\|/\sigma)\varepsilon + \varepsilon^2 \quad (5.2.26)$$

The solution of the problem of average distribution of substances for the period T by application of a statistical model and the solution of nonstationary problem (5.2.11)–(5.2.13) are close enough under assumptions we made.

Notice that solving problems (5.2.4) and (5.2.5) or (5.2.11)–(5.2.13) and averaging the results with (5.2.6) or (5.2.14) respectively, we take into account diffusion of substances resulting from fluctuations of input data.

5.3. Diffusion Approximation. Uniqueness of the Solution

Models of propagation of pollutants in the atmosphere from their sources we discussed in § 5.2 describe the essence of the process, but they idealize to a certain extent real processes which are more complex and richer in their physical content. Assume a case where there is no advective and convective motions in the atmosphere, i. e., $u = v = w = 0$. Then, in compliance with our models the nonstationary problem of substance transport is

$$\frac{\partial \varphi}{\partial t} + \sigma\varphi = f \quad (5.3.1)$$

$$\varphi = \varphi_0 \quad \text{at} \quad t = 0$$

If f does not depend on t the solution is

$$\varphi = \varphi_0 e^{-\sigma t} + \frac{f}{\sigma}(1 - e^{-\sigma t}) \qquad (5.3.2)$$

and becomes the solution of the corresponding stationary problem $\sigma\varphi = f$, i. e., $\varphi = f/\sigma$, at $t \to \infty$.

It is well known that such a simple model cannot describe main features of transport of substances from the source f. As a matter of fact, we know that pollutants spread out in the atmosphere and form a fairly complex distribution of aerosols in far away from the source of emission. No wonder, as even without any wind, the atmosphere is a turbulent medium where small fluctuations, usually vertical, dissipate and generate conditions for new formations.

The spectrum of such fluctuations has been thoroughly studied and largely appears to account for the spread of pollutants in the atmosphere. Since the fluctuations manifest themselves only statistically, they cannot be precomputed or even registered in reality. If it were possible, we could precompute the swell of an aerosol source and propagation of the aerosol on the basis of the model (5.2.4)–(5.2.5) or a more accurate model, (5.2.7), (5.2.12) and (5.2.13). Since there is no such possibility, we have to modify the models, so that they could allow statistically for constantly regenerated atmospheric fluctuations. The physics of fluctuation effects has been well studied now, but their mathematical description is still based mostly on semi-empirical relationships. Let us make a short review of this simple theory.

Assume a function a representable as a sum of an averaged component $\bar{a}$ and a fluctuating component a', i. e., $a = \bar{a} + a'$; assume here that

$$a' \ll \bar{a} \qquad (5.3.3)$$

i. e., fluctuations of the substance a are small. Assume now that a is averaged over a sufficiently large time interval T:

$$\bar{a} = \frac{1}{T} \int_t^{t+T} a \, dt' \qquad (5.3.4)$$

and

$$\bar{a}' = \frac{1}{T} \int\limits_{t}^{t+T} a'\, dt' = 0 \qquad (5.3.5)$$

on this interval.

If the process satisfies conditions (5.3.3)–(5.3.5), take the following possible approach to construction of equations necessary to describe the propagation of substances in different conditions.

Integrate equation (5.1.8) within $t \leq \tau \leq t + T$:

$$\varphi(t+T) - \varphi(t) + \text{div} \int\limits_{t}^{t+T} \mathbf{u}\varphi\, dt' + \sigma \int\limits_{t}^{t+T} \varphi\, dt' = 0 \qquad (5.3.6)$$

Let $\varphi = \bar{\varphi} + \varphi'$, $\mathbf{u} = \bar{\mathbf{u}} + \mathbf{u}'$. Then (5.3.6) transforms into

$$\frac{\varphi(t+T) - \varphi(t)}{T} + \text{div}\, \bar{\mathbf{u}}\bar{\varphi} + \text{div}\, \overline{\mathbf{u}'\varphi'} + \sigma\bar{\varphi} = 0 \qquad (5.3.7)$$

or

$$\frac{\bar{\varphi}(t+T) - \bar{\varphi}(t)}{T} + \text{div}\, \bar{\mathbf{u}}\bar{\varphi} + \text{div}\, \overline{\mathbf{u}'\varphi'} + \sigma\bar{\varphi} = -\frac{\varphi'(t+T) - \varphi'(t)}{T} \qquad (5.3.8)$$

Scale the main quantities. Let $\bar{\varphi} = A\bar{\Phi}$ and $\varphi' = a\Phi'$ where $\bar{\Phi}$ and Φ' are now of the same order. We have $\varphi' \ll \bar{\varphi}$ by assumption, therefore, $a \ll A$. Let $a/A = \varepsilon$. Then rewrite (5.3.8) as

$$\frac{\bar{\Phi}(t+T) - \bar{\Phi}(t)}{T} + \text{div}\, \bar{\mathbf{u}}\bar{\Phi} + \sigma\bar{\Phi} + \varepsilon \text{div}\, \overline{\mathbf{u}'\Phi'} = \frac{\varepsilon}{T}O(1) \qquad (5.3.9)$$

where $O(1)$ means a quantity of an order of Φ'. The right-hand part of (5.3.9) is a small quantity of an order of ε/T and may be disregarded; then obtain

$$\frac{\bar{\varphi}(t+T) - \bar{\varphi}(t)}{T} + \text{div}\, \bar{\mathbf{u}}\bar{\varphi} + \text{div}\, \overline{\mathbf{u}'\varphi'} + \sigma\bar{\varphi} = 0 \qquad (5.3.10)$$

If T is a time interval within which the function $\bar{\varphi}(t)$ changes little, substitute approximately the derivative $\partial\bar{\varphi}/\partial t$ for $[\bar{\varphi}(t+T) - \bar{\varphi}(t)]/T$ and obtain the equation for an averaged component:

$$\frac{\partial\bar{\varphi}}{\partial t} + \text{div}\, \bar{\mathbf{u}}\bar{\varphi} + \text{div}\, \overline{\mathbf{u}'\varphi'} + \sigma\bar{\varphi} = 0 \qquad (5.3.11)$$

which differs from equation (5.1.3) by the fluctuation moment div $\overline{\mathbf{u}'\varphi'}$. It is this term that accounts for the spread of air masses flow, which carries the pollutants. It has been established that the components of the vector $\overline{\mathbf{u}'\varphi'}$ may be described by the following semi-empirical representation through averaged fields of substances:

$$\overline{u'\varphi'} = -\mu\frac{\partial\bar{\varphi}}{\partial x}, \quad \overline{v'\varphi'} = -\mu\frac{\partial\bar{\varphi}}{\partial y}, \quad \overline{w'\varphi'} = -\nu\frac{\partial\bar{\varphi}}{\partial z} \tag{5.3.12}$$

Here $\mu \geq 0$ and $\nu \geq 0$ are respectively horizontal and vertical diffusion coefficients determined experimentally.

Allow for (5.3.12) in (5.3.11) to obtain the diffusion approximation of the equation of propagation of substances in the atmosphere:

$$\frac{\partial\bar{\varphi}}{\partial t} + \operatorname{div}\bar{\mathbf{u}}\bar{\varphi} + \sigma\bar{\varphi} = K\bar{\varphi} \tag{5.3.13}$$

where

$$K\bar{\varphi} = \frac{\partial}{\partial x}\mu\frac{\partial\bar{\varphi}}{\partial x} + \frac{\partial}{\partial y}\mu\frac{\partial\bar{\varphi}}{\partial y} + \frac{\partial}{\partial z}\nu\frac{\partial\bar{\varphi}}{\partial z} \tag{5.3.14}$$

The continuity relationship must be joined to equation (5.3.13):

$$\operatorname{div}\bar{\mathbf{u}} = 0 \tag{5.3.15}$$

and the initial data

$$\bar{\varphi} = \bar{\varphi}_0 \quad \text{at} \quad t = 0 \tag{5.3.16}$$

As concerns boundary conditions, a set of these will be available in checking the uniqueness of the solution.

Omit, for the sake of convenience, later the bars over the functions φ, assuming that we are dealing with an averaged component. Multiply equation (5.3.13) by φ and integrate the result both in time over the interval $0 \leq t \leq T$, and space:

$$\int_\Omega \frac{\varphi_T^2}{2}d\Omega - \int_\Omega \frac{\varphi_0^2}{2}d\Omega + \int_0^T dt \int_S \frac{u_n\varphi^2}{2}dS + \sigma\int_0^T dt \int_\Omega \varphi^2 d\Omega$$

$$= -\int_0^T dt \int_\Omega \left\{\mu\left[\left(\frac{\partial\varphi}{\partial x}\right)^2 + \left(\frac{\partial\varphi}{\partial y}\right)^2\right] + \nu\left(\frac{\partial\varphi}{\partial z}\right)^2\right\}d\Omega$$

$$+ \int_0^T dt \left[\mu\int_\Sigma \varphi\frac{\partial\varphi}{\partial n}d\Sigma + \nu\left(\int_{\Sigma_H} \varphi\frac{\partial\varphi}{\partial z}d\Sigma - \int_{\Sigma_0} \varphi\frac{\partial\varphi}{\partial z}d\Sigma\right)\right] \tag{5.3.17}$$

Here $\varphi_T = \varphi|_{t=T}$, $\varphi_0 = \varphi|_{t=0}$, $\partial\varphi/\partial n$ is a derivative in external normal to Σ. Remember that S is the whole surface of the domain Ω, Σ is the cylindrical side surface, Σ_H is a cross-section of the cylindrical surface at the level of $z = H$, and Σ_0 at the level of $z = 0$. Rewrite (5.3.17) as

$$\int_\Omega \frac{\varphi_T^2}{2} d\Omega + \int_0^T dt \int_S \frac{u_n^+ \varphi^2}{2} dS + \int_0^T dt \int_\Omega \left\{ \mu \left[\left(\frac{\partial\varphi}{\partial x}\right)^2 + \left(\frac{\partial\varphi}{\partial y}\right)^2 \right] + \nu \left(\frac{\partial\varphi}{\partial z}\right)^2 \right\} d\Omega$$

$$+ \sigma \int_0^T dt \int_\Omega \varphi^2 d\Omega = \int_\Omega \frac{\varphi_0^2}{2} d\Omega - \int_0^T dt \int_S \frac{u_n^- \varphi^2}{2} dS$$

$$+ \int_0^T dt \left[\mu \int_\Sigma \varphi \frac{\partial\varphi}{\partial n} d\Sigma + \nu \left(\int_{\Sigma_H} \varphi \frac{\partial\varphi}{\partial z} d\Sigma - \int_{\Sigma_0} \varphi \frac{\partial\varphi}{\partial z} d\Sigma \right) \right] \quad (5.3.18)$$

Prescribe the following boundary conditions:

$$\begin{array}{llll} \varphi = \varphi_s & \text{on} \quad \Sigma & \text{at} & u_n < 0 \\ \partial\varphi/\partial n = 0 & \text{on} \quad \Sigma & \text{at} & u_n \geq 0 \\ \partial\varphi/\partial z = \alpha\varphi & \text{on} \quad \Sigma_0 & & \\ \partial\varphi/\partial z = 0 & \text{on} \quad \Sigma_H & & \end{array} \quad (5.3.19)$$

here $\alpha \geq 0$ is a function which characterizes interaction of pollutants and the underlying surface. Besides,

$$w = 0 \quad \text{at} \quad z = 0, \quad z = H \quad (5.3.20)$$

Prove that conditions (5.3.19) and (5.3.20), conditions of smoothness of the solution and input data and initial conditions (5.3.16) provide a unique solution to this problem. Use (5.3.19) to obtain the main relationship coming from S to Σ by using (5.3.20):

$$\int_\Omega \frac{\varphi_T^2}{2} d\Omega + \int_0^T dt \int_\Sigma \frac{u_n^+ \varphi^2}{2} d\Sigma + \int_0^T dt \int_\Omega \left\{ \mu \left[\left(\frac{\partial\varphi}{\partial x}\right)^2 + \left(\frac{\partial\varphi}{\partial y}\right)^2 \right] + \nu \left(\frac{\partial\varphi}{\partial z}\right)^2 \right\} d\Omega$$

$$+ \sigma \int_0^T dt \int_\Omega \varphi^2 d\Omega + \nu \int_0^T dt \int_{\Sigma_0} \alpha\varphi^2 d\Sigma$$

$$= \int_\Omega \frac{\varphi_0^2}{2} d\Omega - \int_0^T dt \int_\Sigma \frac{u_n^- \varphi_s^2}{2} d\Sigma + \mu \int_0^T dt \int_\Sigma \varphi_s \frac{\partial\varphi}{\partial n} d\Sigma \quad (5.3.21)$$

Prove now the uniqueness of the solution. Assume, like in § 5.2, that there are two solutions φ_1 and φ_2 satisfying equation (5.3.13), initial data (5.3.16), boundary conditions (5.3.19) and additional conditions (5.3.15) and (5.3.20), and obtain an equation for the difference $\omega = \varphi_1 - \varphi_2$

$$\frac{\partial \omega}{\partial t} + \operatorname{div} \mathbf{u}\omega + \sigma\omega = K\omega \tag{5.3.22}$$

initial data

$$\omega = 0 \quad \text{at} \quad t = 0 \tag{5.3.23}$$

and boundary conditions

$$
\begin{array}{llll}
\omega = 0 & \text{on} & \Sigma & \text{at} \quad u_n < 0 \\
\partial \omega / \partial n = 0 & \text{on} & \Sigma & \text{at} \quad u_n \geq 0 \\
\partial \omega / \partial z = \alpha\omega & \text{on} & \Sigma_0 & \\
\partial \omega / \partial z = 0 & \text{on} & \Sigma_H &
\end{array}
\tag{5.3.24}
$$

Such a statement will produce the following (5.3.21)-type of relationship:

$$
\int_{\Omega} \frac{\varphi_T^2}{2} d\Omega + \int_0^T dt \int_{\Sigma} \frac{u_n^+ \omega^2}{2} d\Sigma
$$

$$
+ \int_0^T dt \int_{\Omega} \left\{ \mu \left[\left(\frac{\partial \omega}{\partial x} \right)^2 + \left(\frac{\partial \omega}{\partial y} \right)^2 \right] + \nu \left(\frac{\partial \omega}{\partial z} \right)^2 \right\} d\Omega
$$

$$
+ \sigma \int_0^T dt \int_{\Omega} \omega^2 d\Omega + \nu \int_0^T dt \int_{\Sigma_0} \alpha \omega^2 d\Sigma = 0 \tag{5.3.25}
$$

Since the quantities u_n^+, μ, ν, σ, and α are not negative in (5.3.25), this relationship equals to zero just if $\omega = 0$, i. e., if $\varphi_1 = \varphi_2$. This proves the uniqueness of the solution. To make it simple, we assumed $f = 0$ in the statement of the problem. The source can be taken into account like in § 5.1.

Taking into account this generalization, it is possible to pass to the following statement of problem, which secures the uniqueness of the olution in diffusion approximation if input data are smooth:

$$\frac{\partial \varphi}{\partial t} + \text{div } \mathbf{u}\varphi + \sigma\varphi - K\varphi = f$$

$$\varphi = \varphi_0 \quad \text{at} \quad t = 0$$

$\varphi = \varphi_s$	on	Σ	at	$u_n < 0$
$\partial\varphi/\partial n = 0$	on	Σ	at	$u_n \geq 0$
$\partial\varphi/\partial z = \alpha\varphi$	on	Σ_0		
$\partial\varphi/\partial z = 0$	on	Σ_H		

(5.3.26)

Assume here also that

$$\text{div } \mathbf{u} = 0$$

$$w = 0 \quad \text{at} \quad z = 0, \quad z = H$$

Suppose that the solution $\varphi(x, y, z, t)$ of problem (5.3.26) is contin-
uous in $\Omega \times (0, T)$ and has a derivative $\partial\varphi/\partial t$ quadratically summable
over $\Omega \times (0, T)$. Besides, let the function $\varphi(x, y, z, t)$ belong to the
set of functions $D(A)$ in the real Hilbert space $L_2(\Omega)$, continuous and
differentiable over Ω, at every t, so that $\text{div } \mathbf{u}\varphi + \sigma\varphi - K\varphi \in L_2(\Omega)$.

Given these assumptions, the first equation in (5.3.26) may be written as

$$\frac{\partial \varphi}{\partial t} + A\varphi = f$$

where the operator A with the definition domain $D(A)$ operates in the
Hilbert space $L_2(\Omega)$ and is assigned by the equality

$$A\varphi = \text{div } \mathbf{u}\varphi + \sigma\varphi - \frac{\partial}{\partial x}\mu\frac{\partial\varphi}{\partial x} - \frac{\partial}{\partial y}\mu\frac{\partial\varphi}{\partial y} - \frac{\partial}{\partial z}\nu\frac{\partial\varphi}{\partial z}$$

Notice that the following statement frequently used in computations
allows for a unique solution like (5.3.26):

$$\frac{\partial \varphi}{\partial t} + \text{div } \mathbf{u}\varphi + \sigma\varphi - K\varphi = f$$

$\varphi = \varphi_0$	at	$t = 0$	
$\varphi = \varphi_s$	on	Σ	
$\partial\varphi/\partial z = \alpha\varphi$	on	Σ_0	
$\partial\varphi/\partial z = 0$	on	Σ_H	

(5.3.27)

Other statements of problems providing the unique solution are possible, of course.

The operator K introduced by relationship (5.3.14) may be conveniently represented as a sum of two operators:

$$K = \mu \left(\frac{\partial^2}{\partial x^2} + \frac{\partial^2}{\partial y^2} \right) + \frac{\partial}{\partial z} \nu \frac{\partial}{\partial z} \equiv \mu \Delta + \frac{\partial}{\partial z} \nu \frac{\partial}{\partial z}$$

To make it simple, consider the diffusion coefficient μ in this representation to be independent of time and spatial coordinates.

We have so far discussed a general three-dimensional statement of a problem, though many cases justify two-dimensional (x, y)-approximations which may be statement (5.3.26) or (5.3.27). Examine, for example, (5.3.27). Integrate the diffusion equation in z to obtain

$$\frac{\partial}{\partial t} \int_0^H \varphi dz + \int_0^H \operatorname{div} \mathbf{u}\varphi dz + \sigma \int_0^H \varphi dz$$

$$- \int_0^H \frac{\partial}{\partial z} \nu \frac{\partial \varphi}{\partial z} dz - \mu \Delta \int_0^H \varphi dz = \int_0^H f dz \quad (5.3.28)$$

Deploy the second term in the left-hand part and the first term in the right-hand part. Assume that the components u and v of the velocity vector do not change with altitude in the active layer of the substance transport and diffusion and obtain

$$\int_0^H \operatorname{div} \mathbf{u}\varphi = \frac{\partial}{\partial x} \left(u \int_0^H \varphi \, dz \right) + \frac{\partial}{\partial y} \left(v \int_0^H \varphi \, dz \right) + w\varphi \Big|_{z=0}^{z=H} \quad (5.3.29)$$

Since $w = 0$ at $z = 0$ and at $z = H$, the last term is equal to zero and, consequently

$$\int_0^H \operatorname{div} \mathbf{u}\varphi = \frac{\partial}{\partial x} \left(u \int_0^H \varphi \, dz \right) + \frac{\partial}{\partial y} \left(v \int_0^H \varphi \, dz \right) \quad (5.3.30)$$

Examine now equalities

$$\int_0^H \frac{\partial}{\partial z} \nu \frac{\partial \varphi}{\partial z} dz = \nu \frac{\partial \varphi}{\partial z} \Big|_{z=0}^{z=H} = - \nu \frac{\partial \varphi}{\partial z} \Big|_{z=0} \quad (5.3.31)$$

Simplify (5.3.31) allowing for the boundary condition $\partial\varphi/\partial z = \alpha\varphi$ at $z = 0$:

$$\int_0^H \frac{\partial}{\partial z}\nu\frac{\partial\varphi}{\partial z}dz = -\alpha\nu\varphi|_{z=0}$$

Assuming that the equality

$$\varphi|_{z=0} = \frac{1}{H}\int_0^H \varphi dz$$

approximately holds, obtain finally

$$\int_0^H \frac{\partial}{\partial z}\nu\frac{\partial\varphi}{\partial z}dz = -\frac{\alpha\nu}{H}\int_0^H \varphi dz \qquad (5.3.32)$$

Introduce now integral intensity of the aerosols and the source

$$\bar\varphi = \int_0^H \varphi dz, \quad \bar f = \int_0^H f dz$$

Relationship (5.3.28) now will be written in the form

$$\frac{\partial\bar\varphi}{\partial t} + \frac{\partial u\bar\varphi}{\partial x} + \frac{\partial v\bar\varphi}{\partial y} + \bar\sigma\bar\varphi - \mu\Delta\bar\varphi = \bar f \qquad (5.3.33)$$

where $\bar\sigma = \sigma + \alpha\nu/H$. Notice that if $\sigma\bar\varphi$ is the decaying quantity of the migrant aerosol, integral over the altitude, $(\alpha\nu/H)\bar\varphi$ is the fallout quantity. To make it simple, drop the bars over the solution φ, function f, and quantity σ and obtain a two-dimensional statement of the problem:

$$\frac{\partial\varphi}{\partial t} + \frac{\partial u\varphi}{\partial x} + \frac{\partial v\varphi}{\partial y} + \sigma\varphi - \mu\Delta\varphi = f,$$

$$\varphi = \varphi_0 \quad \text{at} \quad t = 0$$

$$\varphi = \varphi_s \quad \text{on} \quad \Sigma \quad \text{at} \quad u_n < 0 \qquad (5.3.34)$$

$$\frac{\partial\varphi}{\partial n} = 0 \quad \text{on} \quad \Sigma \quad \text{at} \quad u_n \geq 0$$

Another statement is possible:

$$\frac{\partial \varphi}{\partial t} + \frac{\partial u\varphi}{\partial x} + \frac{\partial v\varphi}{\partial y} + \sigma\varphi - \mu\Delta\varphi = f$$

$$\varphi = \varphi_0 \quad \text{at} \quad t = 0 \tag{5.3.35}$$

$$\varphi = \varphi_s \quad \text{on} \quad \Sigma$$

One more important remark: all the conclusions in § 5.2 hold for the diffusion approximation, too. This also applies to the stationary solution of the problem when the solution of the main problem is found by averaging over an ensemble of partial problems of the type

$$\text{div } \mathbf{u}_i\varphi_i + \sigma\varphi - K\varphi_i = f, \quad \text{div } \mathbf{u}_i = 0 \tag{5.3.36}$$

with conditions

$$\begin{array}{llll}
\varphi_i = \varphi_{is} & \text{on} & \Sigma & \text{at} \quad u_{in} < 0 \\
\partial\varphi_i/\partial n = 0 & \text{on} & \Sigma & \text{at} \quad u_{in} \geq 0 \\
\partial\varphi_i/\partial z = \alpha_i\varphi_i & \text{on} & \Sigma_0 & \\
\partial\varphi_i/\partial z = 0 & \text{on} & \Sigma_H &
\end{array} \tag{5.3.37}$$

The solution of the problem of distribution averaged over the period T is

$$\tilde{\varphi} = \frac{1}{T}\sum_{i=1}^{n}\varphi_i\Delta t_i \tag{5.3.38}$$

where $T = \sum_{i=1}^{n}\Delta t_i$, and Δt_i is the time of the stable mode of motion of air masses of a given type.

5.4. The Simple Diffusion Equation

It is convenient to examine transport and diffusion of a substance in simple examples of one-dimensional problems first and then gradually make their mathematical statements more complicated. Therefore examine first a simple diffusional statement of a problem[1]

$$\sigma\varphi - \mu\frac{d^2\varphi}{dx^2} = Q\delta(x - x_0) \tag{5.4.1}$$

---

[1]Hereafter see the footnote at page 190 concerning the δ-function.

in an infinite medium $-\infty < x < \infty$, where Q is a power of the source emitting aerosols into the atmosphere. The boundary conditions are assumed as a boundness of solution over all the definition domain in this case. Notice, that the function f is concrete in equation (5.4.1) and represented characteristically of problems the type we are studying. For convenience, reduce problem (5.4.1) to its equivalent with no δ-function: integrate equation (5.4.1) at the point $x = x_0$:

$$\sigma \int_{x_0-\varepsilon/2}^{x_0+\varepsilon/2} \varphi \, dx = \mu \frac{\partial \varphi}{\partial x}\bigg|_{x_0+\varepsilon/2} - \mu \frac{\partial \varphi}{\partial x}\bigg|_{x_0-\varepsilon/2} + Q$$

Pass to a limit for $\varepsilon \to 0$ and obtain an important relationship:

$$\mu \frac{\partial \varphi}{\partial x}\bigg|_{x_0+} - \mu \frac{\partial \varphi}{\partial x}\bigg|_{x_0-} + Q = 0 \tag{5.4.2}$$

Examine now two regions: $-\infty < x \leq x_0$, $x_0 \leq x < \infty$ and denote solutions φ_- and φ_+ respectively, i. e., examine two problems:

$$\mu \frac{d^2 \varphi_+}{dx^2} - \sigma \varphi_+ = 0$$

$$\varphi_+ = 0 \quad \text{at} \quad x \to \infty \tag{5.4.3}$$

$$\mu \frac{d^2 \varphi_-}{dx^2} - \sigma \varphi_- = 0$$

$$\varphi_- = 0 \quad \text{at} \quad x \to -\infty \tag{5.4.4}$$

The solutions of problems (5.4.3) and (5.4.4) are linked by

$$\mu \frac{d\varphi_+}{dx} - \mu \frac{d\varphi_-}{dx} + Q = 0 \quad \text{at} \quad x = x_0 \tag{5.4.5}$$

Assuming that the solution is regular at all points of the region, including the point $x = x_0$, obtain the second condition:

$$\varphi_+ = \varphi_- \quad \text{at} \quad x = x_0 \tag{5.4.6}$$

It is easy to see that the solutions of problems (5.4.3) and (5.4.4) are

$$\varphi_+ = c_+ \exp\{-\sqrt{\sigma/\mu}(x - x_0)\}$$

$$\varphi_- = c_- \exp\{-\sqrt{\sigma/\mu}(x_0 - x)\} \tag{5.4.7}$$

Substitute (5.4.7) into (5.4.5) and (5.4.6) and solve the linear equations for c_+ and c_- to obtain

$$c_+ = c_- = \frac{Q}{2\sqrt{\mu\sigma}}.$$

The solution of problem (5.4.1) is thus

$$\varphi(x) = \frac{Q}{2\sqrt{\mu\sigma}} \begin{cases} \exp\{-\sqrt{\sigma/\mu}(x - x_0)\} & \text{for} \quad x \geq x_0 \\[2mm] \exp\{-\sqrt{\sigma/\mu}(x_0 - x)\} & \text{for} \quad x \leq x_0 \end{cases} \tag{5.4.8}$$

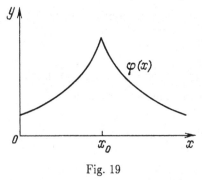

Fig. 19 shows a diagram of the function $\varphi(x)$. The diffusion visibly results in a solution decreasing exponentially and symmetrically in both directions from the point $x = x_0$. It is easy to check that

$$\int_{-\infty}^{\infty} \varphi(x)dx = \frac{Q}{\sigma}$$

Fig. 19

Examine now a more interesting problem where the velocity of the air masses flow differs from zero. Let it be constant and positive. Then

$$u\frac{d\varphi}{dx} + \sigma\varphi - \mu\frac{d^2\varphi}{dx^2} = Q\delta(x - x_0) \tag{5.4.9}$$

on the straight line $-\infty < x < \infty$. Reduce equation (5.4.9) in the same way to two problems:

$$\mu\frac{d^2\varphi_+}{dx^2} - u\frac{d\varphi_+}{dx} - \sigma\varphi_+ = 0$$

$$\varphi_+ = 0 \quad \text{at} \quad x \to \infty \tag{5.4.10}$$

$$\mu\frac{d^2\varphi_-}{dx^2} - u\frac{d\varphi_-}{dx} - \sigma\varphi_- = 0$$

$$\varphi_- = 0 \quad \text{at} \quad x \to -\infty \tag{5.4.11}$$

The solutions of problems (5.4.10) and (5.4.11) are verifiably linked with the relationships

$$\mu \frac{d\varphi_+}{dx} - \mu \frac{d\varphi_-}{dx} + Q = 0, \quad \varphi_+ = \varphi_- \quad \text{at} \quad x = x_0 \tag{5.4.12}$$

Represent the solutions of problems (5.4.10) and (5.4.11) as

$$\varphi_+ = c_+ \exp\left\{ -\left(\sqrt{\frac{\sigma}{\mu} + \frac{u^2}{4\mu^2}} - \frac{u}{2\mu} \right)(x - x_0) \right\} \quad \text{at} \quad x \ge x_0 \tag{5.4.13}$$

$$\varphi_- = c_- \exp\left\{ -\left(\sqrt{\frac{\sigma}{\mu} + \frac{u^2}{4\mu^2}} + \frac{u}{2\mu} \right)(x_0 - x) \right\} \quad \text{at} \quad x \le x_0$$

Substitute (5.4.13) into (5.4.12) to obtain $c_+ = c_- = c$ and

$$c_+ = c_- = c = Q / \sqrt{4\sigma\mu + u^2}.$$

As a result, the solution of problem is

$$\varphi(x) = \frac{Q}{\sqrt{4\sigma\mu + u^2}} \begin{cases} \exp\left\{ -\left(\sqrt{\frac{\sigma}{\mu} + \frac{u^2}{4\mu^2}} - \frac{u}{2\mu} \right)(x - x_0) \right\}, & x \ge x_0 \\[2mm] \exp\left\{ -\left(\sqrt{\frac{\sigma}{\mu} + \frac{u^2}{4\mu^2}} + \frac{u}{2\mu} \right)(x_0 - x) \right\}, & x \le x_0 \end{cases} \tag{5.4.14}$$

Fig. 20 represents the diagram of the function $\varphi(x)$. It is visibly clear that if $u > 0$ the left (from $x = x_0$) part of the exponent tends toward $x = x_0$, while the right part, on the contrary, spreads, which characterizes the transport of a substance by wind and simultaneous diffusion.

Examine now a more complex problem: wind has been blowing steadily for a long time towards positive values of x ($u_1 > 0$) and then backs or veers to change the direction for quite the opposite, negative direction $u_2 < 0$. Then there are two solutions:

$$\varphi_1(x) = \frac{Q}{\sqrt{4\sigma\mu + u_1^2}} \begin{cases} \exp\left\{ -\left(\sqrt{\frac{\sigma}{\mu} + \frac{u_1^2}{4\mu^2}} - \frac{|u_1|}{2\mu} \right)(x - x_0) \right\}, & x \ge x_0 \\[2mm] \exp\left\{ -\left(\sqrt{\frac{\sigma}{\mu} + \frac{u_1^2}{4\mu^2}} + \frac{|u_1|}{2\mu} \right)(x_0 - x) \right\}, & x \le x_0 \end{cases} \tag{5.4.15}$$

$$\varphi_2(x) = \frac{Q}{\sqrt{4\sigma\mu + u_2^2}} \begin{cases} \exp\left\{-\left(\sqrt{\dfrac{\sigma}{\mu} + \dfrac{u_2^2}{4\mu^2}} + \dfrac{|u_2|}{2\mu}\right)(x - x_0)\right\}, & x \geq x_0 \\[4mm] \exp\left\{-\left(\sqrt{\dfrac{\sigma}{\mu} + \dfrac{u_2^2}{4\mu^2}} - \dfrac{|u_2|}{2\mu}\right)(x_0 - x)\right\}, & x \leq x_0 \end{cases} \quad (5.4.16)$$

If the wind had blown toward positive values of x ($u_1 > 0$) for Δt_1 days, and toward negative values of x ($u_2 < 0$) for Δt_2 days, the average value of a substance concentration is found with the formula

$$\varphi(x) = \frac{\Delta t_1}{\Delta t_1 + \Delta t_2}\varphi_1(x) + \frac{\Delta t_2}{\Delta t_1 + \Delta t_2}\varphi_2(x) \qquad (5.4.17)$$

Fig. 21 shows graphically solution (5.4.17). Notice that we used in this statement the method of direct modeling with no allowance for transition processes.

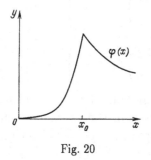

Fig. 20

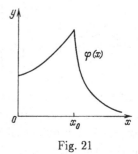

Fig. 21

Examine finally a statistical model in which wind is assigned statistically. Let

$$u(\xi) = \bar{u}p(\xi) \qquad (5.4.18)$$

where ξ is a random quantity within the interval $0 \leq \xi \leq 1$, and $p(\xi)$ is a probability density normed to unit, i. e., $\int_0^1 p(\xi)d\xi = 1$. If currents adjust to wind changes instantly, the solution of problem (5.4.9) with condition (5.4.15) will be, like in the previous problem,

$$\varphi(x) = \frac{Q}{2\mu}\int_0^1 \frac{w(x - x_0, u(\xi))}{\sqrt{\dfrac{\sigma}{\mu} + \dfrac{u^2(\xi)}{4\mu^2}}}d\xi \qquad (5.4.19)$$

where

$$
w(x-x_0, u(\xi)) = \begin{cases} \exp\left\{-\left(\sqrt{\dfrac{\sigma}{\mu} + \dfrac{u^2(\xi)}{4\mu^2}} - \dfrac{u(\xi)}{2\mu}\right)(x - x_0)\right\}, & x \geq x_0 \\[4mm] \exp\left\{-\left(\sqrt{\dfrac{\sigma}{\mu} + \dfrac{u^2(\xi)}{4\mu^2}} + \dfrac{u(\xi)}{2\mu}\right)(x_0 - x)\right\}, & x \leq x_0 \end{cases} \qquad (5.4.20)
$$

Integration in (5.4.19) at every fixed value of x is carried out by the method of statistical tests (the Monte Carlo method).

5.5. Transport and Diffusion of Heavy Aerosols

Heavy aerosols are of a special interest in studies of problems of local environmental pollution. As it propagates in the atmosphere, such an aerosol diffuses and generates a fallout. The fallout velocity may be precomputed by solving the Stokes problem and is a constant directed downwards. Therefore if the modulus of the vertical velocity of fallout particles is denoted as $\bar{w}_g$, a new term, $\bar{w}_g \partial \varphi / \partial z$, will appear in the aerosol transport equations and the problem of aerosol transport and diffusion, (5.3.27), becomes

$$
\frac{\partial \varphi}{\partial t} + \frac{\partial u\varphi}{\partial x} + \frac{\partial v\varphi}{\partial y} + \frac{\partial (w - \bar{w}_g)\varphi}{\partial z} + \sigma\varphi - \frac{\partial}{\partial z}\nu\frac{\partial \varphi}{\partial z} - \mu\Delta\varphi = f
$$

$$
\begin{aligned}
\varphi &= \varphi_0 & \text{at} \quad & t = 0 \\
\varphi &= 0 & \text{on} \quad & \Sigma \\
\frac{\partial \varphi}{\partial z} &= \alpha\varphi & \text{on} \quad & \Sigma_0 \\
\varphi &= 0 & \text{on} \quad & \Sigma_H
\end{aligned} \qquad (5.5.1)
$$

Determine aerosol fallout on an area $\Sigma_i \subset \Sigma_0$ within a time interval $0 \leq t \leq T$ on a plane $z = 0$. Integrate equation (5.5.1) over z within the limits $0 \leq z \leq H$. Let $\int_0^H \varphi \, dz = \bar{\varphi}$, $\int_0^H f \, dz = F$ and u and v be independent of z in the "acting zone," and obtain

$$
\frac{\partial \bar{\varphi}}{\partial t} + \frac{\partial u\bar{\varphi}}{\partial x} + \frac{\partial v\bar{\varphi}}{\partial y} + \sigma\bar{\varphi} = \mu\Delta\bar{\varphi} - (\bar{w}_g + \nu\alpha)\varphi_g + F \qquad (5.5.2)
$$

where $\varphi_g = \varphi|_{z=0}$. As we derived (5.5.2) we used conditions

$$w = 0 \quad \text{at} \quad z = 0, \quad z = H$$

$$\frac{\partial \varphi}{\partial z} = \alpha \varphi \quad \text{at} \quad z = 0$$

and a condition, natural in this case:

$$\varphi \to 0 \quad \text{at} \quad z \to H$$

From equation (5.5.2), an amount of aerosol in atmosphere decreases over the point (x, y) by the quantity $(\bar{w}_g + \alpha \nu)\varphi_g$ in any period of time. Here $\bar{w}_g \varphi_g$ is fallout and $\nu \alpha \varphi_g$ is the portion of aerosols involved in the turbulent exchange in the boundary layer near the Earth's surface. Notice, that if $\bar{w}_g \ll \nu \alpha$, $\bar{w}_g$ may be disregarded in equation (5.5.1) in solving the problem of aerosol fallout on a surface $\Sigma_i \subset \Sigma_0$. If $\bar{w}_g$ is comparable with $\nu \alpha$ or exceeds it, problem (5.5.1) should be taken instead of (5.3.27).

As concerns the basic functionals of the problems, these functionals are usually either the total mass of aerosols in a given domain $\Omega_i \subset \Omega$:

$$J_i = \frac{1}{T} \int_0^T dt \int_{\Omega_i} \varphi d\Omega_i \qquad (5.5.3)$$

or a total mass of the fallout on the area Σ_i of the cylindrical domain Ω:

$$J_i = a \int_0^T dt \int_{\Sigma_i} \varphi_g d\Sigma_i \qquad (5.5.4)$$

The constant a is linked with the gravity force and diffusion mechanisms of aerosol fallout. Given all this,

$$a = \bar{w}_g + \nu \alpha \qquad (5.5.5)$$

and the basic functionals are determined.

We will not specially refer to heavy aerosols later on. However, if gravity proves to be essential, the statement should be corrected to allow for the effects discussed in this section. Therefore we will not examine specifically this particular case, since necessary generalizations, if needed, are trivial.

5.6. The Structure and Modeling of Turbulent Motions in the Atmosphere

Adjustment of advective and convective transport processes to diffusion is very important in studying propagation of passive pollutants in the atmosphere. Correct simultaneous modeling of transport and diffusion is essential for adequate description of propagation of pollutants in the atmosphere. Incorrect (or, rather, disbalanced) modeling results in gross errors. Let us examine in detail adjustment of advective-convective transport and diffusion.

Assume that we are dealing with mesoscale transport processes with characteristic time of several hours. This means that the meteorological variations within these hours may be disregarded and they may be considered in the first approximation as constant; in our case these values are u, v, w. Solving of a problem of substance propagation from a source without diffusion will produce a single trajectory the pollutant travels along. This statement is immediately in conflict with the logic of a physical process, since a substance always spreads as it travels after emission. Therefore we will deal not with a single trajectory, but with an area of non-zero pollutant densities. What determines the characteristics of this spread first of all is all the small-scale wind velocity fluctuations inherent to the statistical nature of atmospheric motions. Measuring the divergence cone and using model problems produce diffusion coefficients related to motions of a given spatial-temporal scale. Examine an example.

Assume a two-dimensional substance transport process described by the equation

$$A\varphi \equiv u\frac{\partial \varphi}{\partial x} + v\frac{\partial \varphi}{\partial y} - \mu\Delta\varphi = Q\delta(\mathbf{r} - \mathbf{r}_0) \qquad (5.6.1)$$

where u, v are given velocity components considered to be constant for simplicity, and μ is a diffusion coefficient still unknown. The solution to equation (5.6.1) for infinite domain is

$$\varphi = \frac{Q}{2\pi\mu}\exp\left\{\frac{u(x - x_0) + v(y - y_0)}{2\mu}\right\}K_0\left(\frac{\sqrt{u^2 + v^2}}{2\mu}|\mathbf{r} - \mathbf{r}_0|\right) \qquad (5.6.2)$$

where $K_0(x)$ is the McDonald's function:

$$K_0(x) = \int\limits_0^\infty e^{-x \operatorname{ch} y} dy, \quad x > 0$$

Transform the coordinates so that the beginning $x = 0$ is at the point $\mathbf{r}_0$, and the direction of the new axis x is parallel to the velocity vector $\mathbf{u}$. Then formula (5.6.2) becomes

$$\varphi = \frac{Q}{2\pi\mu} \exp\left\{\frac{\tilde{u}x}{2\mu}\right\} K_0\left(\frac{\tilde{u}}{2\mu}|\mathbf{r}|\right) \tag{5.6.3}$$

where $\tilde{u} = \sqrt{u^2 + v^2}$.

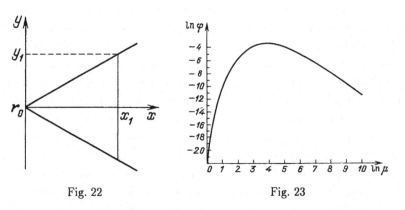

Fig. 22   Fig. 23

Assume now such a divergence cone of the modeled pollutant density that at the distance x_1 from the emission point its width is equal to $2y_1$ (Fig. 22). If x_1 is sufficient for asymptotic representation of the function $K_0(x)$, and the accuracy of pollutant density measurement ε is within the divergence cone, obtain the equation to determine μ

$$\varphi(x_1, y_1, \mu) = \varepsilon \tag{5.6.4}$$

which may be deployed into

$$\frac{Q}{2\sqrt{\pi\mu(x_1^2 + y_1^2)}} \exp\left\{\frac{\tilde{u}}{2\mu}\left(x_1 - \sqrt{x_1^2 + y_1^2}\right)\right\} = \varepsilon \tag{5.6.5}$$

Fig. 23 shows the graphic solution of problem (5.6.5).

If we deal with a transport process with a time scale of an order of several days, we obtain an averaged trajectory by the same reasoning and averaging meteorological elements with regard to their typification over this characteristic time-scale, and disregarding diffusion. This solution will by no means reflect real physical propagation of pollutants, since the average interval will consist of a set of aggregates of transport mesoprocesses with a characteristic time of several hours within which they may be considered as little variable. However, the set of those mesoprocesses determines their variability for several days.

If we have information on the types of wind variability in mesoprocesses for several days, solve a set of corresponding mesoscale transport problems, obtain divergence cones of polluting substances, and statistically average over this aggregate, we may obtain substance density in the process of transport with a characteristic time of several days. This synthesized problem may be solved and the solution checked against the real form of an experimental aerosol cloud. We may thus judge how adequately a set of typical mesoprocesses reflects the physics of a modeled phenomenon.

It is clear, that there is no need for a coefficient of macrodiffusion in this case, and information on the statistical nature of mesoprocesses contributing to the general transport process, will suffice. It is possible, however, to use macrodiffusion approximation in certain cases. Let us apply formally the diffusion equation to describe the process of propagation of a pollutant:

$$\frac{\partial \varphi}{\partial t} + \operatorname{div} \mathbf{u}\varphi + \sigma\varphi - \mu\Delta\varphi = f \tag{5.6.6}$$

It is possible to apply perturbation theory methods to find μ providing, in a certain sense, the best value of a functional, using the vector $\mathbf{u}$ in this model as the average for several days, and assuming that μ is unknown. It may however turn out in solving problem (5.6.6) with chosen μ, that the differential field of solutions may differ from the real (or modeled on the basis of a set of mesoscale problems), though the chosen functional was correct. Therefore, if we are interested in a unique assigned functional, the model of substance transport with a prescribed diffusion coefficient will make a good tool for research.

If we are interested in more detailed characteristics of the pollution field or various functionals, modeling a process with a set of smaller-scale models proves better. Therefore we will proceed from the assumption of possibility to use models in macro-diffusion approximation as we solve optimization problems. But a more detailed study and checking the reliability of results requires a solution of a set of problems corresponding to smaller-scale fluctuations.

This formalism is reduced mathematically to the following: examine the equation of substance transport and diffusion

$$\frac{\partial \varphi}{\partial t} + u\frac{\partial \varphi}{\partial x} + v\frac{\partial \varphi}{\partial y} + w\frac{\partial \varphi}{\partial z} - \mu'\Delta\varphi - \nu'\frac{\partial^2 \varphi}{\partial z^2} = Q\delta(\mathbf{r} - \mathbf{r}_0) \qquad (5.6.7)$$

Assume that u, v, and w are functions

$$u = \bar{u} + u', \quad v = \bar{v} + v', \quad w = \bar{w} + w' \qquad (5.6.8)$$

where u', v', and w' are components of the velocity vector varying within a few hours, which are deviations from the main flow with corresponding velocity vector components $\bar{u}$, $\bar{v}$, and $\bar{w}$ respectively with the norm of deviation known a priori; μ' and ν' are diffusion coefficients corresponding to mesoprocesses. The problem is considered as periodic with a period T inherent to the characteristic time of the process and with boundary conditions

$$\frac{\partial \varphi}{\partial z} = 0 \qquad \text{at} \qquad z = H$$

$$\frac{\partial \varphi}{\partial z} = \alpha\varphi \qquad \text{at} \qquad z = 0 \qquad\qquad (5.6.9)$$

$$\varphi \to 0 \qquad \text{at} \qquad x, y \to \pm\infty$$

The solution of the set of problems so formulated takes into account all the statistical peculiarities of mesoprocesses. Therefore, the functions u', v', and w' will have to be modeled. Since the process is linear here, use the recurrence of these quantities as this corresponds to their action on intervals of time longer than T. The final solution of this problem must be averaged over the interval $[0, T]$.

All the models discussed in this book are thus generalized in a natural manner for more accurate consideration of the statistical structure of input data.

Let us exemplify it. If the substance transport process is described, like in the case discussed above, by equation (5.6.1) with solution (5.6.2), represent the functions u and v as

$$u = \bar{u} + u'(\alpha), \quad v = \bar{v} + v'(\alpha) \qquad (5.6.10)$$

where the deviations u' and v' from the main flow $\bar{u}$ and $\bar{v}$, depending on α, are considered as random fluctuations of a quantity bounded a priori. Substitute (5.6.10) into (5.6.2) to obtain

$$\varphi(x, y, \alpha) = \frac{Q}{2\pi\mu} \exp\left\{ \frac{\bar{u}(x - x_0) + \bar{v}(y - y_0)}{2\mu} \right\}$$

$$\times \exp\left\{ \frac{u'(\alpha)(x - x_0) + v'(\alpha)(y - y_0)}{2\mu} \right\}$$

$$\times K_0 \left(\frac{\sqrt{(\bar{u} + u'(\alpha))^2 + (\bar{v} + v'(\alpha))^2}}{2\mu} |\mathbf{r} - \mathbf{r}_0| \right) \quad (5.6.11)$$

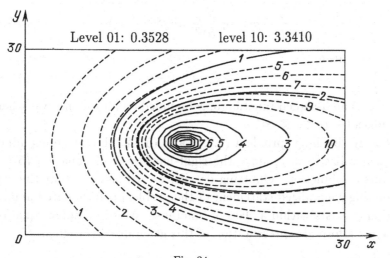

Level 01: 0.3528   level 10: 3.3410

Fig. 24

The function $\varphi(x, y, \alpha)$ represents a set of possible realizations of the aerosol cloud depending on the statistical structure of the fluctuations. Fig. 24 shows isograms of the function φ averaged over the parameter α in assumption of the normal law of fluctuation distribution. Solid isograms of the function φ are presented at $u' = v' = 0$ for comparison.

In studies of global processes forming for weeks and months, the methods discussed above are applicable for construction of an hierarchy of models, which might be identified by space photography.

5.7. Adjoint Equation for a Simple Diffusion Equation

Adjoint equations as applied to various functionals in mathematical physics problems have long interested researchers but were rarely used in solving problems. The perturbation theory gave rise to construction of adjoint problems, as it used both the main and adjoint problems. In solving problems involving finding linear functionals, the adjoint statement which reflects the principle of duality, allows to formulate a number of algorithms which prove the best for analysis of problems and their solution in various conditions.

Examine a simple diffusion equation

$$\frac{\partial \varphi}{\partial t} + \sigma \varphi - \mu \frac{\partial^2 \varphi}{\partial x^2} = Q\delta(x - x_0) \qquad (5.7.1)$$

with a condition

$$\varphi = \varphi_0 \quad \text{at} \quad t = 0 \qquad (5.7.2)$$

where φ is an assigned function of x bounded on all the interval where x changes $(-\infty < x < \infty)$.

Strictly speaking, problem (5.7.1) and (5.7.2) should be examined in a generalized sense, since the right-hand part of equation (5.7.1) contains a δ-function. Omitting details, we will assume that the corresponding generalized statement of the problem is formulated and its solution φ exists within a class of functions. We will assume equation (5.7.1) as a formal representation of this generalized statement.[2]

---

[2]See the footnote at page 40.

Proceed to the adjoint problem: multiply equation (5.7.1) by φ^* (the properties of this function will be clarified later) and integrate the result over time and space:

$$\int\limits_0^T dt \int\limits_{-\infty}^{\infty} \varphi^* \left(\frac{\partial \varphi}{\partial t} + \sigma\varphi - \mu\frac{\partial^2 \varphi}{\partial x^2} \right) dx = Q \int\limits_0^T dt \int\limits_{-\infty}^{\infty} \varphi^* \delta(x - x_0) dx \quad (5.7.3)$$

Assume now that the functions φ and φ^* provide sense to integrals in (5.7.3) and all subsequent transformations. Notice, that

$$Q \int\limits_0^T dt \int\limits_{-\infty}^{\infty} \varphi^* \delta(x - x_0) dx = Q \int\limits_0^T \varphi^*(x_0, t) dt \quad (5.7.4)$$

Modify the left-hand part of identity (5.7.3) so that the function φ is under the integral outside the parentheses and the differential relationship containing the function φ^* is inside the parentheses. To do so, integrate by parts:

$$\int\limits_0^T dt \int\limits_{-\infty}^{\infty} \varphi^* \frac{\partial \varphi}{\partial t} dx = \int\limits_{-\infty}^{\infty} \varphi\varphi^* \Big|_{t=0}^{t=T} dx - \int\limits_0^T dt \int\limits_{-\infty}^{\infty} \varphi\frac{\partial \varphi^*}{\partial t} dx \quad (5.7.5)$$

$$\int\limits_0^T dt \int\limits_{-\infty}^{\infty} \varphi^* \frac{\partial^2 \varphi}{\partial x^2} dx$$

$$= \int\limits_0^T \left(\varphi^*\frac{\partial \varphi}{\partial x} - \varphi\frac{\partial \varphi^*}{\partial x} \right) \Big|_{x=-\infty}^{x=\infty} dt + \int\limits_0^T dt \int\limits_{-\infty}^{\infty} \varphi\frac{\partial^2 \varphi^*}{\partial x^2} dx \quad (5.7.6)$$

Substitute (5.7.5) and (5.7.6) into (5.7.3) to obtain

$$\int\limits_0^T dt \int\limits_{-\infty}^{\infty} \varphi \left(-\frac{\partial \varphi^*}{\partial t} + \sigma\varphi^* - \mu\frac{\partial^2 \varphi^*}{\partial x^2} \right) dx + \int\limits_{-\infty}^{\infty} \varphi\varphi^* \Big|_{t=0}^{t=T} dx$$

$$- \mu \int\limits_0^T \left(\varphi^*\frac{\partial \varphi}{\partial x} - \varphi\frac{\partial \varphi^*}{\partial x} \right) \Big|_{x=-\infty}^{x=\infty} dt = Q \int\limits_0^T \varphi^*(x_0, t) dt \quad (5.7.7)$$

Assume that

$$\varphi^* = 0 \quad \text{at} \quad x \to \pm\infty \quad (5.7.8)$$

Then relationship (5.7.7) will be simpler:

$$\int_0^T dt \int_{-\infty}^{\infty} \varphi \left(-\frac{\partial \varphi^*}{\partial t} + \sigma \varphi^* - \mu \frac{\partial^2 \varphi^*}{\partial x^2} \right) dx + \int_{-\infty}^{\infty} (\varphi_T \varphi_T^* - \varphi_0 \varphi_0^*) dx$$

$$= Q \int_0^T \varphi^*(x_0, t) dt \quad (5.7.9)$$

Assume now that φ^* satisfies the equation

$$-\frac{\partial \varphi^*}{\partial t} + \sigma \varphi^* - \mu \frac{\partial^2 \varphi^*}{\partial x^2} = p \qquad (5.7.10)$$

with the initial data

$$\varphi^* = 0 \quad \text{at} \quad t = T \qquad (5.7.11)$$

and boundary conditions (5.7.8). Here p is an undetermined yet function of x and t. We will refer to this problem as adjoint. With regard to (5.7.10), reduce (5.7.9) to

$$\int_0^T dt \int_{-\infty}^{\infty} p \varphi \, dx = Q \int_0^T \varphi^*(x_0, t) dt + \int_{-\infty}^{\infty} \varphi(x, 0) \varphi^*(x, 0) dx \qquad (5.7.12)$$

Let

$$J = \int_0^T dt \int_{-\infty}^{\infty} p \varphi \, dx \qquad (5.7.13)$$

be a linear functional of φ to be computed by solving the main problem, (5.7.1) and (5.7.2). It follows from (5.7.12) that this functional emerges in solving the adjoint problem, (5.7.10) and (5.7.11):

$$J = Q \int_0^T \varphi^*(x_0, t) dt + \int_{-\infty}^{\infty} \varphi(x, 0) \varphi^*(x, 0) dx \qquad (5.7.14)$$

This is the essence of the principle of duality.

Examine now the problem of functionals. Functional (5.7.13) accepts various physical content. Let

$$p(x, t) = \delta(x - \xi)\delta(t - \tau) \qquad (5.7.15)$$

Substitute (5.7.15) into (5.7.13) to obtain

$$J = \varphi(\xi, \tau) \qquad (5.7.16)$$

i. e., a value of the solution at the point $x = \xi$, $t = \tau$. Notice that (5.7.14) produces the same result, if expression (5.7.15) stands for p in adjoint problem (5.7.10) and (5.7.11).

Examine another case. Suppose the integral quantity of a substance is sought at $a \leq x \leq b$. Choose the function $p(x,t)$ as

$$p(x,t) = \begin{cases} 1, & x \in [a,b] \\ 0, & x \notin [a,b] \end{cases} \qquad (5.7.17)$$

Substitute (5.7.17) into (5.7.13) to obtain

$$J = \int_0^T dt \int_a^b \varphi \, dx \qquad (5.7.18)$$

Formula (5.7.14) produces the same functional.

Assume now that the substance φ integral content is being measured at $[a,b]$ within a time interval $[\tau_1, \tau_2]$ with resolution of the instruments depending on x and t, i. e., the instrument characteristic is described by the function $X = V(t)\chi(x)$. The measured functional must be checked against the calculated. Select the function $p(x,t)$ as

$$p(x,t) = \begin{cases} V(t)\chi(x), & x \in [a,b] \quad \text{and} \quad t \in [\tau_1, \tau_2] \\ 0, & x \notin [a,b] \quad \text{or} \quad t \notin [\tau_1, \tau_2] \end{cases} \qquad (5.7.19)$$

to obtain the functional

$$J = \int_{\tau_1}^{\tau_2} dt \int_a^b \chi(x)V(t)\varphi \, dx \qquad (5.7.20)$$

A set of possible functionals may be expanded. Notice that any linear functional of the solution may be represented in the same manner and thus an adjoint problem may be stated for every functional.

If we are interested in functionals of the solution and not in the solution itself, an adjoint problem may thus be formulated for every of the functionals. At first sight, the best natural way is seemingly to solve the main problem, and then use formula (5.7.13) to calculate a functional. This approach actually proves the most expedient in some cases. However, adjoint equations are indispensable analytic tools in designing utilities emitting aerosols, estimating functional sensitivity to

changes in environmental parameters and solving other similar problems.

Examine an example. Assume that equation (5.7.1) describes propagation of a pollutant in the area $\Omega = (-\infty, \infty)$, and we need to
find a domain $\omega \subset \Omega$ where the functional of the (5.7.16) type (i. e.,
concentration of the pollutant at the point $x = \xi_1$ at the moment of
time $t = \tau_1$) does not exceed a certain assigned given constant c, while
the pollutant emission source with an intensity of Q is located at the
point $x_0 \in \omega$. The initial value of the function φ may be chosen equal
to zero:

$$\varphi = 0 \quad \text{at} \quad t = 0 \qquad (5.7.21)$$

The problem may be solved at least in two ways.

The first is repeated solving equation (5.7.1) at different values of
$x_0 \in \Omega$, determining the values of functional (5.7.16) and isolating
the sought zone ω. This technique implies solution of a considerable
number of problems of the (5.7.1) type and is obviously unacceptable
practically.

The second approach proceeds from dual representation of functional
(5.7.16) and solution of the adjoint equation. The representation for
this problem is

$$J = Q \int_0^T \varphi^*(x_0, t) dt \qquad (5.7.22)$$

where φ^* is the solution of the equation

$$-\frac{\partial \varphi^*}{\partial t} + \sigma \varphi^* - \mu \frac{\partial^2 \varphi^*}{\partial x^2} = \delta(x - \xi_1)\delta(t - \tau_1) \qquad (5.7.23)$$

$$\varphi^* = 0 \quad \text{at} \quad t = T \qquad (5.7.24)$$

Allocation of the zone $\omega \subset \Omega$ for the source of pollution on the basis
of dual representation (5.7.22) of functional (5.7.16) requires thus only
one-time solving the problem adjoint to (5.7.1).

Determine the function φ^* satisfying (5.7.23) and (5.7.24): introduce
a new variable

$$t_1 = T - t, \quad t_1 \in [0, T] \qquad (5.7.25)$$

Then problem (5.7.23)–(5.7.24) becomes

$$\frac{\partial \varphi^*}{\partial t_1} + \sigma \varphi^* - \mu \frac{\partial^2 \varphi^*}{\partial x^2} = \delta(x - \xi_1)\delta(T - t_1 - \tau_1)$$

(5.7.26)

$$\varphi^* = 0 \quad \text{at} \quad t_1 = 0.$$

Notice that the operator in problem (5.7.26) formally coincides with that in problem (5.7.1). The solution to problem (5.7.26) is provided by the formula

$$\varphi^*(x, t_1) = \int_0^{t_1} \int_{-\infty}^{\infty} \tilde{\varphi}(x - \xi, t_1 - \tau)\delta(\xi - \xi_1)\delta(T - t_1 - \tau_1)d\xi d\tau \quad (5.7.27)$$

where $\tilde{\varphi}(x, t)$ is the fundamental solution of the operator of problem (5.7.26):

$$\tilde{\varphi}(x, t) = \frac{\theta(t)}{2\sqrt{\mu\pi t}} \exp\left\{ -\left(\sigma t + \frac{x^2}{4\mu t} \right) \right\}$$

θ is the Heaviside's function:

$$\theta(t) = \begin{cases} 0, & t \leq 0 \\ 1, & t > 0 \end{cases}$$

Turn to the old variables to obtain

$$\varphi^*(x, t)$$
$$= \begin{cases} \dfrac{1}{2\sqrt{\mu\pi(\tau_1 - t)}} \exp\left\{ -\left[\sigma(\tau_1 - t) + \dfrac{(x - \xi)^2}{4\mu(\tau_1 - t)} \right] \right\}, & \text{at } t \in [0, \tau_1) \\ \\ 0, & \text{at } t \in [\tau_1, T] \end{cases}$$

(5.7.28)

Dual representation (5.7.22) of functional (5.7.16) is

$$J = \frac{Q}{2\sqrt{\mu\pi}} \int_0^{\tau_1} \frac{\exp\left\{ -\left[\sigma(\tau_1 - t) + \frac{(x_0 - \xi_1)^2}{4\mu(\tau_1 - t)} \right] \right\}}{\sqrt{\tau_1 - t}} dt \quad (5.7.29)$$

in compliance with formula (5.7.27), or

$$J = \frac{Q}{2\sqrt{\mu\pi}} \sum_{j=0}^{K-1} \frac{\exp\left\{ -\left[\sigma(\tau_1 - t_j) + \frac{(x_0 - \xi_1)^2}{4\mu(\tau_1 - t_j)} \right] \right\}}{\sqrt{\tau_1 - t_j}} \Delta t + O(\Delta t) \quad (5.7.30)$$

where $t_j = j\Delta t,\ \Delta t = \tau_1/K$.

Examine now functional (5.7.29) as a function of $J(x_0)$ at $x_0 \in \Omega$ and draw a diagram of this function. The area ω allocable for the source of pollutant will be determined by the inequality $J(x) < c$. Fig. 25 shows the graphical solution of the examined example; the area ω on the x-axis is marked with double lines.

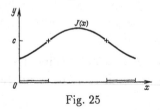

Fig. 25

Find the distribution of the pollutant at every specific $x_0 \in \omega$ by solving problem (5.7.1) and (5.7.2). Rewrite this, in view of the considerations stated above, as

$$\varphi(x,t) = \frac{Q}{2\sqrt{\mu\pi}} \sum_{j=0}^{K-1} \frac{\exp\left\{-\left[\sigma(t-t_j) + \frac{(x-x_0)^2}{4\mu(t-t_j)}\right]\right\}}{\sqrt{t-t_j}} \Delta t + O(\Delta t) \quad (5.7.31)$$

We have just examined a nonstationary case. If the main problem is stationary, like

$$\mu\frac{d^2\varphi}{dx^2} - \sigma\varphi = Q\delta(x - x_0)$$
$$\varphi = 0 \quad \text{at} \quad x \to \pm\infty$$
$$(5.7.32)$$

the adjoint problem is also stationary:

$$\mu\frac{d^2\varphi^*}{dx^2} - \sigma\varphi^* = p(x)$$
$$\varphi^* = 0 \quad \text{at} \quad x \to \pm\infty$$
$$(5.7.33)$$

It is clear that the linear functional J is then

$$J = \int_{-\infty}^{\infty} p(x)\varphi dx \quad (5.7.34)$$

or

$$J = Q\varphi^*(x_0) \quad (5.7.35)$$

If we choose

$$p(x) = \delta(x - \xi) \quad (5.7.36)$$

as the function $p(x)$, the solution to adjoint problem (5.7.33) is similar to that to (5.4.8):

$$\varphi^*(x,\xi) = \frac{1}{2\mu\sqrt{\sigma/\mu}} \begin{cases} \exp\left\{-\sqrt{\frac{\sigma}{\mu}}(x-\xi)\right\}, & x \geq \xi \\ \\ \exp\left\{-\sqrt{\frac{\sigma}{\mu}}(\xi-x)\right\}, & x \leq \xi \end{cases} \qquad (5.7.37)$$

Assume that another solution is sought with $p(x)$ different from (5.7.36). Then

$$\varphi^*(x) = \int_{-\infty}^{\infty} p(x)\varphi^*(x,\xi)d\xi \qquad (5.7.38)$$

where $\varphi^*(x,\xi)$ is the fundamental solution (5.7.37). If, in partic-ular, (5.7.17) is chosen as $p(x)$,

$$\varphi^*(x) = \int_a^b \varphi^*(x,\xi)d\xi$$

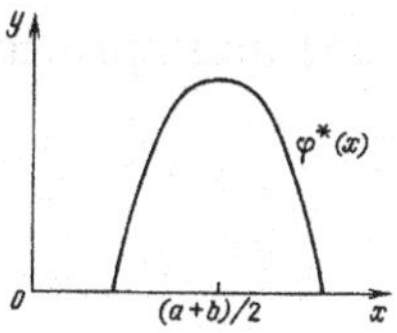

Fig. 26

Substitute (5.7.37) into this expression to obtain

$$\varphi^*(x) = \frac{1}{2\sigma}\left(2 - \exp\{\sqrt{\sigma/\mu}(b-x)\} - \exp\{-\sqrt{\sigma/\mu}(x-a)\}\right) \quad (5.7.39)$$

Fig. 26 shows a diagram of this function.

Examine transport of a substance by diffusion with advection taken into account. The main problem is then

$$\frac{\partial\varphi}{\partial t} + u\frac{\partial\varphi}{\partial x} + \sigma\varphi - \mu\frac{\partial^2\varphi}{\partial x^2} = Q\delta(x - x_0)$$

$$\varphi = \varphi_0 \quad \text{at} \quad t = 0$$
$$\varphi = 0 \quad \text{at} \quad x \to \pm\infty$$

$$(5.7.40)$$

As above, the adjoint one is

$$-\frac{\partial\varphi^*}{\partial t} - u\frac{\partial\varphi^*}{\partial x} + \sigma\varphi^* - \mu\frac{\partial^2\varphi^*}{\partial x^2} = p$$

$$\varphi^* = 0 \quad \text{at} \quad t = T$$
$$\varphi^* = 0 \quad \text{at} \quad x \to \pm\infty$$

$$(5.7.41)$$

The main functional of the problem is

$$J = \int\limits_0^T dt \int\limits_{-\infty}^{\infty} p\varphi \, dx \qquad (5.7.42)$$

and its dual representation is

$$J = Q \int\limits_0^T \varphi^*(x_0, t) dt + \int\limits_{-\infty}^{\infty} \varphi_0 \varphi^*(x, 0) dx \qquad (5.7.43)$$

Examine another example connected with construction planning. Assume that (5.7.40) describes propagation of a pollutant φ in an area Ω at $u = \text{const} > 0$ and $\varphi_0 = 0$. Location of the emission source (the point x_0) must provide that

$$\varphi(\xi_1, \tau_1) < c \qquad (5.7.44)$$

$$|x_0 - \xi_1| = \min_{x_0 \in \Omega} \qquad (5.7.45)$$

where ξ_1, τ_1, and c are prescribed constants.

Solving this problem requires again dual representation of functional (5.7.16):

$$J = Q \int\limits_0^T \varphi^*(x_0, t) dt \qquad (5.7.46)$$

Here φ^* is the solution of equation (5.7.41), where the right-hand part is (5.7.15). Find the fundamental solution of the operator of problem (5.7.41): apply the Fourier transformation to the equation

$$\frac{\partial \varphi}{\partial t_1} - u\frac{\partial \varphi}{\partial x} + \sigma\varphi - \mu\frac{\partial^2 \varphi}{\partial x^2} = \delta(t_1)\delta(x) \qquad (5.7.47)$$

where $t_1 = T - t$, and obtain

$$\frac{\partial}{\partial t_1}F[\varphi] + (iu\xi + \sigma + \mu\xi^2)F[\varphi] = \delta(t_1) \qquad (5.7.48)$$

where

$$F[\varphi] = \int\limits_{-\infty}^{\infty} \varphi(x, t)e^{i\xi x} \, dx$$

The solution to equation (5.7.48) is provided by the function

$$F[\varphi](\xi, t_1) = \theta(t_1) \exp\{-(iu\xi + \sigma + \mu\xi^2)t_1\}$$

Use the inverse Fourier transformation to find the sought fundamental solution:

$$\varphi(x, t_1) = \frac{\theta(t_1)}{2\sqrt{\mu\pi t_1}} \exp\left\{-\left[\sigma t_1 + \frac{(x + ut_1)^2}{4\mu t_1}\right]\right\} \qquad (5.7.49)$$

Use (5.7.27) and the old variables to solve the problem adjoint to (5.7.40) and having in the right-hand part $p = \delta(x - \xi_1)\delta(t - \tau_1)$:

$$\varphi^*(x, t) =$$

$$\begin{cases} \dfrac{1}{2\sqrt{\mu\pi(\tau_1 - t)}} \exp\left\{-\left[\sigma(\tau_1 - t) + \dfrac{(x - \xi_1 + u(\tau_1 - t))^2}{4\mu(\tau_1 - t)}\right]\right\}, & t \in [0, \tau_1) \\[4mm] & \qquad\qquad (5.7.50) \\ 0, & t \in [\tau_1, T] \end{cases}$$

Substitute (5.7.50) into (5.7.46) to obtain an expression for the functional:

$$J = \frac{Q}{2\sqrt{\mu\pi}} \sum_{j=0}^{K-1} \frac{\exp\left\{-\left[\sigma(\tau_1 - t_j) + \frac{(x_0 - \xi_1 + u(\tau_1 - t_j))^2}{4\mu(\tau_1 - t_j)}\right]\right\}}{\sqrt{\tau_1 - t_j}} \Delta t$$

$$+ O(\Delta t) \quad (5.7.51)$$

where $t_j = j\Delta t$, $\Delta t = \tau_1/K$.

Just as done before, determine the area $\omega \subset \Omega$ where inequality (5.7.44) holds. The sought point x_0 is found from condition (5.7.45). Fig. 27 shows the graphical solution to this problem.

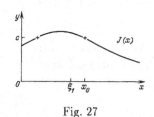

Fig. 27

The formula

$$\varphi(x, t) = \frac{Q}{2\sqrt{\mu\pi}} \sum_{j=1}^{K-1} \frac{\exp\left\{-\left[\sigma(t - t_j) + \frac{(x - x_0 - u(t - t_j))^2}{4\mu(t - t_j)}\right]\right\}}{\sqrt{t - t_j}} \Delta t$$

$$+ O(\Delta t) \quad (5.7.52)$$

where $K = [t/\Delta t]$, may provide an analytical expression for the function $\varphi(x,t)$. This is the solution to direct problem (5.7.40) which takes into account coincidence of the operators in problems (5.7.40) and (5.7.47).

If problem (5.7.40) is stationary:

$$u\frac{\partial\varphi}{\partial x} + \sigma\varphi - \mu\frac{\partial^2\varphi}{\partial x^2} = Q\delta(x - x_0)$$
(5.7.53)

$$\varphi = 0 \quad \text{at} \quad x \to \pm\infty$$

it generates a stationary adjoint problem:

$$-u\frac{\partial\varphi^*}{\partial x} + \sigma\varphi^* - \mu\frac{\partial^2\varphi^*}{\partial x^2} = p(x)$$
(5.7.53')

$$\varphi^* = 0 \quad \text{at} \quad x \to \pm\infty$$

The main problem, (5.7.53), is solved in § 5.4 (see (5.4.14)). Solve now the adjoint problem: find at first the fundamental solution of the operator of (5.7.53') at $p(x) = \delta(x - \xi)$. Similarly to (5.4.14), obtain

$$\varphi^*(x,\xi) = \frac{1}{2\mu\sqrt{\frac{\sigma}{\mu} + \frac{u^2}{4\mu^2}}} \begin{cases} \exp\left\{-\left(\sqrt{\frac{\sigma}{\mu} + \frac{u^2}{4\mu^2}} + \frac{u}{2\mu}\right)(x - \xi)\right\}, & x \geq \xi \\ \\ \exp\left\{-\left(\sqrt{\frac{\sigma}{\mu} + \frac{u^2}{4\mu^2}} - \frac{u}{2\mu}\right)(\xi - x)\right\}, & x \leq \xi \end{cases}$$
(5.7.54)

Any another solution to problem (5.7.53') may be obtained by the fundamental solution:

$$\varphi^*(x) = \int\limits_{-\infty}^{\infty} p(\xi)\varphi^*(x,\xi)\,d\xi$$
(5.7.55)

If, in particular, (5.7.17) is $p(x)$, then

$$\varphi^*(x) = \int\limits_{a}^{b} \varphi^*(x,\xi)\,d\xi$$
(5.7.56)

Substitute (5.7.54) into (5.7.56) to obtain

$$\varphi^*(x) = \frac{1}{2\mu\sqrt{\frac{\sigma}{\mu} + \frac{u^2}{4\mu^2}}} \left[\frac{1}{\sqrt{\frac{\sigma}{\mu} + \frac{u^2}{4\mu^2}} - \frac{u}{2\mu}} \right.$$

$$\times \left(1 - \exp\left\{ -\left(\sqrt{\frac{\sigma}{\mu} + \frac{u^2}{4\mu^2}} - \frac{u}{2\mu} \right)(x - a) \right\} \right)$$

$$+ \frac{1}{\sqrt{\frac{\sigma}{\mu} + \frac{u^2}{4\mu^2}} + \frac{u}{2\mu}} \left(1 - \exp\left\{ -\left(\sqrt{\frac{\sigma}{\mu} + \frac{u^2}{4\mu^2}} + \frac{u}{2\mu} \right)(b - x) \right\} \right) \right] \qquad (5.7.57)$$

Fig. 28 shows a diagram of the function $\varphi^*(x)$.

A set of problems must be adjoint to a set of stationary problems related to different types of motion at different velocities of air mass flows. The functional is found either with a help of the main equations or the adjoint equations.

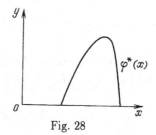

Fig. 28

5.8. General Case of Adjoint Problem for Three-Dimensional Domain

Examine a general three-dimensional problem

$$\frac{\partial \varphi}{\partial t} + \operatorname{div} \mathbf{u}\varphi + \sigma\varphi - \frac{\partial}{\partial z}\nu\frac{\partial \varphi}{\partial z} - \mu\Delta\varphi = f$$

$$\varphi = 0 \quad \text{on} \quad \Sigma \quad \text{at} \quad u_n < 0$$

$$\frac{\partial \varphi}{\partial n} = 0 \quad \text{on} \quad \Sigma \quad \text{at} \quad u_n \geq 0$$

$$\frac{\partial \varphi}{\partial z} = \alpha\varphi \quad \text{on} \quad \Sigma_0 \qquad\qquad (5.8.1)$$

$$\frac{\partial \varphi}{\partial z} = 0 \quad \text{on} \quad \Sigma_H$$

Assume that the solution $\varphi(x, y, z, t)$ of problem (5.8.1) is a function continuous within $\Omega \times (0, T)$, differentiable and periodic in t with a period T). Let $\varphi(x, y, z, t)$ belong to the set $D(A)$ of functions continuous and differentiable in Ω in the real Hilbert space $L_2(\Omega)$ at every t, so that $-\partial/\partial z (\nu \partial \varphi / \partial z) - \mu \Delta \varphi \in L_2(\Omega)$. Every function of $D(A)$ satisfies the boundary conditions (5.8.1). Assume that the components of the vectoral coefficient $\mathbf{u}$ are continuous and differentiable, with div $\mathbf{u} = 0$, and the other functions are sufficiently smooth and provide the unique solution φ.

Then the main problem, (5.8.1), may be written as

$$\frac{\partial \varphi}{\partial t} + A\varphi = f$$

where the operator A with its definition domain $D(A)$ operates in the Hilbert space $L_2(\Omega)$ and is assigned by the equality

$$A\varphi = \operatorname{div} \mathbf{u}\varphi + \sigma\varphi - \frac{\partial}{\partial z} \nu \frac{\partial \varphi}{\partial z} - \mu \Delta \varphi$$

while the function φ is assumed as periodic in t.

Construct now an adjoint problem.

Multiply equation (5.8.1) by a function φ^* and integrate the result over all the definition domain of the solution to obtain

$$\int\limits_0^T dt \int\limits_\Omega \varphi^* \frac{\partial \varphi}{\partial t} d\Omega + \int\limits_0^T dt \int\limits_\Omega \varphi^* \operatorname{div} \mathbf{u}\varphi d\Omega + \sigma \int\limits_0^T dt \int\limits_\Omega \varphi\varphi^* d\Omega$$

$$- \int\limits_0^T dt \int\limits_\Omega \varphi^* \frac{\partial}{\partial z} \nu \frac{\partial \varphi}{\partial z} d\Omega - \mu \int\limits_0^T dt \int\limits_\Omega \varphi^* \Delta \varphi d\Omega = \int\limits_0^T dt \int\limits_\Omega \varphi^* f \, d\Omega \quad (5.8.2)$$

Use the Ostrogradski–Gauss formula and integrate the isolated expressions in (5.8.2) (with regard to the fact that div $\mathbf{u} = 0$) to bring them to

$$\int\limits_0^T dt \int\limits_\Omega \varphi^* \frac{\partial \varphi}{\partial t} d\Omega = \int\limits_\Omega \varphi^*\varphi d\Omega \bigg|_{t=0}^{t=T} - \int\limits_0^T dt \int\limits_\Omega \varphi \frac{\partial \varphi^*}{\partial t} d\Omega \quad (5.8.3)$$

$$\int\limits_0^T dt \int\limits_\Omega \varphi^* \operatorname{div} \mathbf{u}\varphi d\Omega = \int\limits_0^T dt \int\limits_S u_n \varphi\varphi^* dS - \int\limits_0^T dt \int\limits_\Omega \varphi \operatorname{div} \mathbf{u}\varphi^* d\Omega \quad (5.8.4)$$

$$\int\limits_0^T dt \int\limits_\Omega \varphi^* \frac{\partial}{\partial z} \nu \frac{\partial \varphi}{\partial t} d\Omega = \int\limits_0^T dt \int\limits_{\Sigma_H} \nu \left(\varphi^* \frac{\partial \varphi}{\partial z} - \varphi \frac{\partial \varphi^*}{\partial z} \right) d\Sigma$$

$$- \int\limits_0^T dt \int\limits_{\Sigma_0} \nu \left(\varphi^* \frac{\partial \varphi}{\partial z} - \varphi \frac{\partial \varphi^*}{\partial z} \right) d\Sigma + \int\limits_0^T dt \int\limits_\Omega \varphi \frac{\partial}{\partial z} \nu \frac{\partial \varphi^*}{\partial z} d\Omega \quad (5.8.5)$$

$$\mu \int\limits_0^T dt \int\limits_\Omega \varphi^* \Delta \varphi d\Omega$$

$$= \mu \int\limits_0^T dt \int\limits_\Sigma \left(\varphi^* \frac{\partial \varphi}{\partial n} - \varphi \frac{\partial \varphi^*}{\partial n} \right) d\Sigma + \int\limits_0^T dt \int\limits_\Omega \varphi \Delta \varphi^* d\Omega \quad (5.8.6)$$

Substitute (5.8.3)–(5.8.6) into (5.8.2) to obtain

$$\int\limits_0^T dt \int\limits_\Omega \varphi \left(-\frac{\partial \varphi^*}{\partial t} - \operatorname{div} \mathbf{u} \varphi^* + \sigma \varphi^* - \frac{\partial}{\partial z} \nu \frac{\partial \varphi^*}{\partial s} - \mu \Delta \varphi^* \right) d\Omega$$

$$= \int\limits_0^T dt \int\limits_\Omega \varphi^* f dt - \int\limits_\Omega \varphi_T \varphi_T^* d\Omega + dt \int\limits_\Omega \varphi_0 \varphi_0^* d\Omega - \int\limits_0^T dt \int\limits_S u_n \varphi \varphi^* dS$$

$$+ \int\limits_0^T dt \int\limits_{\Sigma_H} \nu \left(\varphi^* \frac{\partial \varphi}{\partial z} - \varphi \frac{\partial \varphi^*}{\partial z} \right) d\Sigma - \int\limits_0^T dt \int\limits_{\Sigma_0} \nu \left(\varphi^* \frac{\partial \varphi}{\partial z} - \varphi \frac{\partial \varphi^*}{\partial z} \right) d\Sigma$$

$$+ \mu \int\limits_0^T dt \int\limits_\Sigma \left(\varphi^* \frac{\partial \varphi}{\partial n} - \varphi \frac{\partial \varphi^*}{\partial n} \right) d\Sigma \quad (5.8.7)$$

Assume now that φ^* satisfies the equation

$$-\frac{\partial \varphi^*}{\partial t} - \operatorname{div} \mathbf{u} \varphi^* + \sigma \varphi^* - \frac{\partial}{\partial z} \nu \frac{\partial \varphi^*}{\partial z} - \mu \Delta \varphi^* = p \quad (5.8.8)$$

Use the boundary conditions of the function φ in (5.8.1) and the periodicity condition of the function φ to transform the right-hand part of equality (5.8.7). Assume also that the function φ^* is periodic in time with the period T. Then

$$-\int\limits_\Omega \varphi_T \varphi_T^* d\Omega + \int\limits_\Omega \varphi_0 \varphi_0^* d\Omega = 0 \quad (5.8.9)$$

Since
$$\varphi = 0 \quad \text{on} \quad S \quad \text{at} \quad u_n < 0$$

as assumed,
$$\int_0^T dt \int_S u_n \varphi \varphi^* dS = \int_0^T dt \int_\Sigma u_n^+ \varphi \varphi^* d\Sigma \qquad (5.8.10)$$

We used also the condition $w = 0$ at $z = 0$, $z = H$ here.

Next,

$$\int_0^T dt \int_{\Sigma_H} \nu \left(\varphi^* \frac{\partial \varphi}{\partial z} - \varphi \frac{\partial \varphi^*}{\partial z} \right) d\Sigma = - \int_0^T dt \int_{\Sigma_H} \nu \varphi \frac{\partial \varphi^*}{\partial z} d\Sigma, \qquad (5.8.11)$$

$$\int_0^T dt \int_{\Sigma_0} \nu \left(\varphi^* \frac{\partial \varphi}{\partial z} - \varphi \frac{\partial \varphi^*}{\partial z} \right) d\Sigma = - \int_0^T dt \int_{\Sigma_0} \nu \varphi \left(\frac{\partial \varphi^*}{\partial z} - \alpha \varphi^* \right) d\Sigma \qquad (5.8.12)$$

We used the boundary conditions in (5.8.1) in (5.8.11) and (5.8.12) respectively at $z = H$ and $z = 0$. Finally, use the conditions $\varphi = 0$ on Σ at $u_n < 0$ and $\partial\varphi/\partial n = 0$ on Σ at $u_n \geq 0$ to obtain

$$\mu \int_0^T dt \int_\Sigma \left(\varphi \frac{\partial \varphi^*}{\partial n} - \varphi^* \frac{\partial \varphi}{\partial n} \right) d\Sigma$$

$$= \mu \int_0^T dt \int_{\Sigma^+} \varphi \frac{\partial \varphi^*}{\partial n} d\Sigma - \mu \int_0^T dt \int_{\Sigma^-} \varphi^* \frac{\partial \varphi}{\partial n} d\Sigma \qquad (5.8.13)$$

where
$$\Sigma^+ = \{ (x,y,z) \in \Sigma \ : \ u_n \geq 0 \}, \quad \Sigma^- = \{ (x,y,z) \in \Sigma \ : \ u_n < 0 \}$$

Take now into account (5.8.8)–(5.8.13) to obtain

$$\int_0^T dt \int_\Omega p \varphi d\Omega = \int_0^T dt \int_\Omega \varphi^* f d\Omega - \int_0^T dt \int_\Sigma u_n^+ \varphi \varphi^* d\Sigma$$

$$- \int_0^T dt \int_{\Sigma_H} \nu \varphi \frac{\partial \varphi^*}{\partial z} d\Sigma + \int_0^T dt \int_{\Sigma_H} \nu \varphi \left(\frac{\partial \varphi^*}{\partial z} - \alpha \varphi^* \right) d\Sigma$$

$$- \mu \int_0^T dt \int_{\Sigma^+} \varphi \frac{\partial \varphi^*}{\partial n} d\Sigma + \mu \int_0^T dt \int_{\Sigma^-} \varphi^* \frac{\partial \varphi}{\partial n} d\Sigma \qquad (5.8.14)$$

Boundary conditions for the solution of the adjoint problem have not been yet fixed. Now assume, along with equation (5.8.8), that

$$-\frac{\partial \varphi^*}{\partial t} - \text{div } \mathbf{u}\varphi^* + \sigma\varphi^* - \frac{\partial}{\partial z}\nu\frac{\partial \varphi^*}{\partial z} - \mu\Delta\varphi^* = p$$

$$
\begin{aligned}
\mu\frac{\partial \varphi^*}{\partial n} + u\varphi^* = 0 \quad &\text{on } \Sigma \quad &\text{at } u_n \geq 0 \\
\varphi^* = 0 \quad &\text{on } \Sigma \quad &\text{at } u_n < 0 \\
\partial\varphi^*/\partial z = \alpha\varphi^* \quad &\text{on } \Sigma_0 \\
\partial\varphi^*/\partial z = 0 \quad &\text{on } \Sigma_H
\end{aligned}
\qquad (5.8.15)
$$

The dual relationship will then become

$$\int\limits_0^T dt \int\limits_\Omega p\varphi d\Omega = \int\limits_0^T dt \int\limits_\Omega \varphi^* f d\Omega \qquad (5.8.16)$$

If the main functional is determined by the equality

$$J = \int\limits_0^T dt \int\limits_\Omega p\varphi d\Omega \qquad (5.8.17)$$

its identical dual representation is

$$J = \int\limits_0^T dt \int\limits_\Omega \varphi^* f d\Omega \qquad (5.8.18)$$

Choose various functions p to obtain various functionals and corresponding adjoint equations.

Rewrite the adjoint problem, (5.8.15), as

$$-\frac{\partial \varphi^*}{\partial t} + A^*\varphi^* = p$$

where A^* is an operator adjoint to A; it operates in the real Hilbert space $L_2(\Omega)$ and is defined by the equality

$$A^*\varphi^* = -\text{div } \mathbf{u}\varphi^* + \sigma\varphi^* - \frac{\partial}{\partial z}\nu\frac{\partial \varphi^*}{\partial z} - \mu\Delta\varphi^*$$

The set of functions φ^* of $L_2(\Omega)$ may be regarded as the definition domain $D(A^*)$ of the operator A^*. These functions are continuous and differentiable in Ω, so that $-\partial/\partial z(\nu\partial\varphi^*/\partial z) - \mu\Delta\varphi^* \in L_2(\Omega)$. Every function in $D(A^*)$ satisfies the boundary conditions in (5.8.15).

The solution $\varphi^*(x, y, z, t)$ of the adjoint problem is a function assumed as differentiable and periodic in t.

Notice that the main equations in the form of (5.3.27) produce a slightly different adjoint problem if the method discussed above is applied:

$$-\frac{\partial\varphi^*}{\partial t} - \operatorname{div}\mathbf{u}\varphi^* + \sigma\varphi^* = K\varphi^* + p$$

$$\varphi^* = 0 \quad \text{at} \quad t = T$$
$$\varphi^* = 0 \quad \text{on} \quad \Sigma$$

$$\frac{\partial\varphi^*}{\partial z} = \alpha\varphi^* \quad \text{on} \quad \Sigma_0$$

$$\frac{\partial\varphi^*}{\partial z} = 0 \quad \text{on} \quad \Sigma_H$$

The conditions $\operatorname{div}\mathbf{u} = 0$ and $w = 0$ at $z = 0$, $z = H$ are complementary conditions for the statement of the problem, as was indicated.

Notice that the initial problem was homogeneous in its boundary conditions and by initial data. The analysis made may be easily generalized for a main problem with heterogeneous conditions. Only the form of functional (5.8.18) will change in the process as more terms will appear due to the heterogeneity.

Notice also peculiar two-dimensional statements of the problem. Since two-dimensional statements are already determined as (5.3.34) and (5.3.35), the methods discussed above may produce corresponding adjoint problems. Adjoint to problem (5.3.34) will then be

$$-\frac{\partial\varphi^*}{\partial t} - \frac{\partial u\varphi^*}{\partial x} - \frac{\partial v\varphi^*}{\partial y} + \sigma\varphi^* - \mu\Delta\varphi^* = p$$

$$\frac{\partial\varphi^*}{\partial n} + u_n\varphi^* = 0 \quad \text{on} \quad \Sigma \quad \text{at} \quad u_n \geq 0$$

$$\varphi^* = 0 \quad \text{on} \quad \Sigma \quad \text{at} \quad u_n < 0 \qquad (5.8.19)$$

$$\varphi^* = \varphi_T^* \quad \text{at} \quad t = T$$

and the problem

$$-\frac{\partial \varphi^*}{\partial t} - \frac{\partial u\varphi^*}{\partial x} - \frac{\partial v\varphi^*}{\partial y} + \sigma\varphi^* - \mu\Delta\varphi^* = p$$

$$\varphi^* = \varphi_T^* \quad \text{at} \quad t = T \tag{5.8.20}$$

$$\varphi^* = 0 \quad \text{at} \quad \Sigma$$

is adjoint to (5.3.35). Their functionals are

$$J = \int_0^T dt \int_\Omega p\varphi \, d\Omega \tag{5.8.21}$$

$$J = \int_0^T dt \int_\Omega f\varphi^* \, d\Omega \tag{5.8.22}$$

To conclude, examine the problem of determination of pollutant fallout on an underlying surface. As was shown above, this problem may be solved by a two-dimensional model, on the assumption that

$$\varphi \approx \frac{1}{H} \int_0^H \varphi \, dz \tag{5.8.23}$$

where H is the altitude of the cylindrical domain.

Formulate the problem as follows: let a single emission φ of a pollutant occur with a strength of Q at a point $\mathbf{r}_0 \in \Omega$ at a moment of time $\tau_0 \in [0, T]$. Find the total fallout on the underlying surface at the point $\mathbf{r}_1 \in \Omega$ at the moment of time $\tau_1 \in [\tau_0, T]$. According to the results of the previous chapter and the statement of the problem, functional (5.8.21) is

$$J = \frac{\bar{w}_g + \alpha\nu}{H} \varphi(\mathbf{r}_1, \tau_1) \tag{5.8.24}$$

where $\bar{w}_g$ is the absolute value of the fallout downward velocity. Its identical dual representation is

$$J = Q\varphi^*(\mathbf{r}_0, \tau_0) \tag{5.8.25}$$

Assume that the boundary of the area Ω is far away from the points $\mathbf{r}_0$ and $\mathbf{r}_1$ and the impact of the emission on the neighborhood of the boundary may be affordably disregarded. Then the model is

$$\frac{\partial \varphi}{\partial t} + \frac{\partial u\varphi}{\partial x} + \frac{\partial v\varphi}{\partial y} + (\sigma + \bar{\sigma})\varphi - \mu\Delta\varphi = Q\delta(\mathbf{r} - \mathbf{r}_0)\delta(t - \tau_0)$$

$$\varphi = 0 \quad \text{at} \quad t = 0 \qquad\qquad (5.8.26)$$

$$\varphi \to 0 \quad \text{at} \quad |\mathbf{r}| \to \infty$$

$$\bar{\sigma} = (\bar{w}_g + \alpha\nu)/H$$

Write the problem adjoint to (5.8.26) problem as

$$-\frac{\partial \varphi^*}{\partial t} - \frac{\partial u\varphi^*}{\partial x} - \frac{\partial v\varphi^*}{\partial y} + (\sigma + \bar{\sigma})\varphi^* - \mu\Delta\varphi^* = \bar{\sigma}\delta(\mathbf{r} - \mathbf{r}_1)\delta(t - \tau_1)$$

$$\varphi^* = 0 \quad \text{at} \quad t = 0 \qquad\qquad (5.8.27)$$

$$\varphi^* \to 0 \quad \text{at} \quad |\mathbf{r}| \to \infty$$

Solve problems (5.8.26) and (5.8.27) on the basis of the fundamental solutions of corresponding operators. Fundamental solutions to one-dimensional diffusion equations allowing for advective terms were obtained in § 5.7. Similarly, find the fundamental solutions $\tilde{\varphi}$ and $\tilde{\varphi}^*$ of the operators of problems (5.8.26) and (5.8.27):

$$\tilde{\varphi}(x, y, t) = \frac{\theta(t)}{4\mu\pi t} \exp\left\{-\left[(\sigma + \bar{\sigma})t + \frac{(x - ut)^2 + (y - vt)^2}{4\mu t}\right]\right\} \quad (5.8.28)$$

$$\tilde{\varphi}^*(x, y, t_1) = \frac{\theta(t)}{4\mu\pi t_1} \exp\left\{-\left[(\sigma + \bar{\sigma})t + \frac{(x + ut_1)^2 + (y + vt_1)^2}{4\mu t_1}\right]\right\} \quad (5.8.29)$$

where $t_1 = T - t$. We assumed that $u = const > 0$, $v = const > 0$ as we derived (5.8.28) and (5.8.29).

The solutions of the main and adjoint problems, with regards to the latter, are:

$$\varphi(x, y, t) = \begin{cases} \dfrac{Q}{4\pi\mu(t - \tau_0)} \exp\{-\alpha(\mathbf{r} - \mathbf{r}_0, t - \tau_0)\}, & t \in (\tau_0, T] \\[4mm] 0, & t \in [0, \tau_0] \end{cases} \quad (5.8.30)$$

$$\varphi^*(x, y, t) = \begin{cases} \dfrac{\bar{\sigma}}{4\pi\mu(\tau_1 - t)} \exp\{-\beta(\mathbf{r} - \mathbf{r}_1, \tau_1 - t)\}, & t \in [0, \tau_1) \\[4mm] 0, & t \in [\tau_1, T] \end{cases} \quad (5.8.31)$$

where

$$\alpha(\mathbf{r}, t) = (\sigma + \bar{\sigma})t + \frac{(x - ut)^2 + (y - vt)^2}{4\mu t}$$

$$\beta(\mathbf{r}, t) = (\sigma + \bar{\sigma})t + \frac{(x + ut)^2 + (y + vt)^2}{4\mu t}$$

The fallout at a given point may be determined by substituting (5.8.30) into (5.8.24) or (5.8.31) into (5.8.25).

Notice that though the final result does not depend on the choice of a functional – the main or its dual representation – the quality of information provided by the solution of the main and adjoint problems is different. Solving the main problem provides complete information on distribution of the pollutant in time and space when the time and location of a single emission are fixed. Solving the adjoint problem in this case provides information on the polluting fallout at a given point of space at a given time, while the time and location of the emission are arbitrary. Therefore, for example, if the time and location of a solitary emission are not known in advance dual representation (5.8.25) of the functional should be used in solving the problem for the point $(\mathbf{r}_1, \tau_1)$.

5.9. Uniqueness of Solution of Adjoint Problem

Prove now that the solution of adjoint problem (5.8.15) is unique. Assume that $\varphi^*(x, y, z, t)$ is continuous in $\Omega \times [0, T]$, differentiable, and periodic in t. Assume also that the function $\varphi^*(x, y, z, t)$ belongs to a set $D(A^*) \subset L_2(\Omega)$, introduced above, at any t. Multiply (5.8.15) by φ^* and integrate the result over all the solution definition domain. Then

$$-\int_0^T dt \int_\Omega \varphi^* \frac{\partial \varphi^*}{\partial t} d\Omega - \int_0^T dt \int_\Omega \varphi^* \operatorname{div} \mathbf{u}\varphi^* d\Omega + \int_0^T dt \int_\Omega \sigma \varphi^{*2} d\Omega$$

$$-\int_0^T dt \int_\Omega \varphi^* \frac{\partial}{\partial z} \nu \frac{\partial \varphi^*}{\partial z} d\Omega - \mu \int_0^T dt \int_\Omega \varphi^* \Delta \varphi^* d\Omega = \int_0^T dt \int_\Omega p\varphi^* d\Omega \quad (5.9.1)$$

Transform the integrals like in the cases above:

$$\int\limits_0^T dt \int\limits_\Omega \varphi^* \operatorname{div} \mathbf{u}\varphi^* d\Omega = \int\limits_0^T dt \int\limits_\Omega \operatorname{div} \mathbf{u}\frac{\varphi^{*2}}{2} d\Omega = \int\limits_0^T dt \int\limits_\Sigma \frac{u_n \varphi^{*2}}{2} d\Sigma \quad (5.9.2)$$

$$\int\limits_0^T dt \int\limits_\Omega \varphi^* \frac{\partial}{\partial z}\nu\frac{\partial\varphi^*}{\partial z} d\Omega = \int\limits_0^T dt \int\limits_{\Sigma_H} \nu\varphi^* \frac{\partial\varphi^*}{\partial z} d\Sigma$$

$$- \int\limits_0^T dt \int\limits_{\Sigma_0} \nu\varphi^* \frac{\partial\varphi^*}{\partial z} d\Sigma - \int\limits_0^T dt \int\limits_\Omega \nu\left(\frac{\partial\varphi^*}{\partial z}\right)^2 d\Omega \quad (5.9.3)$$

$$\mu\int\limits_0^T dt \int\limits_\Omega \varphi^*\Delta\varphi^* d\Omega = \mu\int\limits_0^T dt \int\limits_\Sigma \varphi^* \frac{\partial\varphi^*}{\partial z} d\Sigma - \mu\int\limits_0^T dt \int\limits_\Omega (\nabla\varphi^*)^2 d\Omega \quad (5.9.4)$$

where

$$\nabla\varphi = \frac{\partial\varphi}{\partial x}\mathbf{i} + \frac{\partial\varphi}{\partial y}\mathbf{j}, \quad (\nabla\varphi)^2 = \left(\frac{\partial\varphi}{\partial x}\right)^2 + \left(\frac{\partial\varphi}{\partial y}\right)^2$$

Use (5.9.2)–(5.9.4) to reduce (5.9.1) to

$$-\int\limits_0^T dt \int\limits_\Omega \frac{1}{2}\frac{\partial\varphi^{*2}}{\partial t} d\Omega + \int\limits_0^T dt \left\{\int\limits_\Omega (\sigma\varphi^{*2} + \nu\left(\frac{\partial\varphi^*}{\partial z}\right)^2 + \mu(\nabla\varphi^*)^2)d\Omega\right.$$

$$- \int\limits_\Sigma \varphi^*\left(-\frac{u_n\varphi^*}{2} + u_n\varphi^* + \mu\frac{\partial\varphi^*}{\partial n}\right) d\Sigma - \int\limits_{\Sigma_H} \nu\varphi^* \frac{\partial\varphi^*}{\partial z} d\Sigma$$

$$\left. + \int\limits_{\Sigma_0} \nu\varphi^* \frac{\partial\varphi^*}{\partial z} d\Sigma\right\} = \int\limits_0^T dt \int\limits_\Omega p\varphi^* d\Omega \quad (5.9.5)$$

Since φ^* is periodic,

$$\int\limits_0^T dt \int\limits_\Omega \frac{\partial\varphi^{*2}}{\partial t} d\Omega = 0 \quad (5.9.6)$$

Next, by virtue of the condition

$$\varphi^* = 0 \quad \text{on} \quad \Sigma \quad \text{at} \quad u_n < 0 \quad (5.9.7)$$

obtain

$$\int\limits_0^T dt \int\limits_{\Sigma^-} \varphi^*\left(-\frac{u_n^- \varphi^*}{2} + u_n^- \varphi^* + \mu\frac{\partial\varphi^*}{\partial n}\right) d\Sigma = 0 \quad (5.9.8)$$

and by virtue of the condition

$$u_n\varphi^* + \mu\frac{\partial\varphi^*}{\partial n} = 0 \quad \text{on} \quad \Sigma \quad \text{at} \quad u_n \geq 0 \qquad (5.9.9)$$

obtain

$$\int_0^T dt \int_{\Sigma+} u_n^+ \varphi^{*2} d\Sigma + \int_0^T dt \int_{\Sigma+} \mu\varphi^*\frac{\partial\varphi^*}{\partial n} d\Sigma = 0 \qquad (5.9.10)$$

As concerns the penultimate terms in the left-hand part of (5.9.5), in compliance with the boundary conditions at Σ_0 and Σ_H,

$$\int_0^T dt \int_{\Sigma_H} \nu\varphi^*\frac{\partial\varphi^*}{\partial z} d\Sigma = 0$$

$$(5.9.11)$$

$$\int_0^T dt \int_{\Sigma_0} \nu\varphi^*\frac{\partial\varphi^*}{\partial z} d\Sigma = \int_0^T dt \int_{\Sigma_0} \nu\alpha\varphi^{*2} d\Sigma$$

With regard to (5.9.7)–(5.9.11), (5.9.5) becomes

$$\int_0^T dt \int_{\Sigma} \frac{u_n^+\varphi^{*2}}{2} d\Sigma + \int_0^T dt \int_{\Omega} \sigma\varphi^{*2} d\Omega + \int_0^T dt \int_{\Omega} \nu\left(\frac{\partial\varphi^*}{\partial z}\right)^2 d\Omega$$

$$+ \mu\int_0^T dt \int_{\Omega} (\nabla\varphi^*)^2 d\Omega + \alpha\nu\int_0^T dt \int_{\Sigma_0} \varphi^{*2} d\Sigma = \int_0^T dt \int_{\Omega} p\varphi^* d\Omega \quad (5.9.12)$$

Relationship (5.9.12) will be the main in the proof of the uniqueness of the solution of problem (5.8.15). Assime that problem (5.8.15) has two different solutions, φ_1^* and φ_2^*. Examine their difference $\omega^* = \varphi_1^* - \varphi_2^*$. Then the homogenous problem for ω^* is

$$-\frac{\partial\omega^*}{\partial t} - \text{div }\mathbf{u}\omega^* + \sigma\omega^* - \frac{\partial}{\partial z}\nu\frac{\partial\omega^*}{\partial z} - \mu\Delta\omega^* = 0$$

$$\mu\frac{\partial\omega^*}{\partial n} + u\omega^* = 0 \quad \text{on} \quad \Sigma \quad \text{at} \quad u_n \geq 0$$

$$\omega^* = 0 \qquad \text{on} \quad \Sigma \quad \text{at} \quad u_n < 0 \qquad (5.9.13)$$

$$\partial\omega^*/\partial z = \alpha\omega^* \quad \text{on} \quad \Sigma_0$$

$$\partial\omega^*/\partial z = 0 \quad \text{on} \quad \Sigma_H$$

A relationship of the (5.9.12) type corresponding to this one is

$$\int\limits_0^T dt \int\limits_\Sigma \frac{u_n^+ \omega^{*2}}{2} d\Sigma + \int\limits_0^T dt \int\limits_\Omega \omega^{*2} d\Omega + \int\limits_0^T dt \int\limits_\Omega \nu \left(\frac{\partial \omega^*}{\partial z}\right)^2 d\Omega$$

$$+ \mu \int\limits_0^T dt \int\limits_\Omega (\nabla \omega^*)^2 d\Omega + \alpha \nu \int\limits_0^T dt \int\limits_{\Sigma_0} \omega^{*2} d\Sigma = 0 \quad (5.9.14)$$

Since σ, ν, μ, $\alpha \geq 0$ in this relationship, it holds only at $\omega^* = 0$, i. e., $\varphi_1^* = \varphi_2^*$. The uniqueness of the solution is thus proven. We naturally assumed implicitly required smoothness to justify the transformations.

Notice that the uniqueness theorem for a mixed problem could be similarly proved, if $\varphi^* = \varphi_T$ at $t = T$ and the problem is solved toward decreasing t. Only then the algorithm produces the right solution.

Notice also, that the main and adjoint problems allow the unique solution also if the conditions change: there may be $\varphi = 0$ at Σ and $\varphi^* = 0$ at Σ at all u_n instead of $\varphi = 0$ at Σ at $u_n < 0$ and $\partial\varphi/\partial n = 0$ at Σ at $u_n \geq 0$ in (5.8.1) and conditions $\varphi^* = 0$ at Σ at $u_n < 0$ and $\mu \partial\varphi^*/\partial n + u_n \varphi^* = 0$ at Σ at $u_n \geq 0$ in (5.8.15).

5.10. Adjoint Equation and Lagrange Identity

This chapter discussed a method of construction of equations adjoint to the equation of transport and diffusion of aerosols. These equations may also be derived from general considerations on the basis of the Lagrange identity. Let us examine the technique of deriving equations adjoint to the general evolutionary equation proposed in § 1.7, in a more formal manner.

Examine the problem

$$\frac{\partial \varphi}{\partial t} + A\varphi = f \qquad (5.10.1)$$

where A is a linear operator in the Hilbert space $H = L_2(\Omega)$ defined on a set of functions $D(A) \in H$ where every element satisfies corresponding smoothness conditions, certain additional conditions (e. g. boundary conditions) and other requirements according to the essence of problem. Ω is the domain of variations of spatial variables. Assume problem (5.10.1) as periodic in t with a period T.

Introduce the Hilbert space $\Phi = L_2(\Omega \times (0,T))$ of real functions periodic in t and summable with square in $\mathbf{r}$ and in t on $\Omega \times (0,T)$ with the inner product

$$(g,h) = \int_0^T dt \int_\Omega gh \, d\mathbf{r}, \quad g,h \in \Phi \qquad (5.10.2)$$

To formalize problem (5.10.1) even more, write it as

$$L\varphi = f \qquad (5.10.3)$$

where L operates in the Hilbert space Φ and is determined by the equality $L = \partial/\partial t + A$.

A set of functions $\varphi \in \Phi$ continuously differentiable in t (so that at every t $\varphi \in D(A)$) may be considered as the definition domain $D(L)$ of the operator L.

Introduce an operator L^* with the definition domain $D(L^*)$ adjoint to the linear operator L on a basis of the Lagrange identity:

$$(Lh,g) = (h, L^*g) \qquad (5.10.4)$$

where the operator L and the functions $h \in D(L)$, $g \in D(L^*)$ are assumed to be real. Assign $h = \varphi$ and $g = \varphi^*$ to obtain

$$(L\varphi, \varphi^*) = (\varphi, L^*\varphi^*) \qquad (5.10.5)$$

Since equation (5.10.3) holds, then with a formal notation

$$L^*\varphi^* = p \qquad (5.10.6)$$

where p is not yet a determined function, relationship (5.10.5) becomes

$$(\varphi^*, f) = (\varphi, p) \qquad (5.10.7)$$

If p is a measurement characteristic (e. g., that of an instrument) or that of a system of measurements (e. g., a sum of measurements), the main functional will be

$$J = (\varphi, p) \qquad (5.10.8)$$

From (5.10.7) follows a dual formula

$$J = (\varphi^*, f) \qquad (5.10.9)$$

Apply this method to a concrete problem

$$\frac{\partial \varphi}{\partial t} + \text{div } \mathbf{u}\varphi + \sigma\varphi - \frac{\partial}{\partial z}\nu\frac{\partial \varphi}{\partial z} - \mu\Delta\varphi = f$$

$$\varphi = 0 \qquad \text{on} \quad \Sigma$$

$$\partial\varphi/\partial z = \alpha\varphi \quad \text{on} \quad \Sigma_0 \qquad\qquad (5.10.10)$$

$$\partial\varphi/\partial z = 0 \quad \text{on} \quad \Sigma_H$$

Assume, as before, that the components u, v and w satisfy the conditions

$$\text{div } \mathbf{u} = 0$$

$$w = 0 \quad \text{at} \quad z = 0, \quad z = H$$

Assume, that the solution $\varphi(x, y, z, t)$ of problem (5.10.10) is continuous in $\Omega \times (0, T)$ and is a function periodic in t. Besides, let the function $\varphi(x, y, z, t)$ belong to the set of functions $D(A)$ in the Hilbert space $H = L_2(\Omega)$ continuous and differentiable in Ω at every t, so that $-\partial/\partial z(\nu\partial\varphi/\partial z) - \mu\Delta\varphi \in L_2(\Omega)$. Every function of $D(A)$ satisfies the boundary conditions (5.10.10) here. Rewrite then the main problem, (5.10.10), as

$$\frac{\partial \varphi}{\partial t} + A\varphi = f \qquad\qquad (5.10.11)$$

where A operates in the Hilbert space $H = L_2(\Omega)$ with the definition domain $D(A)$ and is assigned by the equality

$$A\varphi = \text{div } \mathbf{u}\varphi + \sigma\varphi - \frac{\partial}{\partial z}\nu\frac{\partial \varphi}{\partial z} - \mu\Delta\varphi$$

As was seen before, problem (5.10.11) may be written as (5.10.3):

$$L\varphi = f \qquad\qquad (5.10.12)$$

where L operates in the real Hilbert space $\Phi = L_2(\Omega \times (0, T))$ with the definition domain $D(L)$ and is determined by the equality $L = \partial/\partial t + A$.

Use the Lagrange identity (5.10.4) and expand in explicit form its left-hand part at h, $g \in D(L)$ in order to construct the adjoint operator L^*:

$$(Lh,g) = \int\limits_0^T dt \int\limits_\Omega g\left(\frac{\partial h}{\partial t} + \text{div } \mathbf{u}h + \sigma h - \frac{\partial}{\partial z}\nu\frac{\partial h}{\partial z} - \mu\Delta h\right) d\Omega$$

$$= \int\limits_\Omega g_T h_T d\Omega - \int\limits_\Omega g_0 h_0 d\Omega + \int\limits_0^T dt \int\limits_\Sigma u_n gh\, d\Sigma + \sigma\int\limits_0^T dt \int\limits_\Omega gh\, d\Omega$$

$$- \int\limits_0^T dt \int\limits_{\Sigma_H} \nu g\frac{\partial h}{\partial z} d\Sigma + \int\limits_0^T dt \int\limits_{\Sigma_0} \nu g\frac{\partial h}{\partial z} d\Sigma + \int\limits_0^T dt \int\limits_{\Sigma_H} \nu h\frac{\partial g}{\partial z} d\Sigma$$

$$- \int\limits_0^T dt \int\limits_{\Sigma_0} \nu h\frac{\partial g}{\partial z} d\Sigma - \int\limits_0^T dt \int\limits_\Omega h\frac{\partial}{\partial z}\nu\frac{\partial g}{\partial z} d\Omega - \mu\int\limits_0^T dt \int\limits_\Sigma g\frac{\partial h}{\partial n} d\Sigma$$

$$+ \mu\int\limits_0^T dt \int\limits_\Sigma h\frac{\partial g}{\partial n} d\Sigma - \int\limits_0^T dt \int\limits_\Omega \mu h\Delta g\, d\Omega$$

$$- \int\limits_0^T dt \int\limits_\Omega h\frac{\partial g}{\partial t} d\Omega - \int\limits_0^T dt \int\limits_\Omega h\,\text{div } \mathbf{u}g\, d\Omega \quad (5.10.13)$$

Identity (5.10.13) transforms into

$$(Lh,g) = \int\limits_0^T dt \int\limits_\Omega h\left(-\frac{\partial g}{\partial t} - \text{div } \mathbf{u}g + \sigma g - \frac{\partial}{\partial z}\nu\frac{\partial g}{\partial z} - \mu\Delta g\right) d\Omega \quad (5.10.14)$$

Let us denote the operator affecting the function g in the expression in the round brackets as

$$L^*\cdot = -\frac{\partial\cdot}{\partial t} - \text{div } (\mathbf{u}\cdot) + \sigma\cdot - \frac{\partial}{\partial z}\nu\frac{\partial\cdot}{\partial z} - \mu\Delta\cdot \quad (5.10.15)$$

Then the right-hand part of expression (5.10.14) will be an inner product of the (h, L^*g) type. As a result, obtain, quite naturally, identity (5.10.4) and the adjoint operator (5.10.15) in the Hilbert space $\Phi = L_2(\Omega \times (0,T))$ with the definition domain $D(L^*) = D(L)$.

Define formally the adjoint problem

$$L^*\varphi^* = p \quad (5.10.16)$$

or, if expanded,

$$-\frac{\partial \varphi^*}{\partial t} - \operatorname{div} \mathbf{u}\varphi^* + \sigma\varphi^* - \frac{\partial}{\partial z}\nu\frac{\partial \varphi^*}{\partial z} - \mu\Delta\varphi^* = p$$

$$\varphi^* = 0 \quad \text{on} \quad \Sigma$$

$$\frac{\partial \varphi^*}{\partial z} = \alpha\varphi^* \quad \text{on} \quad \Sigma_0 \tag{5.10.17}$$

$$\frac{\partial \varphi^*}{\partial z} = 0 \quad \text{on} \quad \Sigma_H$$

with the function $\varphi^*(x,y,z,t)$ assumed to be periodic in t with a period T. Then, depending on the chosen p, we obtain one or another form of the functional J. If $h = \varphi$ and $g = \varphi^*$, obtain, in compliance with (5.10.7), functionals (5.10.8) and (5.10.9).

Notice that initial and boundary conditions of the main problem were conveniently homogeneous as we used the Lagrange identity to construct the adjoint operator. The functions of the set $D(L)$, which is linear due to the homogeneity of these conditions, satisfy just these conditions.

CHAPTER 6

Adjoint Equations, Optimization

The pace of economic development today requires increasingly larger and more capacious industrial objects and complexes. Labor distribution patterns usually predetermine location of such objects in densely populated areas or nearby, which implies special restrictions concerning objects emitting into the atmosphere aerosols hazardous for human health and ecological systems historically set in the areas. The problem of optimum location of industries is multi-aspect and algorithmically extremely complex. The solution required of this author development of a mathematical apparatus of adjoint problems. The solutions of these may be termed as the functions of influence of aerosol pollution of environment. This chapter applies the results of research presented in Chapter 5 to a concrete object of study of optimum location of industries, discusses mathematical models of typical situations and methods of solution of optimization problems and interprets the results.

6.1. Statement of the Problem

Let us suppose that a new industry is to be based near populated areas, recreation zones, and other environmentally sensitive areas, provided that the total annual pollution by hazardous industrial emissions does not exceed affordable health standards, and the total pollution load Σ_0 on the area is minimum or within the limits of global health standards.

Assume that the plant in question emits into the atmosphere a hazardous aerosol at a rate of Q per unit of time at an altitude of $z = h$.

The emission is then transported by air masses and diffused by small-scale turbulence. Assume that the source of aerosol emission is located in the neighborhood of the point $\mathbf{r}_0 = (x_0,\ y_0,\ h)^T$. Then the source may be described by the function

$$f(\mathbf{r}) = Q\delta(\mathbf{r} - \mathbf{r}_0) \qquad (6.1.1)$$

which produces an equation[1]

$$\frac{\partial \varphi}{\partial t} + \operatorname{div} \mathbf{u}\varphi + \sigma\varphi - \frac{\partial}{\partial z}\nu\frac{\partial \varphi}{\partial z} - \mu\Delta\varphi = Q\delta(\mathbf{r} - \mathbf{r}_0) \qquad (6.1.2)$$

Assume

$$\varphi = 0 \qquad \text{on} \quad \Sigma$$

$$\frac{\partial \varphi}{\partial z} = \alpha\varphi \quad \text{on} \quad \Sigma_0 \qquad (6.1.3)$$

$$\frac{\partial \varphi}{\partial z} = 0 \quad \text{on} \quad \Sigma_H$$

as the boundary conditions.

Seek the solution to problem (6.1.2) and (6.1.3) within a set of sufficiently smooth periodic functions with period T in the variable t. Then

$$\varphi(\mathbf{r}, T) = \varphi(\mathbf{r}, 0) \qquad (6.1.4)$$

The task is to choose for the plant a location $\omega_\Omega \subset \Omega$ where global and local environmental and health standards are met, both for the whole area Σ_0 and specifically selected localities Σ_k. The components of the velocity vector in planetary boundary layer necessary for solving the problem of transport and diffusion of pollutants are computed by mesometeorology methods. Once all the necessary information about the wind force field is available, the problem of propagation of the industrial aerosol emitted at the given point $\mathbf{r}_0 \in \Omega$ is solved by methods of direct modeling. Take average weekly climatic data on components of the wind force field provided by mesometeorology methods and then solve the problem (6.1.2)–(6.1.4).

Figs. 29–32 show examples of the solution of problem (6.1.2)–(6.1.4) on the basis of statistical model described in Chapter 5 where the iso-

---

[1]See the footnote at page 190 concerning the δ-function.

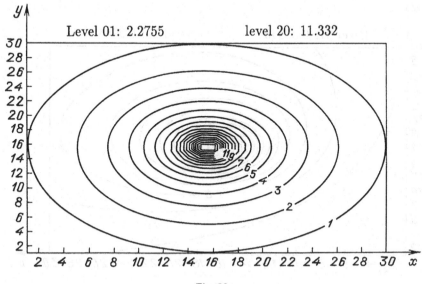

Fig. 29

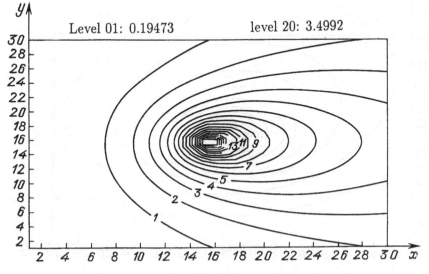

Fig. 30

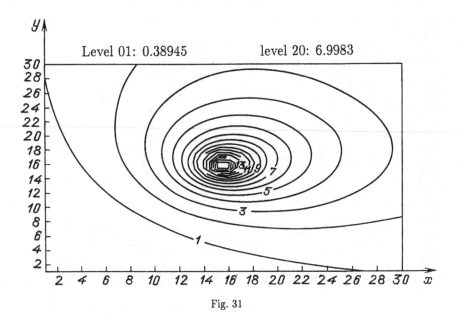

Fig. 31

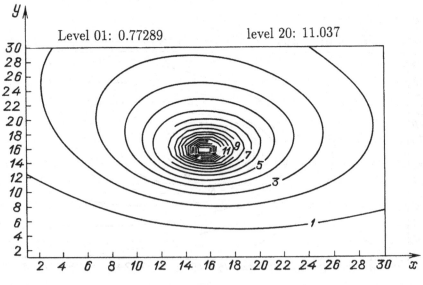

Fig. 32

grams of the functions

$$\bar{\varphi}(x,y) = \frac{1}{TH} \int\limits_0^T \int\limits_0^H \varphi(x,y,z,t)\,dz\,dt$$

are given for different types of motion of air masses. Fig. 29 corresponds to the case where $u = v = w = 0$; Fig. 30 shows the case where $u = 5$ m/s and $v = w = 0$, and Fig. 31 the situation where the velocity vector components are $u = 5$ m/s and $v = w = 0$ in the interval $[0, T/2]$, and $u = 0$, $v = 5$ m/s, $w = 0$ in the interval $[T/2, T]$. Fig. 32 corresponds to the following values of the components of velocity vector:

$$u = \begin{cases} 5, & t \in [0, T/3) \\ 0, & t \in [T/3, 2T/3) \\ -5, & t \in [2T/3, T] \end{cases}$$

$$v = \begin{cases} 0, & t \in [0, T/3) \\ 5, & t \in [T/3, 2T/3) \\ 0, & t \in [2T/3, T] \end{cases}$$

$$w = 0$$

The solution so obtained is integrated within the limits of the annual interval $0 \le t \le T$ and either the average amount of aerosol is calculated in unit cylinder Ω_k over the sensitive area $\Sigma_k \subset \Sigma_0$ for the period T:

$$J_k^B = b \int\limits_0^T dt \int\limits_{\Omega_k} \varphi\,d\Omega_k \tag{6.1.5}$$

or the total fallout in the same area $\Sigma_k \subset \Sigma_0$:

$$J_k^A = a \int\limits_0^T dt \int\limits_{\Sigma_k} \varphi\,d\Sigma_k \tag{6.1.6}$$

Here $b = 1/T$; the constant a reflects a portion of aerosol which falls out, this is primarily heavy aerosols and partially light aerosols falling out by downward diffusion. As was shown in § 5.5, a can be chosen in this case as

$$a = \bar{w}_g + \alpha\nu \tag{6.1.7}$$

Notice that both functionals are important for estimation of the pollution and its impact on environmental conditions of the locality Σ_k. Functional (6.1.5) is important for statistical estimation of the direct effect on oxygen-consuming nature and functional (6.1.6) is essential for assessment of contamination of a soil and water, whose influence on environment may be considerable, but mediated in the framework of corresponding lines of bioceonosis. As was said above, functionals (6.1.5) and (6.1.6) are particular cases of a more common functional

$$J_p = \int_0^T dt \int_\Omega p\varphi d\Omega \qquad (6.1.8)$$

which may be justifiably applied in assessment of pollution in different cases with the same p. If

$$p = \begin{cases} b, & \mathbf{r} \in \Omega_k \\ 0, & \mathbf{r} \notin \Omega_k \end{cases}$$

we come to functional (6.1.5). If

$$p = \begin{cases} a\delta(z), & \mathbf{r} \in \Sigma_k \\ 0, & \mathbf{r} \notin \Sigma_k \end{cases}$$

then we arrive at functional (6.1.6). Remember that $\Sigma_k \in \Sigma_0$ is the base of the cylindrical domain Ω_k on the plane $z = 0$.

If the plant is located at the point $\mathbf{r}_1 \in \Omega$, all the calculations must be remade. It means that solving a problem of optimum location of a plant requires examining a great deal of variants with consequent comparison of the functionals J_k^A and J_k^B, or their linear combinations, with the definite constants a and b:

$$J_k = J_k^A + J_k^B \qquad (6.1.9)$$

Since assessing pollution is linked in the long run with a combination of the functionals J_k^A and J_k^B, introduce a generalized functional

$$J_k = \int_0^T dt \int_{\Omega_k} [b + a\delta(z)]\varphi d\Omega_k \qquad (6.1.10)$$

written as functional (6.1.6), provided that

$$p_k = \begin{cases} b + a\delta(z), & \text{in} \quad \Omega_k \\ 0, & \text{outside} \quad \Omega_k \end{cases}$$

Therefore we will have to deal hereinafter just with the functional (6.1.10).

Introduce another important functional J_{pk}:

$$J_{pk} = \sum_{k=1}^{m} \int_0^T dt \int_{\Omega_k} [b + a\delta(z)]\varphi d\Omega_k \qquad (6.1.11)$$

Assume now that the constants a_k and b_k may be different in different (not overlapping) areas Ω_k and may depend, for example, on the character of the underlying surface. Then functional (6.1.11) may be written again as

$$J_p = \int_0^T dt \int_{\Omega} p\varphi d\Omega \qquad (6.1.12)$$

where

$$p = \begin{cases} b_k + a_k\delta(z), & \text{on} \quad \Omega_k, \ k = 1, ..., m \\ 0, & \text{outside} \quad \bigcup_{k=1}^{m} \Omega_k \end{cases}$$

The physical sense of the functional J_p is: this functional furnishes the effect of pollution all over the sensitive areas Ω_k provided that the source of industrial emissions is located at the point $r_0 \in \Omega$. This functional is in a certain sense global for all the domain Ω and for all the areas Ω_k.

Examine, along with (6.1.11), another global functional:

$$Y_p = \int_0^T dt \int_{\Omega} p_c\varphi d\Omega \qquad (6.1.13)$$

where

$$p_c = \begin{cases} B_k + A_k\delta(z), & r \in \Omega_k, \ k = 1, ..., m \\ 0, & \text{outside} \quad \bigcup_{k=1}^{m} \Omega_k \end{cases}$$

Here A_k and B_k are quantities linked to the health (physiological) effect of industrial aerosols on all the areas $\Omega_k \subset \Omega$ and may be selected in several ways. For instance, the constants B_k may express correlations of the amount of aerosols in Ω_k with its health hazard degree or its direct impact on certain objects (including living objects) in the area Ω_k. The same applies to A_k. The data are obtained experimentally over several years.

Make another generalization. Assume that all the data on the physiological effects are available not only in the sensitive localities Σ_k, but also at all the other points $r \in \Sigma_0$. Then the values A and B become functions: $A = A(x,y)$, $B = B(x,y)$. Make the functional Y_p with the value

$$p_c = B + A\delta(z) \quad \text{in} \quad \Omega \qquad (6.1.14)$$

We will use functionals (6.1.13) and (6.1.14) for assessing the pollution load on an area. The task is to find a point $r_0 \in \Omega$ where

$$Y_p = \min_{r_0 \in \Omega} \qquad (6.1.15)$$

This problem, as was noted above, is solved solving the main problem, (6.1.2)–(6.1.4), by repeated selecting r_0 in the domain Ω and the combinatorial search method. This search requires a great deal of computations and is difficult even for contemporary computers. Therefore a directed search is made in reality with regard to the wind rose and other considerations of a statistical character. We will see later, however, that problem (6.1.15) solves uniquely using only one computation variant of an adjoint problem. This makes one of remarkable features of duality. The next section discusses this algorithm.

6.2. Adjoint Equations and Optimization Problem

Since the main functional of the problem is chosen as

$$Y_p = \int_0^T dt \int_\Omega p_c \varphi \, d\Omega, \quad p_c = B + A\delta(z) \quad \text{in} \quad \Omega \qquad (6.2.1)$$

adjoint to the main problem, according to Chapter 5 is

$$-\frac{\partial \varphi^*}{\partial t} - \operatorname{div} \mathbf{u}\varphi^* + \sigma\varphi^* - \frac{\partial}{\partial z}\nu\frac{\partial \varphi^*}{\partial z} - \mu\Delta\varphi^* = p_c$$

$$\varphi^* = 0 \qquad \text{on} \quad \Sigma$$

$$\frac{\partial \varphi^*}{\partial z} = \alpha \varphi^* \quad \text{on} \quad \Sigma_0 \qquad\qquad (6.2.2)$$

$$\frac{\partial \varphi^*}{\partial z} = 0 \quad \text{on} \quad \Sigma_H$$

$$\varphi^*(\mathbf{r}, T) = \varphi^*(\mathbf{r}, 0)$$

Multiply equation (6.1.2) by φ^* and equation (6.2.2) by φ, integrate the results all over time interval and the domain Ω, subtract from each other, and use the boundary conditions and initial data of (6.1.3), (6.1.4), and (6.2.2). Since the problems are adjoint (see Chapter 5), transform and simplify them to obtain the dual form of the functional Y_p (see (5.8.17) and (5.8.18)):

$$Y_p = \int\limits_0^T dt \int\limits_\Omega p_c \varphi \, d\Omega \qquad\qquad (6.2.3)$$

$$Y_p = Q \int\limits_0^T \varphi^*(\mathbf{r}_0, t) dt \qquad\qquad (6.2.4)$$

Denote Y_p as $Y_p(\mathbf{r}_0)$, since this functional depends on the location $\mathbf{r}_0 \in \Omega$ of the plant.

Assume now that the adjoint problem, (6.2.2), is solved and the function $\varphi^*(\mathbf{r}, t)$ is found. Substitute it into (6.2.4) to obtain

$$Y_p(\mathbf{r}) = Q \int\limits_0^T \varphi^*(\mathbf{r}, t) dt \qquad\qquad (6.2.5)$$

Use the auxiliary function $Y_p(\mathbf{r})$ to find $\mathbf{r}_0$ from the condition

$$Y_p(\mathbf{r}) = \min_{\mathbf{r} \in \Omega} \qquad\qquad (6.2.6)$$

The point $\mathbf{r}_0$ will be just the point minimizing $Y_p(\mathbf{r})$.

Further action is obvious and involves construction of a field of the function $Y_p(x, y, h)$ where h is the altitude of the emission limited by building technology. As a result, a field of isograms $Y_p(x, y, h) = const$ is obtained on the plane (x, y).

The unique solution of the optimization problem, however, is not necessary in many cases, since the final decision requires meeting a number of limitations on the part of the area geology, available labor, water and infrastructure. Therefore it is important to choose a range of affordable values of health and environment standards. Let ω_Ω denote an area satisfying the condition

$$Y_p \leq c \tag{6.2.7}$$

and be the solution of the problem.

Re-examine now the classic solution of the optimization problem. Assume that $\mathbf{r}_0$ is found. Then main problem (6.1.2)–(6.1.4) has to be solved for this point (see Figs. 29–32) and full information be obtained on the fields of pollution in the local sense (and not in the global), i. e., the information on pollution in separate areas. If health and environment standards are met in all the areas, the problem is solved; if not, a more complex and more informative multi-criterial problem has to be solved. This problem is discussed in Section 6.3.

6.3. Multi-Criterial Optimization Problem

Global assessing the pollution of all the area Σ_0 with hazardous industrial emissions may show certain sensitive areas Σ_k as over-polluted beyond affordable health and environment standards. Examine the following multi-criterial problem to find a method of calculation which helps avoid it.

Let Σ_k $(k = 1, 2, \ldots, m)$ be selected sensitive areas on the plane $z = 0$ in the domain Σ_0. These may be populated areas, recreation zones, water reservoirs, etc. Locate a new industrial plant so that pollution of all the m areas Σ_k does not exceed approved standards (if such a location is possible in Σ_0 at all). If the location proves impossible on Σ_0, formulate restrictions on the rate of polluting emissions Q, which will make the location of the plant possible.

Examine at first a simpler problem where the area $\Sigma_k \subset \Sigma_0$ is unique. Require a priori that pollution does not exceed the standard c_k in the area, i. e.,

$$Y_{pk} = \int\limits_0^T dt \int\limits_{\Omega_k} p_{ck}\varphi d\Omega_k \leq c_k \qquad (6.3.1)$$

$$p_{ck} = \begin{cases} b_k + a_k\delta(z), & \text{on} \quad \Omega_k \\ 0, & \text{outside} \quad \Omega_k \end{cases} \qquad (6.3.2)$$

integrate in (6.3.1) not all over the domain Ω like in the case discussed in § 6.2, but only over its part Ω_k. Then instead of (6.2.2) the problem is

$$-\frac{\partial \varphi_k^*}{\partial t} - \operatorname{div} \mathbf{u}\varphi_k^* + \sigma\varphi_k^* - \frac{\partial}{\partial z}\nu\frac{\partial \varphi_k^*}{\partial z} - \mu\Delta\varphi_k^* = p_{ck}$$

$$\varphi_k^* = 0 \qquad \text{on} \quad \Sigma$$

$$\frac{\partial \varphi_k^*}{\partial z} = \alpha\varphi_k^* \quad \text{on} \quad \Sigma_0 \qquad (6.3.3)$$

$$\frac{\partial \varphi_k^*}{\partial z} = 0 \qquad \text{on} \quad \Sigma_H$$

$$\varphi_k^*(\mathbf{r}, T) = \varphi_k^*(\mathbf{r}, 0)$$

where p_{ck} is the function of the (6.3.2) type. Since the principle of duality

$$Y_{pk} = \int\limits_0^T dt \int\limits_{\Omega_k} p_{ck}\varphi \, d\Omega_k, \quad Y_{pk} = Q\int\limits_0^T \varphi_k^*(\mathbf{r}_0, t)dt \qquad (6.3.4)$$

holds, the equivalence condition

$$Y_{pk}(\mathbf{r}_0) = Q\int\limits_0^T \varphi_k^*(\mathbf{r}_0, t)dt \leq c_k \qquad (6.3.5)$$

also holds, as well as (6.3.1). Use the equivalence condition to locate the plant. Assume that problem (6.3.3) is solved and we have $\varphi_k^*(\mathbf{r}, t)$. Find the value $Y_{pk}(\mathbf{r})$ by the formula

$$Y_{pk}(\mathbf{r}) = Q\int\limits_0^T \varphi^*(\mathbf{r}, t)dt \qquad (6.3.6)$$

and draw the isograms of $Y_{pk}(\mathbf{r}) = const$.

Denote the sought location as $\omega_k \subset \Sigma_0$. An area ω_k is thus explicitly found where the location of the plant is possible. Then environmental and other considerations will in turn determine the choice of the best suitable location. If there is perchance no areas ω_k inside Σ_0, it may be established anyway by reducing affordable Q. This will of course impose limits on emissions and possibly on the technology of the plant. Assume now that there are several sensitive areas Σ_k $(k = 1, 2, \ldots, m)$. Then m adjoint problems of the (6.3.3) type have to be solved to obtain $\varphi_1^*, \varphi_2^*, \ldots, \varphi_m^*$. Use these solutions to form m functionals

$$Y_{pk}(\mathbf{r}_0) = Q \int\limits_0^T \varphi_k^*(\mathbf{r}_0, t)dt, \quad k = 1, 2, ..., m \qquad (6.3.7)$$

and obtain respectively m limitations

$$Y_{pk}(\mathbf{r}_0) \le c_k, \quad k = 1, 2, ..., m \qquad (6.3.7')$$

Locate (find ω_k) now the plant in every area Σ_k. The overlapping of all the areas ω_k $(k = 1, 2, \ldots, m)$ will be the location ω sought for the plant, where pollution standards will be met in all the areas of Σ_k. Fig. 33 illustrates the situation; isograms are provided for the case of several sensitive areas.

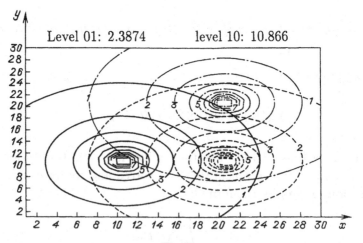

Fig. 33

The location ω is thus found. If it is not possible at a given Q, ocation may always be furnished by reduction of Q.

Summing up the discussion above, we may note, that the problem of rograms of location of atmosphere polluting plants for every ecological rea may now be realistically stated as part of environment protection. Climatic wind filed charts allowing for the impact of the relief must be made for every area Σ_0, adjoint problems solved on the basis of these harts and the functions $\varphi_1^*, \varphi_2^*, \ldots, \varphi_m^*$ be found. The latter are to be used to obtain the functionals $Y_{pk}(\mathbf{r})$ which will determine locations of he plants and limits of their emissions. This job should be done first f all in newly developed areas in industrial planning. It opens a road owards environmentally sound solutions. This criterion will be a top priority in the future.

ⅰ.4. Problem of Minimax

The principles formulated in § 6.3 provide an approach to the problem f minimax which may be described as follows. Assume Ω as a closed egion of space with a boundary $S = \Sigma \cup \Sigma_0 \cup \Sigma_H$. Inside, there are m environmentally sensitive areas on Σ_0 requiring special protection against industrial pollution; denote them as $\Sigma_1, \Sigma_2, \ldots, \Sigma_m$. Use the solution f problem (6.3.3) to construct the functionals $Y_{p1}, Y_{p2}, \ldots, Y_{pm}$ for very area. Examine now the functional

$$Y_p(\mathbf{r}) = \max_k Y_{pk}(\mathbf{r}) \qquad (6.4.1)$$

or a possible location $M(\mathbf{r}_0) \in \Omega$ of a plant emitting an aerosol at rate of Q per unit of time at an altitude of h. Then the minimum pollution for all the Ω_k areas is only possible if the point $M(\mathbf{r})$ is chosen n condition that

$$\max_k Y_{pk}(\mathbf{r}) = \min_{M \in \omega} \qquad (6.4.2)$$

'ind $M_0(\mathbf{r}_0)$ where condition (6.4.2) is satisfied, by combinatorial earch among the values of the functional $\max_k Y_{pk}(\mathbf{r})$ at all the points ı the region ω. We suppose here that ω has no common points with he region's boundary Σ. This is a very rare case when the problem of minimax is solved within a complex problem of mathematical physics explicitly as a result of one search of the functionals. Fig. 34 shows isograms of the functional $Y_p(x, y)$ which qualitatively characterizes the case with three sensitive areas. The types of wind circulations are here the same as in the Fig. 33.

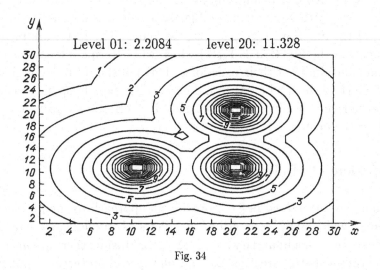

Fig. 34

6.5. Generalized Optimization Problem of Plant Location

We have thus obtained two statements of the problem of possible location of an environment-unfriendly plant. The global estimate of pollution in the optimization problem is provided for all the region Σ_0, but it may not meet specific conditions for all the environmentally sensitive areas Σ_k. The multi-criterial problem is to be solved for all sensitive areas, but it does not take into full account potential pollution of other areas of the region, though basically sensitive areas Σ_k may cover all the region Σ_0, so that $\cup \Sigma_k = \Sigma_0$, and adjoint problems may be solved for every area. This is feasible but difficult, since sensitive areas may be very many. Meanwhile, combining both problems may succeed. First the multi-criterial problem is solved and the possible plant location ω_k is found provided the pollution standards are met for all the areas Σ_k. Then the global pollution assessment problem is solved for all the region. This all results in the plant location where the limited-pollution condition is met for all the region Σ_0. Let this area be ω_Ω. Then the intersection of the areas ω_k, ω_Ω is the location of the plant, which satisfies both conditions. Denote it as ω_c.

Fig. 35 illustrates the method discussed above. The functional isograms are obtained by solving adjoint problems with the right-hand parts

$$p_1 = 1 \quad \text{at} \quad (x,p) \in \Omega$$

$$p_2 = 1 \quad \text{at} \quad (x,p) \in \Omega_1$$

where Ω_1 is a protected area in the center of the diagram. Solid isograms correspond to the local criterion and dashes to the global (the 1st level is 1.0602 and the 10th level 2.0584). Components of the velocity vector were assumed $u = 10$ m/s and $v = 0$.

We will come back to the optimization problem now and then in the future with stricter still limits imposed by factors linked with rehabilitation of polluted environment and amortization. These are, however, problems of an economic nature discussed in § 6.7–6.9.

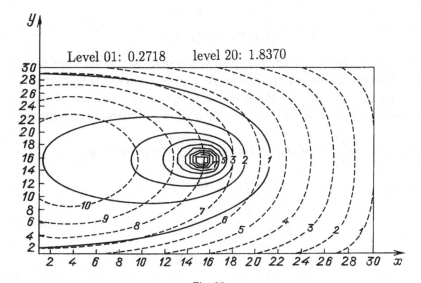

Fig. 35

6.6. Some General Remarks

The density of aerosols has been assumed as equal to zero at the boundary Σ of the domain Ω. This condition is not met in many real cases, since wind may export aerosols from one region to another across the boundary Σ. Therefore pollution imported from plants located in adjacent regions must be taken into account. Plants already operating in a given region may emit into the atmosphere the same aerosol, and their emission may be taken into consideration without any change in the principal scheme of solution of optimization problems discussed in § 6.1–6.5. Examine the problem

$$\frac{\partial \varphi'}{\partial t} + \operatorname{div} \mathbf{u}\varphi' + \sigma\varphi' - \frac{\partial}{\partial z}\nu\frac{\partial \varphi'}{\partial z} - \mu\Delta\varphi' = q + Q\delta(\mathbf{r} - \mathbf{r}_0)$$

$$\varphi' = f_s \qquad \text{on} \quad \Sigma$$

$$\frac{\partial \varphi'}{\partial z} = \alpha\varphi' \quad \text{on} \quad \Sigma_0 \tag{6.6.1}$$

$$\frac{\partial \varphi'}{\partial z} = 0 \quad \text{on} \quad \Sigma_H$$

$$\varphi'(\mathbf{r}, T) = \varphi'(\mathbf{r}, 0)$$

where $q(x, y, z)$ is a source of aerosol emitted by operating plants, and f_s is intensity of aerosol imported across the boundary of Ω from regions adjoining Σ_0. Then, since the problem is linear, the solution of (6.6.1) may be represented as a sum $\varphi' = \varphi^0 + \varphi$ where φ^0 satisfies the problem

$$\frac{\partial \varphi^0}{\partial t} + \operatorname{div} \mathbf{u}\varphi^0 + \sigma\varphi^0 - \frac{\partial}{\partial z}\nu\frac{\partial \varphi^0}{\partial z} - \mu\Delta\varphi^0 = q$$

$$\varphi^0 = f_s \qquad \text{on} \quad \Sigma$$

$$\frac{\partial \varphi^0}{\partial z} = \alpha\varphi^0 \quad \text{on} \quad \Sigma_0$$

$$\frac{\partial \varphi^0}{\partial z} = 0 \quad \text{on} \quad \Sigma_H \tag{6.6.2}$$

$$\varphi^0(\mathbf{r}, T) = \varphi^0(\mathbf{r}, 0)$$

and φ satisfies the problem we already know

$$\frac{\partial \varphi}{\partial t} + \operatorname{div} \mathbf{u}\varphi + \sigma\varphi - \frac{\partial}{\partial z}\nu\frac{\partial \varphi}{\partial z} - \mu\Delta\varphi = Q\delta(\mathbf{r} - \mathbf{r}_0)$$

$$\varphi = 0 \quad \text{on} \quad \Sigma$$

$$\frac{\partial \varphi}{\partial z} = \alpha\varphi \quad \text{on} \quad \Sigma_0 \qquad (6.6.3)$$

$$\frac{\partial \varphi}{\partial z} = 0 \quad \text{on} \quad \Sigma_H$$

$$\varphi(\mathbf{r}, T) = \varphi(\mathbf{r}, 0)$$

Now link the functionals of problem (6.6.1) with similar functionals of problems (6.6.2) and (6.6.3) and examine the functional

$$Y'_{pk} = \int_0^T dt \int_{\Omega_k} p_{ck}\varphi' \, d\Omega_k \qquad (6.6.4)$$

where

$$p_{ck} = \begin{cases} b_k + a_k\delta(z), & \text{on} \quad \Omega_k \\ 0, & \text{outside} \quad \Omega_k \end{cases} \quad k = 1, ..., m$$

Assume the limits on the functionals Y'_{pk}

$$Y'_{pk} \le c_k \qquad (6.6.5)$$

Substitute the sum $\varphi^0 + \varphi$ for φ' in formula (6.6.4) to obtain

$$Y'_{pk} = Y^0_{pk} + Y_{pk} \qquad (6.6.6)$$

where

$$Y^0_{pk} = \int_0^T dt \int_{\Omega_k} p_{ck}\varphi^0 \, d\Omega_k, \quad Y_{pk} = \int_0^T dt \int_{\Omega_k} p_{ck}\varphi \, d\Omega_k \qquad (6.6.7)$$

Rewrite restriction (6.6.5) with regard to (6.6.6) as

$$Y^0_{pk} + Y_{pk} \le c_k \qquad (6.6.8)$$

or $Y_{pk} \leq c_k - Y_{pk}^0$. Assuming $c_k - Y_{pk}^0 = \bar{c}_k$ obtain the sought restriction

$$Y_{pk} \leq \bar{c}_k \tag{6.6.9}$$

both for problem (6.6.3) and for the problem we examined under new restrictions. The functionals Y^0 are the result of solution of problem (6.6.2).

As concerns functionals local in time, notice, that we assumed always that the main functionals are quantities integral all over the time interval $0 \leq t \leq T$, i. e., we assumed that the optimizing functional is linked with the total annual aerosol fallout and dispersed in the zone Σ_k. This assumption, though, disregards possible short but very intensive aerosol emissions due to sudden backing and veering of the wind and change of its force. Such changes may affect significantly pollution in parts of the region Σ_0, though the annual average pollution by that aerosol may remain within established limits. This means that, along with the examined functionals of a total annual standard, functionals local in time may be introduced for periods characteristic of meteorological situations stability.

Assume that j_0 typical meteorological situations occur in a given region with a total annual limit for each equal to Δt_j. Summing all in $j = 1, 2, \ldots, j_0$, obtain

$$\sum_{j=1}^{j_0} \Delta t_j = T$$

If transition processes are disregarded and additivity assumably holds, the whole process may be considered as continuous in time with a random alternation of meteorological situations of various types, and j_0 different main problems may be solved (see Chapter 5):

$$\operatorname{div} \mathbf{u}\varphi_j + \sigma\varphi_j - \frac{\partial}{\partial z}\nu\frac{\partial\varphi_j}{\partial z} - \mu\Delta\varphi_j = Q\delta(\mathbf{r} - \mathbf{r}_0)$$

$$\varphi_j = 0 \qquad \text{on} \quad \Sigma$$

$$\frac{\partial\varphi_j}{\partial z} = \alpha\varphi_j \quad \text{on} \quad \Sigma_0 \tag{6.6.10}$$

$$\frac{\partial\varphi_j}{\partial z} = 0 \qquad \text{on} \quad \Sigma_H$$

Notice that if transition effects are disregarded, problem (6.6.10) may be solved through the non-stationary problem:

$$\frac{\partial \varphi}{\partial t} + \operatorname{div} \mathbf{u}\varphi + \sigma\varphi - \frac{\partial}{\partial z}\nu\frac{\partial \varphi}{\partial z} - \mu\Delta\varphi = Q\delta(\mathbf{r} - \mathbf{r}_0)$$

$$\varphi = 0 \quad \text{on} \quad \Sigma$$

$$\frac{\partial \varphi}{\partial z} = \alpha\varphi \quad \text{on} \quad \Sigma_0 \tag{6.6.11}$$

$$\frac{\partial \varphi}{\partial z} = 0 \quad \text{on} \quad \Sigma_H$$

$$\varphi(\mathbf{r}, t_{j+1}) = \varphi(\mathbf{r}, t_j)$$

where t_j, t_{j+1} are the limits of time interval Δt_j a typical meteorological situation develops within. Then

$$\varphi_j = \frac{1}{\Delta t_j} \int_{t_j}^{t_{j+1}} \varphi \, dt \tag{6.6.12}$$

Examine the problem adjoint to the main, (6.6.10),

$$-\operatorname{div} \mathbf{u}\varphi_{jk}^* + \sigma\varphi_{jk}^* - \frac{\partial}{\partial z}\nu\frac{\partial \varphi_{jk}^*}{\partial z} - \mu\Delta\varphi_{jk}^* = p_{ck}$$

$$\varphi_{jk}^* = 0 \quad \text{on} \quad \Sigma,$$

$$\frac{\partial \varphi_{jk}^*}{\partial z} = \alpha\varphi_{jk}^* \quad \text{on} \quad \Sigma_0 \tag{6.6.13}$$

$$\frac{\partial \varphi_{jk}^*}{\partial z} = 0 \quad \text{on} \quad \Sigma_H$$

where

$$p_{ck} = \begin{cases} b_k + a_k\delta(z), & \text{on} \quad \Omega_k \\ 0, & \text{outside} \quad \Omega_k \end{cases}$$

Problem (6.6.13) also may be solved through the non-stationary problem

$$-\frac{\partial\varphi_k^*}{\partial t} - \operatorname{div}\mathbf{u}\varphi_k^* + \sigma\varphi_k^* - \frac{\partial}{\partial z}\nu\frac{\partial\varphi_k^*}{\partial z} - \mu\Delta\varphi_k^* = p_{ck}$$

$$\varphi_k^* = 0 \qquad \text{on} \quad \Sigma$$

$$\frac{\partial\varphi_k^*}{\partial z} = \alpha\varphi_k^* \quad \text{on} \quad \Sigma_0 \qquad\qquad (6.6.14)$$

$$\frac{\partial\varphi_k^*}{\partial z} = 0 \qquad \text{on} \quad \Sigma_H$$

$$\varphi_k^*(\mathbf{r}, t_{j+1}) = \varphi_k^*(\mathbf{r}, t_j)$$

Then

$$\varphi_{jk}^* = \frac{1}{\Delta t_j}\int\limits_{t_j}^{t_{j+1}}\varphi_k^* dt$$

Introduce functionals

$$Y_{pjk} = \int\limits_{\Omega_k} p_{ck}\varphi_j d\Omega_k, \quad Y_{pjk} = Q\varphi_{jk}^*(\mathbf{r}_0) \qquad (6.6.15)$$

$$j = 1, 2, \ldots, j_0; \quad k = 1, 2, \ldots, m$$

Now that a set of adjoint problems is available, examine, like in the previous case, inequality

$$Y_{pjk} \leq c_{jk} \qquad\qquad (6.6.16)$$

where c_{jk} are admissible pollution standards. As a result, the problem reduces to location ω_{kj} of the plant meeting standards in respect to the area Σ_k. Denote the intersection of the regions ω_{kj} $(j = 1, 2, \ldots, j_0)$ as $\bar{\omega}_k$. This area will meet all the requirements concerning all types of meteorological processes involved into transport and diffusion of the aerosol.

Hence the intersection of all $\bar{\omega}_k$ $(k = 1, 2, \ldots, m)$ is the possible location $\bar{\omega}_c$ of the plant which would safely meet pollution limits with regard to typical meteorological processes in the area.

6.7. Estimation of Biospherical Losses to Environmental Pollution by Industrial Emissions

The previous sections discussed the problem on location of industrial plants so that they would meet pollution standards established for a given area. In this section, we will introduce a new functional linked to the cost for rehabilitation of polluted environment. This functional and others examined above, give an accurate idea of potential consequences of the biospheric pollution and the cost of environmental rehabilitation.

Since industrial emissions almost always depress all animal and vegetable life, a certain integral estimate of emission-inflicted losses over the region Σ_0 would be only too natural.

To obtain such an estimate, introduce differential characteristics describing the quantity of biomass of a given lth component lost to pollution of jth aerosol per unit of area per unit of time. Denote this quantity as $n_l b_{jl}$ $(j = 1, 2, \ldots, m; \; l = 1, 2, \ldots, s)$ where $n_l(x, y)$ is the density of the lth population in the area Σ_0 and b_{jl} is the specific loss of biomass of this population per unit of pollution density. Then the total losses of the lth component of life lost to pollution by aerosol with a concentration φ_j in the area Σ_0 per year is determined by the formula

$$\beta_{lj} = \int\limits_0^T dt \int\limits_{\Sigma_0} n_l b_{jl} \varphi_j d\Sigma_0 \qquad (6.7.1)$$

Denote the price of a unit of the biomass component as β_l. Then the total price (or pecuniary loss of biomass to pollution) is

$$c_{lj} = \int\limits_0^T dt \int\limits_{\Sigma_0} \beta_l n_l b_{jl} \varphi_j d\Sigma_0 \qquad (6.7.2)$$

Sum c_{lj} in all the components:

$$c_j = \int\limits_0^T dt \int\limits_{\Sigma_0} p_0^j \varphi_j d\Sigma_0 \qquad (6.7.3)$$

where

$$p_0^j = \sum_{l=1}^s n_l \beta_l b_{jl} \qquad (6.7.4)$$

The quantities b_{jl} which characterize the estimate of depression of biospheric components by aerosols of a certain type, are obtained in experiments. Notice that functions b_{jl} become nonlinear at large concentrations of pollutants. Sum the results in all the components of aerosols emitted by the plant to obtain the overall price of loses of the biosphere in the region Σ_0:

$$c = \sum_{j=1}^{m} \int_0^T dt \int_{\Sigma_0} p_0^j \varphi_j d\Sigma_0 \qquad (6.7.5)$$

Formulate now the optimization problem. Examine m problems corresponding to main emission components:

$$\frac{\partial \varphi_j}{\partial t} + \operatorname{div} \mathbf{u}\varphi_j + \sigma_j\varphi_j - \frac{\partial}{\partial z}\nu\frac{\partial \varphi_j}{\partial z} - \mu\Delta\varphi_j = Q\delta(\mathbf{r} - \mathbf{r}_0)$$

$$\varphi_j = 0 \qquad \text{at} \quad \Sigma$$

$$\frac{\partial \varphi_j}{\partial z} = \alpha\varphi_j \quad \text{at} \quad \Sigma_0 \qquad (6.7.6)$$

$$\frac{\partial \varphi_j}{\partial z} = 0 \qquad \text{at} \quad \Sigma_H$$

$$\varphi_j(\mathbf{r}, T) = \varphi_j(\mathbf{r}, 0), \quad j = 1, 2, ...m$$

and m adjoint problems:

$$-\frac{\partial \varphi_j^*}{\partial t} - \operatorname{div} \mathbf{u}\varphi_j^* + \sigma_j\varphi_j^* - \frac{\partial}{\partial z}\nu\frac{\partial \varphi_j^*}{\partial z} - \mu\Delta\varphi_j^* = p_0^j\delta(z)$$

$$\varphi_j^* = 0 \qquad \text{at} \quad \Sigma$$

$$\frac{\partial \varphi_j^*}{\partial z} = \alpha\varphi_j^* \quad \text{at} \quad \Sigma_0 \qquad (6.7.7)$$

$$\frac{\partial \varphi_j^*}{\partial z} = 0 \qquad \text{at} \quad \Sigma_H$$

$$\varphi_j^*(\mathbf{r}, T) = \varphi_0^*(\mathbf{r}, 0), \quad j = 1, 2, ...m$$

Assume problems (6.7.6) and (6.7.7) are solved, and examine the functional

$$I_j = \int_0^T dt \int_{\Sigma_0} p_0^j \varphi_j d\Sigma_0 \qquad (6.7.8)$$

Write it in a dual form found by solving the adjoint problem:

$$I_j = Q_j \int_0^T \varphi_j^*(\mathbf{r}_0, t) dt \qquad (6.7.9)$$

Notice that problem (6.7.7) is solved only once for this functional provided there is one determined aerosol. The function $\varphi^*(\mathbf{r}, t)$ is then computed and the area ω_B is found where the loss of biomass is minimum. There appears thus an area ω_B where the condition of an acceptable price of losses of biomass in the region Σ_0 is met, along with the region ω_c where pollution standards are met in sensitive areas (this region was introduced in the previous sections). The intersection of these regions (see Fig. 36) provides the location of the new plant.

Examine now a multi-component mixture of passive aerosols which does not transform chemically as it propagates, and therefore $\sigma_j = 0$. Assume that the aerosol is light and consists of oxides, which means that the mixture propagates according to one law. The problems (6.7.6) and (6.7.7) may be formulated immediately for the mixture density.

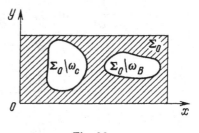

Fig. 36

Assume that the output of sources Q_j for given aerosol components is known and

$$Q_j/Q = \gamma_j, \quad j = 1, ..., m \qquad (6.7.10)$$

Assume also that

$$\varphi_j = \gamma_j \varphi, \quad \varphi_j^* = \gamma_j \varphi^* \qquad (6.7.11)$$

where

$$\sum_{j=1}^m \gamma_j = 1 \qquad (6.7.12)$$

It follows from (6.7.10), (6.7.11) with regard to (6.7.12) that

$$\varphi = \sum_{j=1}^{m} \varphi_j, \quad \varphi^* = \sum_{j=1}^{m} \varphi_j^*, \quad Q = \sum_{j=1}^{m} Q_j \qquad (6.7.13)$$

Sum now every relationship in (6.7.6) in j using (6.7.4) to obtain the problem of propagation of the mix of aerosols:

$$\frac{\partial \varphi}{\partial t} + \operatorname{div} \mathbf{u}\varphi - \frac{\partial}{\partial z}\nu\frac{\partial \varphi}{\partial z} - \mu\Delta\varphi = Q\delta(\mathbf{r} - \mathbf{r}_0)$$

$$\varphi = 0 \qquad \text{on} \quad \Sigma$$

$$\frac{\partial \varphi}{\partial z} = \alpha\varphi \quad \text{on} \quad \Sigma_0 \qquad (6.7.14)$$

$$\frac{\partial \varphi}{\partial z} = 0 \qquad \text{on} \quad \Sigma_H$$

$$\varphi(\mathbf{r}, T) = \varphi(\mathbf{r}, 0)$$

Similarly, obtain the adjoint problem for the same mix:

$$-\frac{\partial \varphi^*}{\partial t} - \operatorname{div} \mathbf{u}\varphi^* - \frac{\partial}{\partial z}\nu\frac{\partial \varphi^*}{\partial z} - \mu\Delta\varphi^* = p_0\delta(z)$$

$$\varphi^* = 0 \qquad \text{on} \quad \Sigma$$

$$\frac{\partial \varphi^*}{\partial z} = \alpha\varphi^* \quad \text{on} \quad \Sigma_0 \qquad (6.7.15)$$

$$\frac{\partial \varphi^*}{\partial z} = 0 \qquad \text{on} \quad \Sigma_H$$

$$\varphi^*(\mathbf{r}, T) = \varphi^*(\mathbf{r}, 0)$$

where

$$p_0 = \sum_{j=1}^{m} p_0^j$$

The functional, i. e., the price of biospheric losses in the region Σ_0, is

$$I = \int_0^T dt \int_{\Sigma_0} p_0\varphi d\Sigma_0 \qquad (6.7.16)$$

or

$$I = Q \int_0^T \varphi^*(\mathbf{r}_0, t)dt \qquad (6.7.17)$$

where

$$p_0 = \sum_{j=1}^m \sum_{l=1}^s n_l \beta_l b_{jl} = \sum_{l=1}^s n_l \beta_l b_l, \quad b_l = \sum_{j=1}^m b_{jl} \qquad (6.7.18)$$

After problem (6.7.15) is solved and the plant location determined, find the solution φ of problem (6.7.14) and use (6.7.11) to compute component densities $\varphi_j = \gamma_j \varphi$ $(j = 1, 2, \ldots, m)$. Simultaneously, the differential distribution of aerosol density is refined in the examined region Σ_0. If the aerosols propagate according to different laws, computation of the total pollution c will require solving m problems of this type.

Notice that determining coefficients b_{jl} is essential for the computations. This problem is not trivial, since it requires either simultaneous monitoring all the components of the biosphere exposed to pollution for many years, or building complex mathematical models of bioceonosis describing as fully as possible interaction between biospheric components. Examine an example. Assume that the aerosol with concentration φ_j proved lethal for insects. It will entail a sharp decline in the population of birds feeding on insects, which will in turn affect that of predatory birds also keeping down rodents. As a result, rodent populations will soar and ravage crops, and so on. Therefore, the problem of determining coefficients b_{jl} should be taken very seriously since they are important for industrial planning and forecasting industrial effects on environment.

6.8. Economics of Natural Resources

The price of losses in various components of the biosphere to environmental pollution we have just discussed is just only one side of the problem. The other is the cost of rehabilitation, improvement or at least conservation of the present environmental conditions in given areas. If, for example, pollution of water decreases spawning and thus reproduction of fish, fish farms have to be set up to breed fry to rebuild the fish population to the optimum size determined by food resources. If crops decline, they have to be rehabilitated by better cultivation, application of organic and chemical fertilizers and melioration. If a game population drops, there must be provided extra food, care, restrictions on hunting and so on.

Damage pollution does to nature (and especially to wildlife) must be compensated for at the expense of a pre-calculated environment (pollution) tax on plants as mandatory as their own capital depreciation instalments.

Formulate a mathematical model to estimate such extra cost of plant operation. Denote the cost of measures taken to rehabilitate a unit of mass of the lth component of the biosphere depressed by pollution to the original level, as p_l. Compute the loss of j-th component of a biomass:

$$\Delta M = \int_0^T dt \int_{\Sigma_0} n_l b_{jl} \varphi_j d\Sigma_0 \qquad (6.8.1)$$

Then a price of rehabilitation is

$$R_{jl} = \int_0^T dt \int_{\Sigma_0} p_l n_l b_{jl} \varphi_j d\Sigma_0 \qquad (6.8.2)$$

Sum (6.8.2) in l and j to obtain

$$R_g = \sum_{j=1}^m \sum_{l=1}^s \int_0^T dt \int_{\Sigma_0} p_l n_l b_{jl} \varphi_j d\Sigma_0 \qquad (6.8.3)$$

Rewrite it as

$$R_g = \sum_{j=1}^m \int_0^T dt \int_{\Sigma_0} \xi_j \varphi_j d\Sigma_0 \qquad (6.8.4)$$

where

$$\xi_j = \sum_{l=1}^s p_l n_l b_{jl} \qquad (6.8.5)$$

If the pollutants are light and decay quickly, $\sigma_j = 0$ $(j = 1, 2, \ldots, m)$. Then write with regard to (6.7.11)

$$R_g = \int_0^T dt \int_{\Sigma_0} \xi \varphi d\Sigma_0 \qquad (6.8.6)$$

where

$$\xi = \sum_{j=1}^{m} \gamma_j \xi_j \tag{6.8.7}$$

Formulate the adjoint problem with respect to the functional R_g:

$$-\frac{\partial \varphi^*}{\partial t} - \operatorname{div} \mathbf{u}\varphi^* - \frac{\partial}{\partial z}\nu\frac{\partial \varphi^*}{\partial z} - \mu\Delta\varphi^* = \xi\delta(z)$$

$$\begin{aligned}
\varphi^* &= 0 && \text{at} \quad \Sigma \\
\frac{\partial \varphi^*}{\partial z} &= \alpha\varphi^* && \text{at} \quad \Sigma_0 \\
\frac{\partial \varphi^*}{\partial z} &= 0 && \text{at} \quad \Sigma_H
\end{aligned} \tag{6.8.8}$$

$$\varphi^*(\mathbf{r},T) = \varphi^*(\mathbf{r},0)$$

Then, on the basis of the general theory, obtain two equalities:

$$R_g = \int_0^T dt \int_{\Sigma_0} \xi\varphi d\Sigma_0, \quad R_g = Q\int_0^T \varphi^*(\mathbf{r}_0,t)dt \tag{6.8.9}$$

Fig. 37 shows a diagram of the function $R_g(\mathbf{r}_0)$ corresponding to the solution $\varphi^*(\mathbf{r},t)$.

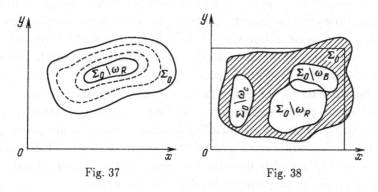

Fig. 37   Fig. 38

Use this algorithm to locate the area ω_R where the plant operation will involve environment rehabilitation costs below a certain prescribed standard, $R_g \le B_R$.

Planning and location of an industrial plant emitting hazardous pollutants must therefore meet three criteria:

1. Pollution of the environmentally sensitive areas must not exceed established health and conservancy standards, which takes us to the area $\omega_c \subset \Sigma_0$.

2. The price of losses of biological resources must be possibly minimum, which gives us the area $\omega_B \subset \Sigma_0$.

3. The cost of rehabilitation of biological resources lost through environmental pollution must be minimum, which happens to be in the area $\omega_R \subset \Sigma_0$.

The overlap area of the three regions (crosshatched in Fig. 38) is the likeliest location for the plant. If regions do not overlap, economic criteria should be changed.

We examined a case where all the aerosol components behave as their whole mixture does. If certain components propagate differently, all the problems have to be solved for every differing component j. As a result we face a more complex, but a more general problem of intersection of $3j$ separate regions.

6.9. Common Economic Criterion

All the areas meeting optimization criteria may sometimes provide in an assigned region Σ_0 an empty intersection. Then some of restrictions on the functionals should be slackened or the area should be expanded. Such modifications in the optimization problem are basically heuristic and are not strictly determined process.

On the other hand, the optimization criteria formulated in the preceding chapters produce definitely prohibited areas for location of industrial plants. Economically justified locations require more information. This section will discuss a common economic criterion of the total cost of rehabilitation of polluted environment given optimum location of plants in relation to environmentally sensitive areas.

The problem of optimum location includes considerable number of factors linked to the cost of the project on a particular site, the cost of infrastructure and telecommunications in local conditions and prospec-

tive development of the whole area. The problem may be methodolo-
gically classified as follows:

– planning of plant locations in the area in agreement with the existing environmental structure (e. g., of a large cluster of plants outside
a city);

– forming an environmental structure around plants (e. g., building housing and services for future plants based on local resources or
minerals and therefore tied to the specific location); and

– simultaneous planning the location of plants and environmentally
sensitive areas.

Formulate a common economic criterion for a situation of the first
type. Examine the cost of rehabilitation of environment, including
the cost of health protection in the area, more better-quality food,
recreation facilities and hospitals.

Denote the annual per capita per unit of jth aerosol concentration
cost of health protection in the sensitive area No. k as a_{jk} with a total
population of N_k. The cost of health protection is then

$$c_{jk} = \int\limits_0^T dt \int\limits_{\Sigma_k} a_{jk} N_k \varphi_j d\Sigma_k \qquad (6.9.1)$$

Sum this expression over all the environmentally sensitive areas to
obtain

$$R_{pj} = \sum_{k=1}^n c_{jk} = \sum_{k=1}^n \int\limits_0^T dt \int\limits_{\Sigma_k} a_{jk} N_k \varphi_j d\Sigma_k \qquad (6.9.2)$$

or

$$R_{pj} = \int\limits_0^T dt \int\limits_{\Sigma_0} P_p^j \varphi_j d\Sigma_0 \qquad (6.9.3)$$

where

$$P_p^j = \begin{cases} \sum\limits_{k=1}^n a_{jk} N_k, & \mathbf{r} \in \bigcup\limits_{k=1}^n \Omega_k \\ 0, & \text{outside the domain} \end{cases}$$

Then there is the price of the loss of biomass of all the components
of environment (animal and vegetable life, etc.) because of the shrink
of the productive area. Denote that price as R_{bj}. Use the results of

§ 6.7 to write

$$R_{bj} = \int\limits_{0}^{T} dt \int\limits_{\Sigma_0} P_b^j \varphi_j d\Sigma_0 \qquad (6.9.4)$$

where

$$P_b^j = \sum_{l=1}^{s} n_l \beta_l b_{jl}$$

The third part of the cost is that of rehabilitation and maintenance of biological resources productivity at a prescribed level

$$R_{gj} = \int\limits_{0}^{T} dt \int\limits_{\Sigma_0} P_g^j \varphi_j d\Sigma_0 \qquad (6.9.5)$$

where

$$P_g^j = \sum_{l=1}^{s} p_l n_l b_{jl}$$

Obtain m functionals for m different pollutants:

$$R_j = R_{pj} + R_{bj} + R_{gj} \qquad (6.9.6)$$

$$R = \sum_{j=1}^{m} R_j \qquad (6.9.7)$$

Denote the cost of construction of plant No. i at point $\mathbf{r}$ of area Σ_0 as $c_i(\mathbf{r})$, that of construction and operation of infrastructure per unit of the shortest distance between the ith plant's location $\mathbf{r}$ and the kth protected area (including the average cost of freight and commuter transport) as $c_{ik}(\mathbf{r})$, and a set of these distances as $r_{ik}(\mathbf{r})$. Examine the functional

$$E_{ik}(\mathbf{r}) = c_i(\mathbf{r}) + c_{ik}(\mathbf{r}) r_{ik}(\mathbf{r}) \qquad (6.9.8)$$

which describes the cost of the ith enterprise and infrastructural connection with the kth sensitive area. The value of the functional increases with the distance between an enterprise site and protected zone. Sum (6.9.8) over all k with regard to functional (6.9.7) to obtain the single functional

$$I_i(\mathbf{r}_0) = R(\mathbf{r}_0) + \sum_{k=1}^{n} E_{ik}(\mathbf{r}_0) \qquad (6.9.9)$$

Its $I_i(\mathbf{r}) = const$ level isograms provide the best location of the plant.

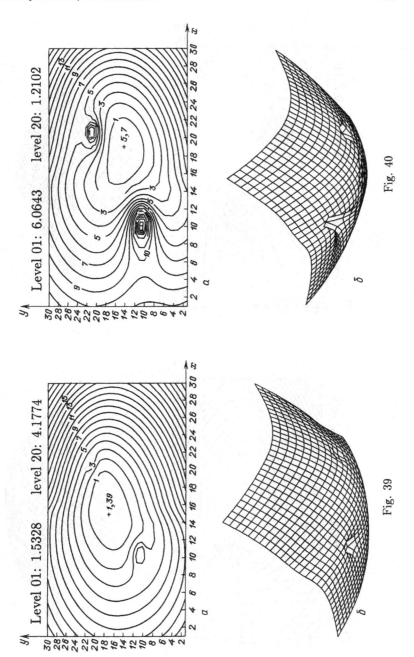

Fig. 40

Fig. 39

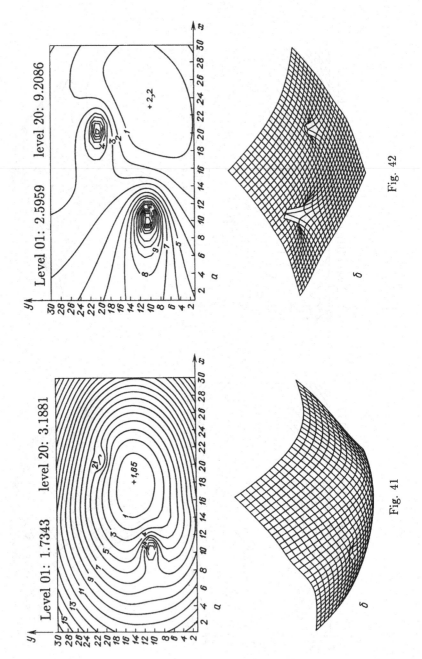

Fig. 42

Fig. 41

Examine a model situation, where of two populated areas located at (x_1, y_1) and (x_2, y_2) the first is twice as large. The prevailing wind direction coincides with the x axis over a time interval $[0, T]$. Point (x_2, y_1) marks the location of a plant with a minimum cost. The cost rises with distance from this point in proportion to a function of the type $\Psi(x, y) = 2 - \exp\{-\alpha[(x - x_2)^2 + (y - y_1)^2]\}$. The cost of infrastructure in the area $x > (x_1 + x_2)/2$ is double that in $[0, (x_1 + x_2)/2]$.

Figs. 39–42 show isograms of a functional of the type (6.9.9). Different isograms correspond to domination of different costs discussed above. Fig. 39,a reflects domination of the cost of infrastructure, i. e.,

$$I_i(\mathbf{r}_0) = R(\mathbf{r}_0) + \sum_{k=1}^{n} [2c_i(\mathbf{r}_0) + 20c_{ik}(\mathbf{r}_0)r_{ik}(\mathbf{r}_0)]$$

Fig. 39,b shows a surface generated by this functional. Fig. 40,a corresponds to larger share of the construction cost of the plant with the cost of infrastructure comparable with that of the building, i. e.,

$$I_i(\mathbf{r}_0) = R(\mathbf{r}_0) + \sum_{k=1}^{n} [10c_i(\mathbf{r}_0) + 8c_{ik}(\mathbf{r}_0)r_{ik}(\mathbf{r}_0)]$$

Fig. 40,b shows a surface corresponding to this functional.

Fig. 41,a illustrates approximate coincidence of the sizes of the cost of infrastructure and other costs, i. e.,

$$I_i(\mathbf{r}_0) = R(\mathbf{r}_0) + \sum_{k=1}^{n} [2c_i(\mathbf{r}_0) + 4c_{ik}(\mathbf{r}_0)r_{ik}(\mathbf{r}_0)]$$

and Fig. 41,b shows the respective surface. Finally, Fig. 42,a belongs to a case where all the costs are of the same order, and Fig. 42,b shows the surface of the respective functional. Crosses mark locations of plants where the total costs (in chosen units) are minimum.

6.10. Mathematical Problems of Optimization of Emissions at Operating Industrial Plants

Protection of environment from pollution by plants is becoming one of the most urgent problems sciences and technology face. Preceding sections discussed the aspect related to location of new environment polluting plants with the minimum damage to environmentally sensitive areas. This section deals with another aspect of the problem. Assume, there already exist aerosol-emitting plants in a certain area. The task is to find limits of emissions for every plant so that the sum would not exceed established standards. At the same time, the total emission cannot be reduced considerably, since it would result in a lower efficiency of the plants. Therefore limits must both provide maximum efficiency and meet assigned restrictions.

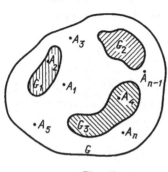

Fig. 43

Assume n plants A_i are located at points $\mathbf{r}_i$ $(i = 1, 2, \ldots, n)$ of assigned area Ω with a boundary S. The plants emit Q_i $(i = 1, 2, \ldots, n)$ of aerosols per unit of time. To simplify, assume the emission composition as the same (see Fig. 43). Single out m sensitive areas Ω_k $(k = 1, 2, \ldots, m)$ in the region Ω and assign a limit for aerosol fallout concentration per time interval $[0, T]$. The statement of the problem is as follows.

There is the equation of diffusion of substances emitted by n industrial objects:

$$\frac{\partial \varphi}{\partial t} + \operatorname{div} \mathbf{u}\varphi - \frac{\partial}{\partial z}\nu\frac{\partial \varphi}{\partial z} - \mu\Delta\varphi = \sum_{i=1}^{n} Q_i\delta(\mathbf{r} - \mathbf{r}_i) \qquad (6.10.1)$$

with conditions

$$\begin{aligned}
\varphi &= f_s \quad \text{on} \ \Sigma \\
\frac{\partial \varphi}{\partial z} &= \alpha\varphi \quad \text{on} \ \Sigma_0 \\
\frac{\partial \varphi}{\partial z} &= 0 \quad \text{on} \ \Sigma_H
\end{aligned} \qquad (6.10.2)$$

Since the problem is climatically periodic (the period is one year), we have

$$\varphi(\mathbf{r}, T) = \varphi(\mathbf{r}, 0) \qquad (6.10.3)$$

The components of the velocity vector $\mathbf{u}$ are linked at every moment of time with the continuity relationship

$$\frac{\partial u}{\partial x} + \frac{\partial v}{\partial y} + \frac{\partial w}{\partial z} = 0$$

where $w = 0$ at $z = 0$, $z = z_H$; ν and μ are coefficients of vertical and horizontal turbulent exchange, $\mathbf{r}_i = (x_i, y_i, z_i)^T$. The coefficient α characterizes the probability of fallout taking off again into the atmosphere, and f_s are the aerosol sources at Σ.

Examine the functional

$$Y_k = \int\limits_0^T dt \int\limits_{\Omega_k} p_c \varphi \, d\Omega_k \qquad (6.10.4)$$

which characterizes aerosol fallout on the Earth's surface ($z = 0$) in the sensitive area Ω_k. The task is to find a set of envisageable emissions Q_i which would meet the annual average limit of pollution

$$Y_k \le c_k, \quad k = 1, 2, ..., m \qquad (6.10.5)$$

with the minimum cost of reconstruction of the plants allowing to maintain the assigned production level with the assigned reduction in emissions.

Add to restrictions (6.10.5) a minimizing functional, e. g.,

$$I = \sum_{i=1}^n \xi_i(\bar{Q}_i - Q_i) \qquad (6.10.6)$$

where $\bar{Q}_i$ is initial and Q_i is affordable planned emission rate, the coefficient ξ_i determines investment into technology to keep up the same production level and reduce emissions per unit of emission rate. The functional I represents the total cost of improvement of technology at all the enterprises A_i to cut emissions from $\bar{Q}_i$ to the established discharges Q_i. As a result, the task is to find in (6.10.1)–(6.10.3) emissions Q_i satisfying the conditions

$$\begin{aligned} I &= \sum_{i=1}^n \xi_i(\bar{Q}_i - Q_i) = \min, \\ Y_k &\le c_k, \quad k = 1, 2, 3, ..., m \end{aligned} \qquad (6.10.7)$$

Problem (6.10.1)–(6.10.3) and (6.10.7) may be reduced to a problem of linear programming. Two different approaches are possible: main and adjoint equations.

First optimize using the main equations. Represent the solution to problem (4.10.1)–(4.10.3) as a superposition of solutions of elementary problems:

$$\varphi = \sum_{i=1}^{n} Q_i \varphi_i(\mathbf{r}, t) + \varphi_s \qquad (6.10.8)$$

where $\varphi_i(\mathbf{r}, t)$ is a solution to the problem

$$\frac{\partial \varphi_i}{\partial t} + \operatorname{div} \mathbf{u}\varphi_i + \sigma\varphi_i - \frac{\partial}{\partial z}\nu\frac{\partial \varphi_i}{\partial z} - \mu\Delta\varphi_i = \delta(\mathbf{r} - \mathbf{r}_i) \qquad (6.10.9)$$

with boundary conditions

$$\varphi_i = 0 \quad \text{on } \Sigma$$
$$\frac{\partial \varphi_i}{\partial z} = \alpha\varphi_i \quad \text{on } \Sigma_0 \qquad (6.10.10)$$
$$\frac{\partial \varphi_i}{\partial z} = 0 \quad \text{on } \Sigma_H$$

and condition

$$\varphi_i(\mathbf{r}, T) = \varphi_i(\mathbf{r}, 0) \qquad (6.10.11)$$

Add to (6.10.9)–(6.10.11) at $i = 1, 2, \ldots, n$ another problem to be solved to determine the background aerosols imported into Ω across the boundary S:

$$\frac{\partial \varphi_s}{\partial t} + \operatorname{div} \mathbf{u}\varphi_s + \sigma\varphi_s - \frac{\partial}{\partial z}\nu\frac{\partial \varphi_s}{\partial z} - \mu\Delta\varphi_s = 0 \qquad (6.10.12)$$

with conditions

$$\varphi_s = f_s \quad \text{on } \Sigma$$
$$\frac{\partial \varphi_s}{\partial z} = \alpha\varphi_s \quad \text{on } \Sigma_0 \qquad (6.10.13)$$
$$\frac{\partial \varphi_s}{\partial z} = 0 \quad \text{on } \Sigma_H$$
$$\varphi_s(\mathbf{r}, T) = \varphi_s(\mathbf{r}, 0) \qquad (6.10.14)$$

Assume that every problem in (6.10.9)–(6.10.11) at $i = 1, 2, \ldots, n$ and problem (6.10.12)–(6.10.14) are solved. Then representation

(6.10.8) holds; use it to compute the functionals Y_k. Substitute (6.10.8) into (6.10.4) to obtain

$$Y_k = \sum_{i=1}^{n} Q_i a_{ik} + b_k \qquad (6.10.15)$$

where

$$a_{ik} = \int_0^T dt \int_{\Omega_k} p_c \varphi_i(\mathbf{r}, t)\, d\Omega_k, \quad b_k = \int_0^T dt \int_{\Omega_k} p_c \varphi_s(\mathbf{r}, t)\, d\Omega_k$$

$$i = 1, 2, ..., n, \quad k = 1, 2, ..., m$$

Now a_{ik} and b_k are known constants. Combine (6.10.14) and (6.10.15) to arrive at

$$\sum_{i=1}^{n} \xi_i(\bar{Q}_i - Q_i) = \min$$
$$\sum_{i=1}^{n} Q_i a_{ik} + b_k \leq c_k, \quad k = 1, 2, 3, ..., m \qquad (6.10.16)$$

Pass from Q_i to $q_i = \bar{Q}_i - Q_i \geq 0$ to obtain a linear programming problem of finding the optimum set through the solution of the problem

$$\sum_{i=1}^{n} \xi_i q_i = \min, \quad \sum_{i=1}^{n} a_{ik} \geq R_k, \quad k = 1, 2, ..., m$$
$$q_i \geq 0, \quad i = 1, 2, ..., n \qquad (6.10.17)$$

where

$$R_k = \sum_{i=1}^{n} a_{ik}\bar{Q}_i + b_k - c_k$$

There may be more restrictions, of course, imposed by social and economic requirements depending on their priority.

Optimize now using the adjoint problem. According to the results of Chapter 5, the problem adjoint to (6.10.9)–(6.10.11) is

$$-\frac{\partial \varphi_k^*}{\partial t} - \operatorname{div}(\mathbf{u}\varphi_k^*) - \frac{\partial}{\partial z}\nu\frac{\partial \varphi_k^*}{\partial z} - \mu\Delta\varphi_k^* = p_k \qquad (6.10.18)$$

$$\varphi_k^* = 0 \text{ at } \Sigma$$
$$\frac{\partial \varphi_k^*}{\partial z} = \alpha\varphi_k^* \text{ at } \Sigma_0 \qquad (6.10.19)$$
$$\frac{\partial \varphi_k^*}{\partial z} = 0 \text{ at } \Sigma_H$$

$$\varphi_k^*(\mathbf{r}, T) = \varphi_k^*(\mathbf{r}, 0)$$

The right-hand part p_k of equation (6.10.18) determines the functional Y_k in (6.10.4) so that

$$p_k = \begin{cases} p_c, & \text{in} \quad \Omega_k \\ 0, & \text{outside} \quad \Omega_k \end{cases}$$

Examine the functional a_{ik} of (6.10.15)

$$a_{ik} = \int\limits_0^T dt \int\limits_{\Omega_k} p_c \varphi_i(\mathbf{r}, t) d\Omega_k \tag{6.10.20}$$

where φ_i is the solution to the main problem, (6.10.9)–(6.10.11). Routinely (see Chapter 5), obtain the dual form of functional (6.10.20):

$$a_{ik} = \int\limits_0^T dt \int\limits_{\Omega_k} p_c \varphi_i(\mathbf{r}, t) d\Omega_k = \int\limits_0^T \varphi_k^*(\mathbf{r}_i, t) dt \tag{6.10.21}$$

where φ_k^* is the solution to adjoint problem (6.10.18), (6.10.19). Assume notations

$$a_{ik}^* = \int\limits_0^T \varphi_k^*(\mathbf{r}_i, t) dt, \quad b_k^* = \int\limits_0^T dt \int\limits_{\Omega} p_k \varphi_s(\mathbf{r}, t) d\Omega \tag{6.10.22}$$

Then another formula follows from (6.10.15) to compute the functional Y_k:

$$Y_k = \sum_{i=1}^n Q_i a_{ik}^* + b_k^* \tag{6.10.23}$$

Similarly to the main problem, we thus obtain the optimization problem for adjoint equations:

$$\sum_{i=1}^n \xi_i(\bar{Q}_i - Q_i) = \min, \quad \sum_{i=1}^n Q_i a_{ik}^* + b_k^* \leq c_k, \quad k = 1, 2, 3, ..., m \tag{6.10.24}$$

or, introducing $q_i = \bar{Q}_i - Q_i \geq 0$, transform problem (6.10.24) into

$$\sum_{i=1}^n \xi_i q_i = \min, \quad \sum_{i=1}^n a_{ik}^* q_i \geq R_k^*, \quad k = 1, 2, ..., m$$
$$q_i \geq 0, \quad i = 1, 2, ..., n \tag{6.10.25}$$

where
$$R_k^* = \sum_{i=1}^{n} a_{ik}^* \overline{Q} + b_k^* - c_k$$
This is again a linear programming problem.

It is convenient to formulate the optimization problem using main equations in certain cases and adjoint in other cases. If polluting plants are few and sensitive areas are many, it will be more convenient to use main equations; in the opposite case, adjoint equations are better.

6.11. Perturbation Theory

The value of adjoint problems is not only in the opportunity they offer to formulate independently optimization problems. They also provide, without actually solving the problem, valuable information on the sensitivity of functionals of the Y_k type to parameter deviations from "standard." Examine a simple case of the perturbation theory using variations in emissions of aerosols.

Assume that a plants emits $Q_i' = Q_i + \delta Q_i$ aerosols instead of Q_i. The problem is then

$$\frac{\partial \varphi_i'}{\partial t} + \operatorname{div} \mathbf{u}\varphi_i' + \sigma\varphi_i' - \frac{\partial}{\partial z}\nu\frac{\partial \varphi_i'}{\partial z} - \mu\Delta\varphi_i' = Q_i'\delta(\mathbf{r} - \mathbf{r}_i) \qquad (6.11.1)$$

$$\varphi_i' = 0 \quad \text{on } \Sigma$$
$$\frac{\partial \varphi_i'}{\partial z} = \alpha\varphi_i' \quad \text{on } \Sigma_0 \qquad (6.11.2)$$
$$\frac{\partial \varphi_i'}{\partial z} = 0 \quad \text{on } \Sigma_H$$
$$\varphi_i'(\mathbf{r}, T) = \varphi_i'(\mathbf{r}, 0)$$

where $\varphi_i' = \varphi_i + \delta\varphi_i$.

Examine the functional
$$Y_{ik} = \int_0^T \int_{\Omega_k} p_k\varphi_i d\Omega_k \qquad (6.11.3)$$

For Y_{ik}' we have
$$Y_{ik}' = \int_0^T \int_{\Omega_k} p_k\varphi_i' d\Omega_k = Y_{ik} + \delta Y_{ik} \qquad (6.11.4)$$

where

$$\delta Y_{ik} = \int\limits_0^T \int\limits_{\Omega_k} p_k \delta\varphi_i d\Omega_k \qquad (6.11.5)$$

The functional δY_{ik} can be computed otherwise through the solution φ_k^* of the adjoint problem

$$-\frac{\partial\varphi_k^*}{\partial t} - \operatorname{div}(\mathbf{u}\varphi_k^*) - \frac{\partial}{\partial z}\nu\frac{\partial\varphi_k^*}{\partial z} - \mu\Delta\varphi_k^* = p_k; \qquad (6.11.6)$$

$$\begin{aligned} \varphi_k^* &= 0 \quad \text{on } \Sigma \\ \frac{\partial\varphi_k^*}{\partial z} &= \alpha\varphi_k^* \quad \text{on } \Sigma_0 \\ \frac{\partial\varphi_k^*}{\partial z} &= 0 \quad \text{on } \Sigma_H \\ \varphi_k^*(\mathbf{r}, T) &= \varphi_k^*(\mathbf{r}, 0) \end{aligned} \qquad (6.11.7)$$

By (6.11.1), (6.11.2), (6.11.6.) and (6.11.7) obtain, routinely, the duality relationship

$$(Q_i + \delta Q_i)\int\limits_0^T \varphi_k^* dt - \int\limits_0^T dt \int\limits_{\Omega_k} p_k(\varphi_i + \delta\varphi_i)d\Omega_k = 0 \qquad (6.11.8)$$

Since

$$Q_i\int\limits_0^T \varphi_k^* dt = \int\limits_0^T dt \int\limits_{\Omega_k} p_k\varphi_i d\Omega_k = Y_{ik}, \quad \delta Y_{ik} = \int\limits_0^T dt \int\limits_{\Omega_k} p_k\delta\varphi_i d\Omega_k \quad (6.11.9)$$

and by (6.11.8), obtain the formula of the perturbation theory

$$\delta Y_{ik} = \delta Q_i\int\limits_0^T \varphi_k^* dt, \quad i = 1, 2, ..., n; \quad k = 1, 2, ..., m \qquad (6.11.10)$$

Calculated quantities $\int_0^T \varphi_k^* dt$ and isograms $\int_0^T \varphi_k^* dt = const$ will provide the worst polluted areas, where local plants are the main sources of pollutant aerosols. Pollution limits must thus be calculated primarily for those plants. That does not solve the whole optimization problem, of course, but clarifies the problem to a certain extent.

Let us discuss another important issue. We have so far assumed that the input parameters of the main and adjoint problems are constant. In the meantime, in optimization problems, aerosol concentration over the region Ω does vary, which brings about changes in the local circulation of the atmosphere. This means that the components of the vector $\mathbf{u}$ may vary: $\mathbf{u}' = \mathbf{u} + \delta\mathbf{u}$ and so may the coefficients of the turbulent exchange, $\nu' = \nu + \delta\nu$, $\mu' = \mu + \delta\mu$. Perturbations make the main problem appear as

$$\frac{\partial \varphi_i'}{\partial t} + \operatorname{div}(\mathbf{u}\varphi_i') + \sigma\varphi_i' - \frac{\partial}{\partial z}\nu'\frac{\partial \varphi_i'}{\partial z} - \mu\Delta\varphi_i' = \delta(\mathbf{r} - \mathbf{r}_i) \qquad (6.11.11)$$

$$\varphi_i' = 0 \quad \text{on} \quad \Sigma$$
$$\frac{\partial \varphi_i'}{\partial z} = \alpha\varphi_i' \quad \text{on} \quad \Sigma_0 \qquad (6.11.12)$$
$$\frac{\partial \varphi_i'}{\partial z} = 0 \quad \text{on} \quad \Sigma_H$$
$$\varphi_i'(\mathbf{r}, T) = \varphi_i'(\mathbf{r}, 0)$$

Add to this one unperturbed adjoint problem (6.11.6) and (6.11.7), multiply equation (6.11.11) by φ_k^* and equation (6.11.6) by φ_i', integrate the results all over the definition domain, and subtract from each other. Then use boundary conditions in (6.11.12), (6.11.17) and integrate by parts to obtain

$$\int_0^T dt \int_\Omega \left(\operatorname{div}\delta\mathbf{u}\varphi_i' - \frac{\partial}{\partial z}\delta\nu\frac{\partial \varphi_i'}{\partial z} - \mu\Delta\varphi_i' \right) \varphi_k^* d\Omega$$

$$= \delta Q_i \int_0^T \varphi_k^*(\mathbf{r}_i, t)dt - \delta Y_{ik} \qquad (6.11.13)$$

therefore,

$$\delta Y_{ik} = \delta Q_i \int_0^T \varphi_k^*(\mathbf{r}_i, t)dt$$

$$- \int_0^T dt \int_\Omega \left(\operatorname{div}\delta\mathbf{u}\varphi_i' - \frac{\partial}{\partial z}\delta\nu\frac{\partial \varphi_i'}{\partial z} - \mu\Delta\varphi_i' \right) \varphi_k^* d\Omega \qquad (6.11.14)$$

If the perturbations $\delta\varphi_i$, $\delta\mathbf{u}$, $\delta\nu$, $\delta\mu$ are small, we obtain a formula of the small perturbation theory with accuracy up to small values of the second order:

$$\delta Y_{ik} = \delta Q_i \int_0^T \varphi_k^*(\mathbf{r}_i, t)dt$$

$$-\int_0^T dt \int_\Omega \left(\mathrm{div}\delta\mathbf{u}\varphi_i - \frac{\partial}{\partial z}\delta\nu\frac{\partial\varphi_i}{\partial z} - \mu\Delta\varphi_i\right)\varphi_k^*\,d\Omega \quad (6.11.15)$$

The formula obtained for δY_{ik} makes possible assessment of the feedback in atmospheric processes generated by variations in the aerosol background in Ω.

The optimization problem discussed in § 6.10 and in this section may be generalized for other functionals as done in § 6.1–6.6.

As concerns numerical applications of the algorithms, the main and adjoint problems, linear and periodic in time, may be solved by the periodization method, proceeding from certain initial data until the periodicity is obtained. Two or three annual cycles of calculations usually suffice. Notice that an adjoint problem must be solved down the time to have correct calculations. Linear programming problems are solved by standard techniques. Since $q_i \geq 0$ and all the coefficients a_{ik}, a_{ik}^* are also positive, the solution lies on the sides of polyhedrons built in constructing the restriction region.

6.12. Global Transport of Pollutants

While solving global problems of atmosphere physics and environment protection, detailed description of meteorological fields structure and pollutants propagation processes is needed. This problem became acute in recent years in connection with active interference of the Man with the environment. Growing enterprise activity and realization of huge industrial projects affected the environment on a global scale.

Mathematical models are helpful in the assessment of possible consequences of these activities. They allow to estimate perturbations of the main parameters characterizing the changes of the climate system under the influence of natural and anthropogenic factors. There is one more reason for the modelling: many experiments in the fields of ecology and social economics may bring about the consequences which are

rreversible. Mathematical modelling allows to consider various scena-
·ios of humane activities impact to the environment and estimate the
ιfter-effects. The information obtained in such a modelling allows to
lynamically monitor the behavior of the system under investigation.
)ne of the most important problems is a search for optimal solutions
vhile planning the industrial activity with regard to limiting loads to
.he environment and social and economical criteria. The final goal is
ιo determine the boundaries of stability of the system and the limits of
ιllowed changes. We will follow the work by Marchuk and Aloyan [96]
n this Section.

To estimate the influence regions of remote pollution sources affec-
ing the ecological state of various regions of the Earth the pollutant
ransport equations are used with regard to turbulent exchange and
nteraction of pollutants with underlying surface. We will model here
he global propagation of pollutants in the atmosphere and obtain, as
ι result, the estimation of total amount of pollutants at underlying
urface.

Model computations on pollutant transport are usually performed
or long periods therefore, to make the results more reliable, we used
nformation on wind velocity fields and other meteoelements based on
ιbservation data. Power of the sources and pollutant distribution func-
ion are assigned as the functions of spatial coordinates and time. The
nodel considers the Earth as a sphere and uses the spherical system
·f coordinates (λ, ψ, z) where λ is a longitude, ψ is a supplement to
ιtitude and z is a height counted from underlying surface. Write the
ιain pollutant transport equation as follows:

$$\frac{\partial \varphi}{\partial t} + \frac{u}{a \sin \psi} \frac{\partial \varphi}{\partial \lambda} + \frac{v}{a} \frac{\partial \varphi}{\partial \psi} + (w - w_g) \frac{\partial \varphi}{\partial z}$$

$$-\frac{\partial}{\partial z} \nu \frac{\partial \varphi}{\partial z} - \frac{1}{a^2 \sin^2 \psi} \frac{\partial}{\partial \lambda} \mu \frac{\partial \varphi}{\partial \lambda} - \frac{1}{a^2 \sin \psi} \frac{\partial}{\partial \psi} \mu \sin \psi \frac{\partial \varphi}{\partial \psi} = f \quad (6.12.1)$$

[ere $\varphi = \varphi(\lambda, \psi, z, t)$ is a pollutant concentration, $\mathbf{u} = (u, v, w - w_g)^T$
‐ wind velocity vector with components in the directions λ, ψ, z re-
pectively, w_g – gravitation subsidence velocity, μ, ν – turbulence duf-
ιsion coefficients in horizontal and vertical directions, $f = f(\lambda, \psi, z, t)$

function of allocation and power of sources, a is the average radius of the Earth.

We solve this problem in the domain $\Omega \times (0, T)$, where $\Omega = S \times (b, H)$, $S = \{(\lambda, \psi) : 0 < \lambda < 2\pi, \ 0 < \psi < \pi\}$, $b = b(\lambda, \psi)$ is a function describing the relief of underlying surface, H is an upper boundary of the domain.

Let us formulate a boundary condition of problem (6.12.1) at the level of a height of surface layer of the atmosphere. As is known from the Obukhov–Monin similarity theory [126], turbulent flow of a passive impurity in the atmosphere surface layer may be considered as constant in height. Then the following relationship is valid at $z \leq b + h$:

$$\frac{\partial \varphi}{\partial z} = \frac{\varphi_*}{z} \eta(\xi) \tag{6.12.2}$$

$$\varphi - \varphi_0 = \varphi_* \int_{\xi_0}^{\xi} \frac{\eta(\xi)}{\xi} d\xi \equiv \varphi_* f_\varphi(\xi, \xi_0) \tag{6.12.3}$$

$$\nu(\xi) = \frac{u_* \kappa z}{\eta(\xi)}, \quad u_* = \frac{\kappa |\mathbf{u}|}{f_u(\xi, \xi_0)} \equiv c_u |\mathbf{u}| \tag{6.12.4}$$

where φ_* is a scale for a change of pollutant concentration, η, f_u, f_φ are universal functions (see [126, 127]), $\xi = z/L$ is a dimensionless length characterizing the atmosphere stability, the zero index means that corresponding values are taken at $z = b + z_0$, were z_0 is a roughness parameter, L is a turbulent layer scale, κ is the Karman's constant, u_* is a scale of wind velocity, $c_u = \kappa / f_u(\xi, \xi_0)$. Obtain by (6.12.2) and (6.12.3)

$$\varphi_* = \frac{\varphi - \varphi_0}{f_\varphi(\xi, \xi_0)} \tag{6.12.5}$$

$$\frac{\partial \varphi}{\partial z} = \frac{\eta(\xi)}{z} \frac{\varphi - \varphi_0}{f_\varphi(\xi, \xi_0)} \tag{6.12.6}$$

Multiply equation (6.12.6) by ν to obtain

$$\nu \frac{\partial \varphi}{\partial z} = \alpha(\varphi - \varphi_0)$$

where $\alpha = c_u c_\varphi |\mathbf{u}|$, $c_\varphi = \kappa/f_\varphi(\xi, \xi_0)$. Examine this relationship at $z = b + h$.

$$\nu \frac{\partial \varphi}{\partial z} = \alpha(\varphi - \varphi_0) \quad \text{at} \quad z = b + h \qquad (6.12.7)$$

Assume this condition as a sought boundary condition for the main problem. Notice that the parameter α characterizes interaction of a pollutant with underlying surface (see [11, 138]).

Substitute $\Omega = S \times (b + h, H)$ for the definition domain Ω since the atmosphere surface layer may be excluded through its parameterization. The function φ_0 is thus the only unknown function in boundary condition (6.12.7) and we use the following expedient to determine it. Write the impurity balance equation in the neighborhood of underlying surface

$$-(\nu \frac{\partial \varphi}{\partial z})_0 + (\beta_i - w_g)\varphi_0 = \sum_{k=1}^{K} Q_{0k}\delta(x - x_k)\delta(y - y_k) \qquad (6.12.8)$$

where K is the number of all surface sources with coordinates (x_k, y_k), Q_{0k} is a power of each source i. e., amount of pollutant falling out at a unit area of underlying surface in a unit of time, β_i $(i = 1, 2)$ – coefficient characterizing interaction of a pollutant with underlying surface $\beta_1 = 0.01$ m/s corresponds to land, $\beta_2 = 1$ m/s – to sea surface). As concerns δ-function, see the footnote at page 160.

Equation (6.12.8) is written in Cartesian coordinates x, y for convenience; with regard to (6.12.7) it takes the form

$$\varphi_0 = \frac{\sum\limits_{k=1}^{K} Q_{0k}\delta(x - x_k)\delta(y - y_k) + c_u c_\varphi |\mathbf{u}_{b+h}|\varphi_{b+h}}{\beta_i - w_g + c_u c_\varphi |\mathbf{u}_{b+h}|} \qquad (6.12.9)$$

where $\mathbf{u}_{b+h} = \mathbf{u}|_{z=b+h}$, $\varphi_{b+h} = \varphi|_{z=b+h}$.

There is only one unknown value $\varphi_{b+h} = \varphi|_{z=b+h}$ in formula (6.12.9). Equations (6.12.7) and (6.12.9) compose the system we may find the expression φ_{b+h} from, starting with the first approximation φ^0_{b+h} (for instance, taking it from preceding time step). Then, substituting φ^0_{b+h} into boundary condition (6.12.7) and solving the problem with this approximate boundary condition, obtain new value of φ_{b+h}. The value thus obtained may be again substituted into (6.12.9) and new value of φ_0 will be obtained and so on, forming an iteration process to find φ_{b+h}.

It turns out in practice that for the most of global problems the first approximation provides sufficient accuracy that is, φ_{b+h} may be taken for (6.12.9) from preceding time step. The boundary condition (6.12.7) is thus determined. Though, it must be accentuated that formula (6.12.9) is needed to estimate a concentration of pollutant fallen out at underlying surface.

Let us use the following boundary condition for problem (6.12.1) at the top of the atmosphere at $z = H$:

$$\frac{\partial \varphi}{\partial z} = 0 \quad \text{at} \quad z = H \tag{6.12.10}$$

Choose the initial condition

$$\varphi = \bar{\varphi} \quad \text{at} \quad t = 0 \tag{6.12.11}$$

where $\bar{\varphi} = \bar{\varphi}(\lambda, \psi, z)$ is a background concentration of a pollutant. We assume $\bar{\varphi} = 0$ hereinafter for simplicity. Periodicity conditions are set for all functions on lateral boundaries in horizontal coordinates:

$$\varphi(0, \psi, z, t) = \varphi(2\pi, \psi, z, t)$$

$$\varphi(\lambda, -\psi, z, t) = \varphi(\lambda + \pi, \psi, z, t) \tag{6.12.12}$$

$$\varphi(\lambda, \pi + \psi, z, t) = \varphi(\lambda + \pi, \pi - \psi, z, t)$$

Then assume that the function $\varphi(\lambda, \psi, z, t)$ as a solution of pollutant transport problem in the form (6.12.1), (6.12.7), (6.12.10)–(6.12.12), is continuous in $\Omega \times [0, T]$ and differentiable in t. Besides, let at each t the function $\varphi(\lambda, \psi, z, t)$ belong to the set of functions $D(A)$ from real Hilbert space $L_2(\Omega)$ continuous and differentiable in Ω, so that they satisfy the condition

$$\frac{\partial}{\partial z} \nu \frac{\partial \varphi}{\partial z} + \frac{1}{a^2 \sin^2 \psi} \frac{\partial}{\partial \lambda} \mu \frac{\partial \varphi}{\partial \lambda} + \frac{1}{a^2 \sin \psi} \frac{\partial}{\partial \psi} \mu \sin \psi \frac{\partial \varphi}{\partial \psi} \in L_2(\Omega)$$

Here each function from $D(A)$ satisfies homogeneous conditions (6.12.10), (6.12.12). As concerns the coefficient $\mathbf{u}$ we assume that components of this vector are continuous and differentiable with div $\mathbf{u} = 0$. Assume all the rest of functions and parameters as sufficiently smooth, so that there exists a unique solution to problem (6.12.1), (6.12.7), (6.12.10)–(6.12.12).

One more important question. Since vertical resolution in global models is not sufficient to restore the fields in lower layers of the atmosphere with proper accuracy, we compute meteorological characteristics of boundary layer using the following parameterization of the atmosphere planetary boundary layer discussed in [48, 49]. The following external parameters are determined with a help of model of planetary boundary layer at the nodes of calculation grid in horizontal plane at the first level through known values of fields of velocity and temperature:

$$Ro = \frac{|\mathbf{u}_{g0}|}{l z_0}, \quad S_T = \frac{\beta \delta \hat{\theta}}{l |\mathbf{u}_{g0}|}$$

$$\eta_x = \frac{\kappa^2}{l} \frac{\partial u_g}{\partial z} = -\frac{\beta \kappa^2}{l^2} \frac{\partial \hat{\theta}}{\partial y}, \quad \eta_y = \frac{\kappa^2}{l} \frac{\partial v_g}{\partial z} = \frac{\beta \kappa^2}{l^2} \frac{\partial \hat{\theta}}{\partial x}$$

(6.12.13)

where Ro is the Kibbel–Rossbi number, S_T is a stratification parameter, u_g, v_g are the components of geostrophic wind velocity $\mathbf{u}_g$, $|\mathbf{u}_{g0}|$ is a modulus of geostrophic wind velocity close to underlying surface, l is the Coriolis' parameter, η_x, η_y are baroclinicity parameters, $\hat{\theta}$ is the potential temperature, $\beta = g/\hat{\theta}$ is a buoyancy parameter, g is the gravity acceleration, $\delta\hat{\theta}$ is a difference between the values of potential temperature at the boundary of planetary boundary layer and at underlying surface. By the values of Ro, S_T, η_x, η_y we find the values: $C_g = u_*/|\mathbf{u}_g|$, coefficient of "geostrophic resistance"; α, an angle between turbulent friction stress at underlying surface and $\mathbf{u}_{g0}$; $\mu = h_0/L_0$, dimensionless "internal" stratification parameter.

The following notation is accepted here: $h_0 = \kappa u_*/l$ is "internal" height scale of a boundary layer; $L_0 = -c_p \rho u_*^3/(\kappa \beta q_0)$ is a Monin–Obukhov length scale; c_p is a heat capacity of the air; ρ is the air density; q_0 is a heat flux at the Earth surface. By the values C_g, α, μ the value q_0 is calculated and then by C_g and $\mathbf{u}_g$ the value u_* and the coefficients of turbulent exchange at the heights $z \geq h$ are computed through the formula

$$\nu(z) = \frac{\kappa^2 u_*^2}{l^2} \times \begin{cases} h/(1 + 10\mu h) & \text{for} \quad \mu \geq 0 \\ h & \text{for} \quad -2.33 \leq \mu \leq 0 \\ (-0.07/\mu)^{-1/3} h^{4/3} & \text{for} \quad \mu \leq -2.33 \end{cases} \quad (6.12.14)$$

where h is a height of a surface layer.

The components of a velocity vector may be written as follows:

$$u = |\mathbf{u}_{g0}| \cos\alpha + \frac{u_*}{\kappa}\alpha_1$$

$$(6.12.15)$$

$$v = -|\mathbf{u}_{g0}| \sin\alpha + \frac{u_*}{\kappa}\alpha_2$$

where α_1 and α_2 are dimensionless velocity "defects" (see [96]) calculated through the formulas

$$\alpha_1 = \kappa\frac{u - u_g}{u_*} = -\exp(-\alpha_3)\frac{\cos\alpha_3 - \sin\alpha_3}{\alpha_4}$$

$$(6.12.16)$$

$$\alpha_2 = \kappa\frac{v - v_g}{u_*} = \exp(-\alpha_3)\frac{\cos\alpha_3 + \sin\alpha_3}{\alpha_4}$$

where $\alpha_3 = (z - h)/\alpha_4$, $\alpha_4 = (2\nu)^{1/2}$.

Vertical velocity w at the "top" boundary of atmosphere planetary boundary layer is computed with the method discussed in [48].

Impurities with small velocities of gravitational subsidence are of special interest for global pollutant transport, since they being suspended for a long period are transported by the air masses at large distances. In this case the theory of turbulence, used in general circulation models for description of turbulent exchange with temperature and humidity, is applicable for the pollutants. In particular, the following model may be used to determine horizontal turbulence coefficient as the first approximation [253]:

$$\mu = k_1^2 \Delta S |D_N|, \quad D_N = (D_T^2 + D_S^2)^{1/2}$$

$$(6.12.17)$$

$$D_T = -\frac{1}{a\sin\psi}\frac{\partial u}{\partial\lambda} - \frac{1}{2}\frac{\partial v}{\partial\psi}, \quad D_S = -\frac{1}{a\sin\psi}\frac{\partial v}{\partial\lambda} + \frac{1}{2}\frac{\partial u}{\partial\psi}$$

where ΔS is an area of elementary cell of a grid domain, k_1 is dimensionless parameter.

Numerical algorithms used for solving the problem formulated here are based on the splitting method (see Appendix I). The numerical scheme consists of two stages at every time step: transport of pollutant along the trajectories and turbulent diffusion. At the first stage an explicit monotonous scheme of the second order of approximation in spatial variables and time is used [40, 138, 179], at the second stage an implicit approximation is applied [73].

Examine now the functionals from the solution φ in the form

$$J = \int_0^T dt \int_\Omega p\varphi \, d\Omega \qquad (6.12.18)$$

where $p = p(\lambda, \psi, z, t)$ is an assigned function from $L_2(\Omega \times (0, T))$. If, for example,

$$p = \begin{cases} p_0, & (\lambda, \psi, z) \in \omega \\ 0, & (\lambda, \psi, z) \in \Omega \backslash \omega \end{cases} \qquad p_0 > 0, \quad \omega \subset \Omega \qquad (6.12.19)$$

then the functional J represents total concentration of pollutant in selected subdomain ω of the domain Ω weighted with a weight p_0. The region ω corresponds to the zone where the pollution is estimated. We may obtain different integral characteristics of pollution fields depending on assignment of ω. So, the problem is reduced to the estimation of the functionals of (6.12.18) type defined on a set of state functions satisfying original problem (6.12.1), (6.12.7), (6.12.10)–(6.12.12). In order to estimate the functionals it is necessary to know the fields of function φ, including real information on pollutant sources acting in the domain $\Omega \times (0, T)$ and initial state of atmosphere pollution in the domain Ω at a fixed moment of time $t = 0$.

It is useful to realize an approach based on adjoint equations, which allows to estimate an extent of potential danger of atmosphere pollution in the region ω from all the sources located in the domain $\Omega \times (0, T)$ with prescribed scenarios of meteorological regime of the atmosphere. Construct a problem adjoint to problem (6.12.1), (6.12.7), (6.12.10)–6.12.12), following the theory discussed in preceding sections. It will look out as

$$-\frac{\partial\varphi^*}{\partial t} - \frac{u}{a\sin\psi}\frac{\partial\varphi^*}{\partial\lambda} - \frac{v}{a}\frac{\partial\varphi^*}{\partial\psi} - (w-w_g)\frac{\partial\varphi^*}{\partial z} - \left(\frac{\partial}{\partial z}\nu\frac{\partial\varphi^*}{\partial z}\right.$$

$$\left. + \frac{1}{a^2\sin^2\psi}\frac{\partial}{\partial\lambda}\mu\frac{\partial\varphi^*}{\partial\lambda} + \frac{1}{a^2\sin\psi}\frac{\partial}{\partial\psi}\mu\sin\psi\frac{\partial\varphi^*}{\partial\psi}\right) = p \quad (6.12.20)$$

$$\varphi^* = 0 \quad \text{at} \quad t = T$$

$$\nu\frac{\partial\varphi^*}{\partial z} = \alpha\varphi^* \quad \text{at} \quad z = b+h$$

$$\nu\frac{\partial\varphi^*}{\partial z} = 0 \quad \text{at} \quad z = H \qquad (6.12.21)$$

$$\varphi^*(0,\psi,z,t) = \varphi^*(2\pi,\psi,z,t)$$

$$\varphi^*(\lambda,-\psi,z,t) = \varphi^*(\lambda+\pi,\psi,z,t)$$

$$\varphi^*(\lambda,\pi+\psi,z,t) = \varphi^*(\lambda+\pi,\pi-\psi,z,t)$$

where p is a function determining functional (6.12.18).

Assume that the solution $\varphi^*(\lambda,\psi,z,t)$ of problem (6.12.20)–(6.12.21) is continuous in $\Omega\times[0,T]$ and is a function differentiable in t. Besides, let the function $\varphi^*(\lambda,\psi,z,t)$ belong at each t to the set of functions $D(A^*) = D(A)$ from $L_2(\Omega)$. Assume again that other functions and parameters of the problem are sufficiently smooth which provides for existence of unique solution of problem (6.12.20)–(6.12.21). As follows from general theory, problem (6.12.20)–(6.12.21) is correct while being solved from $t = T$ up the time to $t = 0$. That is why $\varphi^* = 0$ at $t = T$ was chosen as "initial" condition in (6.12.21).

Perform some transformations. Make an inner product of equation (6.2.1) and φ^* in $L_2(\Omega)$ and that of equation (6.12.20) and φ, subtract the results from each other and integrate in t on the interval $[0,T]$. Integrating by parts with regard to boundary conditions we arrive at another representation of the same functional J

$$J = \int_0^T dt \int_\Omega f\varphi^* \, d\Omega + \int_0^T dt \int_0^{2\pi} d\lambda \int_0^\pi \alpha\varphi_0\varphi^* \bigg|_{z=b+H} d\psi \qquad (6.12.22)$$

In many cases the second term in formula (6.12.22) may be neglected. Since we consider this very case, formula (6.12.22) may be written as

$$J = \int\limits_0^T dt \int\limits_\Omega f\varphi^* \, d\Omega \qquad (6.12.23)$$

So, total amount of pollutant in the subdomain ω may be obtained using the solution φ^* of adjoint problem (6.12.20)–(6.12.21). This function is a weight function affecting contribution of every pollution source f to the atmosphere pollution over selected subdomain ω. It characterizes thus an extent of danger of polluting the atmosphere in this subdomain by a source which may be located anywhere in the domain $\Omega \times (0, T)$. In other words, the input of the source to the functional is equal to the product of power of a discharge and the value of φ^* at each t in the domain where the source is located.

More or less dangerous zones may be separated and mapped in the domain $\Omega \times (0, T)$ using the values of the function φ^*, as related to the atmosphere pollution in the subdomain ω. To do so, the function φ^* is normed by its maximum value φ^*_{max}.

The figures 44–51 present, for illustration, some results of numerical experiments with above models. They show two-dimensional cross-sections of influence function φ^* in the plane (λ, ψ). The estimates of pollution danger were computed for four regions of the planet: Arctic (Fig. 44, Fig. 45, $\varphi^*_{max} = 0.276 \cdot 10^{-7}$), Europe (Fig. 46, Fig. 47, $\varphi^*_{max} = 0.843 \cdot 10^{-7}$), USA–Canada (Fig. 48, Fig. 49, $\varphi^*_{max} = 0.681 \cdot 10^{-7}$) and Africa (Fig. 50, Fig. 51, $\varphi^*_{max} = 0.126 \cdot 10^{-6}$). The input parameters: the step in longitude $\Delta\lambda = 5°$, in latitude $\Delta\psi = 2.5°$, 15 levels in vertical coordinate. Time step $\Delta t = 30$ min, $T = 40$ days. The meteorological information "TOGA Data Set" was used in experiments, which represents the results of objective analysis produced in European Centre for Medium Range Weather Forecasts.

The numerical experiments with global transport of pollutant and calculations of sensitivity function were performed for the period from the January 1st to the March 1st of 1986. The sensitivity function φ^* was computed for each of four selected regions and for time intervals 10 days (figures with odd numbers) and 60 days (figures with even numbers). The values of this function characterize the relative contribution of all the surface sources to total concentration of pollutant in lower

1.5 km layer of the atmosphere over selected regions. The functionals characterizing pollution for separate regions are taken in the form of (6.12.18)

$$J_i = \int\limits_0^T dt \int\limits_\Omega p_i \varphi \, d\Omega$$

where $p = p_i$ are the weight functions different for different regions; $\alpha = 0$ for simplicity. Define the functions p_i $(i = \overline{1,4})$ as follows

$$p_i(\lambda, \psi, z, t) = \begin{cases} \dfrac{1}{\Delta \tilde{z} \mathrm{mes}\, \omega_i}, & \text{if } (\lambda, \psi, z) \in \omega_i \\ 0, & \text{otherwise} \end{cases}$$

where $\Delta \tilde{z} = 1500$ m, ω_1 is the Arctic, ω_2 is the Europe, ω_3 is USA and Canada territory, ω_4 is African continent, $\mathrm{mes}\, \omega_i$ is an area of ω_i $(i = \overline{1,4})$. Dashed parts correspond to the regions ω_i at the figures.

It is expedient to estimate the functionals J using the solution φ^* of adjoint problem (6.12.20)–(6.12.21) and formula (6.12.23). The influence function φ^* is drawn at the figures by isolines with numbers 1–9 corresponding to the values $\varphi_j^* = d_j \times \varphi_{\max}^*$ $(j = 1, \ldots, 9)$ where $d_1 = 0.01$, $d_2 = 0.05$, $d_3 = 0.10$, $d_4 = 0.25$, $d_5 = 0.40$, $d_6 = 0.55$, $d_7 = 0.70$, $d_8 = 0.85$, $d_9 = 0.95$.

The Arctic region embraces the area in the interval $77.5° \leq \psi \leq 90°$ N. Fig. 44 and Fig. 45 characterize relative contribution to total pollutant concentration over Arctic region in lower 1500-m atmosphere layer from all surface sources. The analysis of the figures shows that the sources located in Canada, Siberia and in Northern part of USA are the most influential for the Arctic zone.

Isolines of the function φ^* for the European region are presented at the Fig. 46 and Fig. 47. The function reaches the territory of Canada and USA. Fig. 48 and Fig. 49 present the influence function for the region of Northern America.

Isolines of the influence function for Africa (Fig. 50 and Fig. 51), unlike other regions, are located mostly close to this continent. The influence zone expands in the northern and southern regions which is connected with peculiarities of atmospheric circulation affected by extratropical convergence zone.

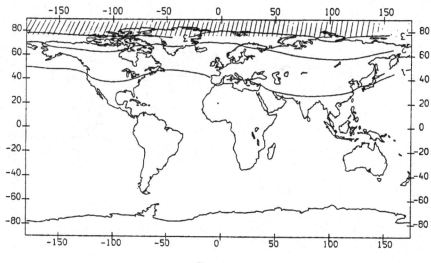

Fig. 44

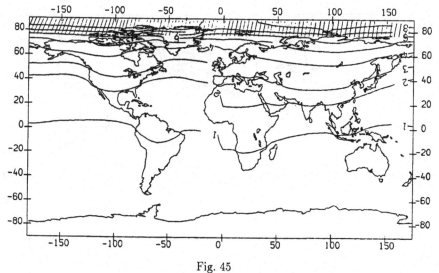

Fig. 45

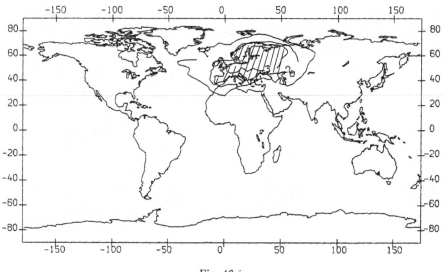

Fig. 46

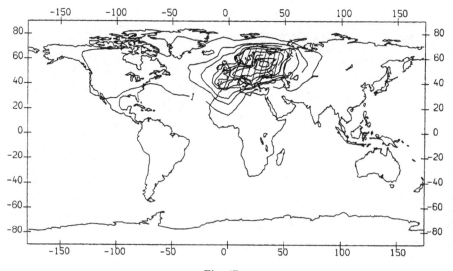

Fig. 47

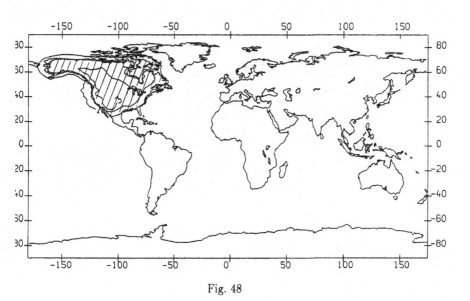

Fig. 48

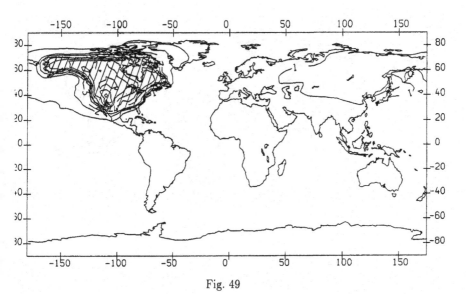

Fig. 49

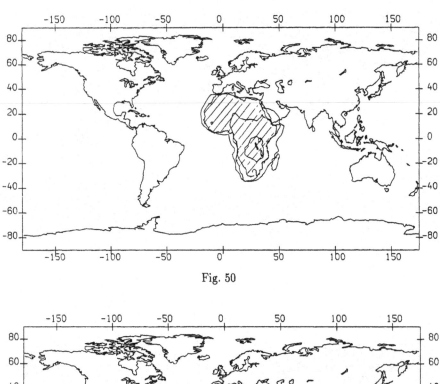

Fig. 50

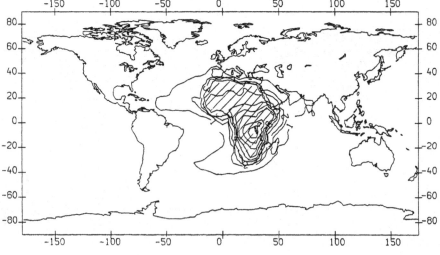

Fig. 51

The function φ^*, as a rule, is maximum within the selected regions. It is natural, since contribution of near sources is always higher than of remote ones, and a danger of pollution decreases with a distance from the region of interest. Since the functional depends on a product of sensitivity function and power of a source, it may happen that weak sources make big contribution to total pollution of a region. Size and configuration of regions covered by the influence function depend on current state of the atmosphere.

So, spatial–temporal structure of solutions of adjoint problems provides an information about the zones which may affect the pollution of a certain region. If a separate state is selected as a region ω, the adjoint functions and the sensitivity functionals may help in the estimation of the trans-boundary transport of pollution into the region ω from all the rest of planet's regions.

CHAPTER 7

Adjoint Equations and Models of General Circulation of Atmosphere and Ocean

Climatic change is one of the most complex problems science is facing nowadays, since all too many processes contribute to climate: hydrodynamical currents, turbulent exchange, interaction of processes in the atmosphere and the ocean, radiation regime of the system, phase transformations of the water releasing or absorbing heat, and many others. Moreover, the hydrothermodynamical processes in the atmosphere and the ocean are utterly nonlinear and unstable, and even small perturbations may cause the loss of stability of problem solutions. Unstable motions usually resolve as wave motions of a cyclone scale in the atmosphere and as mesoscale vortices in the ocean. All told, this makes precalculation of climatic changes an extremely difficult job.

If, however, the task is to examine merely the background climatic changes in the atmosphere and the ocean averaged over large regions – of an order of several thousands of square kilometers – and periods from a month to a season, cyclone-scale atmospheric and typical oceanic vortices may be examined in average statistically. Fast variations in atmospheric processes interacting with the ocean are therefore unlikely to affect appreciably the average background which slow and stable processes largely determine. These components should just be regarded as decisive in climatic changes in the system.

Even in this case, however, the atmosphere-ocean climatic system happens to be still too complex and requires thorough analysis of sensitivity to different affecting factors. Directly computed solution of a forecast problem will not help. Therefore we again, as in environmental protection problems, have to use the new mathematical technique of adjoint equations.

We will discuss the climate sensitivity problem starting with simple models and gradually complicate them to arrive at practical inspiring mathematical descriptions. We will start with a simple diagnostic climatic problem examined in § 7.1 and 7.2. Then this chapter will deal with problems related to the global circulation of the atmosphere and the ocean.

Assessment of man-made impacts on the climate requires mathematical tools which may provide estimates of temperature anomalies and other anomalies in prescribed areas of the Earth on the basis of solution of problems of atmospheric and ocean dynamics and a priori information on the climatic state of the atmosphere and actual information on deviation fields of meteorological elements. The regions should have characteristic sizes adjusted in sensitivity analysis. Notice that local meteorological information is very sensitive to unpredictable "meteorological noises" inherent in any model, even the most "rich" ones. The noises usually result from resolution of physical instabilities which continuously happen in the atmosphere. Some information is lost in inaccurate models used in calculations. This loss may be conventionally attributed to meteorological noises. Therefore, the choice of characteristic scales of the regions is a central problem in the predictability theory.

Another problem is constructing theories helping obtain estimates of global circulation variations following given deviations of functionals of meteorological elements from their climatic values. This problem statement is trivial in linear models. In primitive nonlinear problems, however, it requires a new mathematical technique on the basis of specially defined adjoint equations linked with the problem functionals to be forecast.

This chapter (in § 7.3–7.6) gives a more or less common approach to construction of adjoint equations of the global circulation of the atmosphere and ocean, provides perturbation theory formulas as a basis for the predictability analysis and for estimation of anomalies.

A temperature anomaly over an area of the Earth surface will be regarded as the main functional. It is easy to extend the theory to other linear functionals from fields of meteorological elements. We shall examine both linear and nonlinear statements of the problem.

This chapter discusses rather simple models of the atmosphere and ocean dynamics first and increasingly complex afterwards. They will be more complex and diverse still in real problems, of course. This author believes that it will be easier to grasp general ideas exemplified by simple models, and that more complex and real models will not involve basically new problems, and real models may be examined on the basis of proposed ideas.

This chapter emphasizes the algorithmic aspects of the problem. They may be theoretically substantiated just like in Chapter 1, starting with formation of Hilbert spaces of functions containing solutions of the main and adjoint equations, and smoothness assumptions necessary for transformations discussed in this chapter.

7.1. Temperature Anomalies in the Atmosphere

In this section we will focus the reader's attention on the process of deriving the adjoint equations which describe the anomalies of temperature in assigned region of the planet. Therefore, we examine a simple model of heat transport and diffusion in the atmosphere, assuming that the atmospheric density $\rho = const$ and velocity vector of air masses is determined over all the solution's definition domain, which is bounded by the planes $z = 0$ and $z = h$. Assume the periodicity condition in x-coordinate in the domain $0 \leq x \leq l$ and omit y-coordinate since it provides nothing new in transformations. Then we arrive at the following problem[1] for temperature deviation ϑ from some constant value $\overline{T}$ (i. e., $T = \overline{T} + \vartheta$) for all the domain:

$$L\vartheta \equiv \frac{\partial \vartheta}{\partial t} + u\frac{\partial \vartheta}{\partial x} + w\frac{\partial \vartheta}{\partial z} - \mu\frac{\partial^2 \vartheta}{\partial x^2} - \frac{\partial}{\partial z}\nu\frac{\partial \vartheta}{\partial z} = q \qquad (7.1.1)$$
$$x \in (0,l), \quad z \in (0,h), \quad t \in (0,\bar{t})$$

---

[1] To avoid confusion in notations, we consider hereafter the temporal interval $0 \leq t \leq \bar{t}$ instead of $0 \leq t \leq T$.

with boundary conditions

$$\vartheta(0, z, t) = \vartheta(l, z, t)$$

$$\nu\frac{\partial\vartheta}{\partial z} = f(x, t) \qquad \text{at} \quad z = 0$$

$$\nu\frac{\partial\vartheta}{\partial z} = 0 \qquad \text{at} \quad z = h \tag{7.1.2}$$

and initial data

$$\vartheta(x, z, 0) = g(x, z) \tag{7.1.3}$$

where $q(x, z, t) = \varepsilon/(c_p\rho)$, ε is a heat flux to a unit of volume, c_p is specific heat capacity of a medium. Horizontal and vertical coefficients of turbulent heat conduction $\mu, \nu = const > 0$ are assumed to be known.

Assume that the assigned periodic in x functions u and w with a period l satisfy the conditions

$$\frac{\partial u}{\partial x} + \frac{\partial w}{\partial z} = 0 \quad \text{at any} \quad (x, z, t)$$

$$w(x, 0, t) = w(x, h, t) = 0 \tag{7.1.4}$$

Assume that the functions $q(x, z, t)$, $f(x, t)$, and $g(x, z)$ are assigned, and solution ϑ, functions u, w, q, f, and g are sufficiently smooth, so that problem (7.1.1)–(7.1.4) has a unique solution within a class of periodic in x functions $D(L)$. Properties of their elements will be defined later.

Check first the solution for uniqueness. Assume, as usual, that there are two solutions, ϑ_1 and ϑ_2, satisfying problem (7.1.1)–(7.1.4) and examine their difference

$$\psi = \vartheta_1 - \vartheta_2$$

Then obtain the following homogeneous problem for ψ:

$$\frac{\partial\psi}{\partial t} + u\frac{\partial\psi}{\partial x} + w\frac{\partial\psi}{\partial z} - \mu\frac{\partial^2\psi}{\partial x^2} - \frac{\partial}{\partial z}\nu\frac{\partial\psi}{\partial z} = 0$$

$$\psi(0, z, t) = \psi(l, z, t)$$

$$\nu\frac{\partial\psi}{\partial z} = 0 \qquad \text{at} \quad z = 0, \quad z = h$$

$$\psi = 0 \qquad \text{at} \quad t = 0 \tag{7.1.5}$$

Multiply the first equation in (7.1.5), term by term, by ψ and intraegte the result all over the solution's definition domain:

$$\int\limits_0^{\bar{t}} dt \int\limits_0^l dx \int\limits_0^h \psi \left(\frac{\partial \psi}{\partial t} + u\frac{\partial \psi}{\partial x} + w\frac{\partial \psi}{\partial z} - \mu\frac{\partial^2 \psi}{\partial x^2} - \frac{\partial}{\partial z}\nu\frac{\partial \psi}{\partial z} \right) dz = 0 \quad (7.1.6)$$

Take into account

$$\int\limits_0^l dx \int\limits_0^h dz \int\limits_0^{\bar{t}} \psi\frac{\partial \psi}{\partial t} dt = \frac{1}{2}\int\limits_0^l dx \int\limits_0^h [\psi^2(x,z,\bar{t}) - \psi^2(x,z,0)]dz$$

$$= \frac{1}{2}\int\limits_0^l dx \int\limits_0^h \psi^2(x,z,\bar{t})dz \quad (7.1.7)$$

The initial quantity $\psi = 0$ at $t = 0$ was used here.

Next, examine the expression

$$\int\limits_0^{\bar{t}} dt \int\limits_0^l dx \int\limits_0^h \psi \left(u\frac{\partial \psi}{\partial x} + w\frac{\partial \psi}{\partial z} \right) dz = \frac{1}{2}\int\limits_0^{\bar{t}} dt \int\limits_0^l dx \int\limits_0^h \left(u\frac{\partial \psi^2}{\partial x} + w\frac{\partial \psi^2}{\partial z} \right) dz$$

$$= \frac{1}{2}\int\limits_0^{\bar{t}} dt \int\limits_0^l dx \int\limits_0^h \left[\left(\frac{\partial(u\psi^2)}{\partial x} + \frac{\partial(w\psi^2)}{\partial z} \right) - \psi^2\left(\frac{\partial u}{\partial x} + \frac{\partial w}{\partial x} \right) \right] dz \quad (7.1.8)$$

Since the components u and w of the velocity vector satisfy continuity equation (7.1.4), write (7.1.8) finally as

$$\int\limits_0^{\bar{t}} dt \int\limits_0^l dx \int\limits_0^h \psi \left(u\frac{\partial \psi}{\partial x} + w\frac{\partial \psi}{\partial z} \right) dz = \frac{1}{2}\int\limits_0^{\bar{t}} dt \int\limits_0^l dx \int\limits_0^h \left(\frac{\partial(u\psi^2)}{\partial x} + \frac{\partial(w\psi^2)}{\partial z} \right) dz$$

$$= \frac{1}{2}\int\limits_0^{\bar{t}} dt \int\limits_0^h u\psi^2\Big|_{x=0}^{x=l} dz + \frac{1}{2}\int\limits_0^{\bar{t}} dt \int\limits_0^l w\psi^2\Big|_{z=0}^{z=h} dx \quad (7.1.9)$$

Actually, two last terms in (7.1.9) turn to zero: the first because the integrand is periodic, and the second because $w = 0$ at $z = 0$ and $z = h$. Transform the other of terms in (7.1.6) to obtain

$$\int\limits_0^{\bar{t}} dt \int\limits_0^l dx \int\limits_0^h \psi\frac{\partial^2 \psi}{\partial x^2} dz$$

$$= \int\limits_0^{\bar{t}} dt \int\limits_0^h \psi\frac{\partial \psi}{\partial x}\Big|_{x=0}^{x=l} dz - \int\limits_0^{\bar{t}} dt \int\limits_0^l dx \int\limits_0^h \left(\frac{\partial \psi}{\partial x} \right)^2 dz \quad (7.1.10)$$

The term outside integral in (7.1.10) turns to zero due to its periodicity; therefore,

$$\int_0^{\bar{t}} dt \int_0^l dx \int_0^h \psi \frac{\partial^2 \psi}{\partial x^2} dz = - \int_0^{\bar{t}} dt \int_0^l dx \int_0^h \left(\frac{\partial \psi}{\partial x}\right)^2 dz \qquad (7.1.11)$$

Finally,

$$\int_0^{\bar{t}} dt \int_0^l dx \int_0^h \psi \frac{\partial}{\partial z} \nu \frac{\partial \psi}{\partial z} dz$$

$$= \int_0^{\bar{t}} dt \int_0^l \nu \psi \frac{\partial \psi}{\partial z}\Big|_{z=0}^{z=h} dx - \int_0^{\bar{t}} dt \int_0^l dx \int_0^h \nu \left(\frac{\partial \psi}{\partial x}\right)^2 dz \qquad (7.1.12)$$

The first integral in the right-hand part of (7.1.12) turns to zero due to the condition

$$\nu \frac{\partial \psi}{\partial z} = 0 \quad \text{at} \quad z = 0, \quad z = h$$

As a result,

$$\int_0^{\bar{t}} dt \int_0^l dx \int_0^h \psi \frac{\partial}{\partial z} \nu \frac{\partial \psi}{\partial z} dz = - \int_0^{\bar{t}} dt \int_0^l dx \int_0^h \nu \left(\frac{\partial \psi}{\partial x}\right)^2 dz \qquad (7.1.13)$$

Rewrite relationship (7.1.6) with regard to (7.1.7), (7.1.9), (7.1.11), and (7.1.13) as

$$\frac{1}{2} \int_0^l dx \int_0^h \psi^2(x, z, \bar{t}) dz + \int_0^{\bar{t}} dt \int_0^l dx \int_0^h \left[\mu \left(\frac{\partial \psi}{\partial x}\right)^2 + \nu \left(\frac{\partial \psi}{\partial z}\right)^2\right] dz = 0 \qquad (7.1.14)$$

Since the functions we are dealing with belong to a class of continuous functions quadratically summable together with all their first and second spatial derivatives, it follows from (7.1.14) that the left-hand part of (7.1.14) is equal to zero only if $\psi = 0$, i. e., at $\vartheta_1 = \vartheta_2$ which proves that the solution of the stated problem is unique. At the same time we defined the set of functions $\vartheta \in D(L)$, which is the definition domain of the operator of the problem. Additional conditions on functions from $D(L)$ are determined by the homogeneous boundary conditions in (7.1.12), including the periodicity condition.

Examine again the initial problem, (7.1.1)–(7.1.4). The unique solution to this problem is determined by the initial condition $\vartheta(x, z, 0) = g(x, z)$ and by the constant heat flux at the boundary

$$\nu \frac{\partial \vartheta}{\partial z} = f(x, t) \quad \text{at} \quad z = 0$$

As concerns the initial heat field in the atmosphere, its influence on the thermic regime of the atmosphere is damped in one or two weeks, so it may be disregarded in most cases. This means, that the temperature regime reaches the steady state in the atmosphere in this model mostly as determined by heat fluxes through the surface $z = 0$ and mainly in energetically active zones of the ocean, and by real flows of air masses which transport heat absorbed from the ocean at velocities of atmospheric motions. Since our goal is not forecasting the temperature field in the atmosphere, but analyzing the model sensitivity to various perturbing factors, we pick the components u and w of the air flows velocity vector from real observations. Besides, the ocean exchanges heat with the atmosphere not only through regular currents like the Gulf Stream and Kuroshio which transport heat to higher latitudes where they yield their stored heat to the atmosphere through the unstable stratification of the ocean; storms contribute, too, as the heat flux at the boundary between the atmosphere and the ocean rises steeply; our models must account for this phenomenon. It is well known that the atmosphere derives up to 40 % of all its energy from the ocean in the storm zones.

Construct now the perturbation theory for the selected functionals and estimate the model sensitivity. To do so, define first of all the functional we are interested in, so far in a general form, assuming

$$J = \int\limits_0^l dx \int\limits_0^h g^*(x, z)\vartheta(x, z, \bar{t})dz$$

$$- \int\limits_0^{\bar{t}} dt \int\limits_0^l f^*(x, t)\vartheta(x, 0, t)dx + \int\limits_0^{\bar{t}} dt \int\limits_0^l dx \int\limits_0^h p^*\vartheta dz \quad (7.1.15)$$

where $g^*(x, z)$, $f^*(x, t)$ and $p^*(x, z, t)$ are some yet undetermined functions whose concrete form will be assumed later according to the physical meaning of the measured fields. Until then they will be referred to as the functions linked with characteristics of measurements of the temperature field ϑ.

Constructing the perturbation theory formula for J requires formulation of an adjoint problem. Multiply equation (7.1.1) by the adjoint function $\vartheta^* \in D(L^*)$, where properties of the set $D(L^*)$ will be defined later by further transformations. Assume only that ϑ^* is periodic in x with a period l. Integrate the result all over the definition domain of the functions ϑ and ϑ^* to obtain

$$\int_0^{\bar{t}} dt \int_0^l dx \int_0^h \vartheta^* \left(\frac{\partial \vartheta}{\partial t} + u \frac{\partial \vartheta}{\partial x} + w \frac{\partial \vartheta}{\partial z} - \mu \frac{\partial^2 \vartheta}{\partial x^2} - \frac{\partial}{\partial z} \nu \frac{\partial \vartheta}{\partial z} - q \right) dz = 0 \quad (7.1.16)$$

Examine consequently the terms in formula (7.1.16) and transform them into a form convenient for definition of the adjoint problem:

$$\int_0^{\bar{t}} dt \int_0^l dx \int_0^h \vartheta^* \frac{\partial \vartheta}{\partial t} dz = \int_0^l dt \int_0^h \vartheta^* \vartheta \Big|_{t=0}^{t=\bar{t}} dz - \int_0^{\bar{t}} dt \int_0^l dx \int_0^h \vartheta \frac{\partial \vartheta^*}{\partial t} dz$$

$$= \int_0^l dx \int_0^h g^*(x,z) \vartheta(x,z,\bar{t}) dz - \int_0^l dx \int_0^h g(x,z) \vartheta^*(x,z,0) dz$$

$$- \int_0^{\bar{t}} dt \int_0^l dx \int_0^h \vartheta \frac{\partial \vartheta^*}{\partial t} dz \quad (7.1.17)$$

Here
$$\vartheta^*(x,z,\bar{t}) = g^*(x,z) \quad (7.1.18)$$

Assume g^* as yet undefined function.

Examine another term:

$$\int_0^{\bar{t}} dt \int_0^l dx \int_0^h \vartheta^* \left(u \frac{\partial \vartheta}{\partial x} + w \frac{\partial \vartheta}{\partial z} \right) dz$$

$$= \int_0^{\bar{t}} dt \int_0^l dx \int_0^h \left(\frac{\partial}{\partial x} (u \vartheta \vartheta^*) + \frac{\partial}{\partial z} (w \vartheta \vartheta^*) \right) dz$$

$$- \int_0^{\bar{t}} dt \int_0^l dx \int_0^h \vartheta \left(\frac{\partial u \vartheta^*}{\partial x} + \frac{\partial w \vartheta^*}{\partial z} \right) dz \quad (7.1.19)$$

The first right-hand integral turns to zero after the integration since the functions u, ϑ and ϑ^* are periodic in x and $w = 0$ at $z = 0$ and $z = h$. So

$$\int\limits_0^{\bar{t}} dt \int\limits_0^l dx \int\limits_0^h \vartheta^* \left(u\frac{\partial \vartheta}{\partial x} + w\frac{\partial \vartheta}{\partial z} \right) dz = -\int\limits_0^{\bar{t}} dt \int\limits_0^l dx \int\limits_0^h \vartheta \left(\frac{\partial u\vartheta^*}{\partial x} + \frac{\partial w\vartheta^*}{\partial z} \right) dz$$

$$= -\int\limits_0^{\bar{t}} dt \int\limits_0^l dx \int\limits_0^h \vartheta \left[\left(u\frac{\partial \vartheta^*}{\partial x} + w\frac{\partial \vartheta^*}{\partial z} \right) + \vartheta^* \left(\frac{\partial u}{\partial x} + \frac{\partial w}{\partial z} \right) \right] dz \quad (7.1.20)$$

Remember that $\partial u/\partial x + \partial w/\partial z = 0$, then finally

$$\int\limits_0^{\bar{t}} dt \int\limits_0^l dx \int\limits_0^h \vartheta^* \left(u\frac{\partial \vartheta}{\partial x} + w\frac{\partial \vartheta}{\partial z} \right) dz$$

$$= -\int\limits_0^{\bar{t}} dt \int\limits_0^l dx \int\limits_0^h \vartheta \left(u\frac{\partial \vartheta^*}{\partial x} + w\frac{\partial \vartheta^*}{\partial z} \right) dz \quad (7.1.21)$$

Transform now the next term:

$$\int\limits_0^{\bar{t}} dt \int\limits_0^l dx \int\limits_0^h \vartheta^* \frac{\partial^2 \vartheta}{\partial x^2} dz = \int\limits_0^{bart} dt \int\limits_0^h \vartheta^* \frac{\partial \vartheta}{\partial x} \Bigg|_{x=0}^{x=l} dz$$

$$-\int\limits_0^{\bar{t}} dt \int\limits_0^l dx \int\limits_0^h \frac{\partial \vartheta}{\partial x} \frac{\partial \vartheta^*}{\partial x} dz = \int\limits_0^{\bar{t}} dt \int\limits_0^h \vartheta^* \frac{\partial \vartheta}{\partial x} \Bigg|_{x=0}^{x=l} dz$$

$$-\int\limits_0^{\bar{t}} dt \int\limits_0^h \vartheta \frac{\partial \vartheta^*}{\partial x} \Bigg|_{x=0}^{x=l} dz + \int\limits_0^{\bar{t}} dt \int\limits_0^l dx \int\limits_0^h \vartheta \frac{\partial^2 \vartheta^*}{\partial x^2} dz \quad (7.1.22)$$

The first two integrals in the right-hand part of (7.1.22) turn to zero due to periodicity in x, therefore

$$\int\limits_0^{\bar{t}} dt \int\limits_0^l dx \int\limits_0^h \vartheta^* \frac{\partial^2 \vartheta}{\partial x^2} dz = \int\limits_0^{\bar{t}} dt \int\limits_0^l dx \int\limits_0^h \vartheta \frac{\partial^2 \vartheta^*}{\partial x^2} dz \quad (7.1.23)$$

Finally transform the last term:

$$\int_0^{\bar{t}} dt \int_0^l dx \int_0^h \vartheta^* \frac{\partial}{\partial z} \nu \frac{\partial \vartheta}{\partial z} dz = \int_0^{\bar{t}} dt \int_0^l \vartheta^* \left(\nu \frac{\partial \vartheta}{\partial z} \right) \Big|_{z=0}^{z=h} dx$$

$$- \int_0^{\bar{t}} dt \int_0^l \vartheta \left(\nu \frac{\partial \vartheta^*}{\partial z} \right) \Big|_{z=0}^{z=h} dx + \int_0^{\bar{t}} dt \int_0^l dx \int_0^h \vartheta \frac{\partial}{\partial z} \nu \frac{\partial \vartheta^*}{\partial z} dz \quad (7.1.24)$$

Introduce

$$\nu \frac{\partial \vartheta^*}{\partial z} = f^* \quad \text{at} \quad z = 0 \quad\quad (7.1.25)$$

and require that

$$\nu \frac{\partial \vartheta^*}{\partial z} = 0 \quad \text{at} \quad z = h \quad\quad (7.1.26)$$

Consider the function f^* as arbitrary for the time being. Its concrete form will be determined by the choice of the functional we are interested in. Therefore

$$\int_0^{\bar{t}} dt \int_0^l dx \int_0^h \vartheta^* \frac{\partial}{\partial z} \nu \frac{\partial \vartheta}{\partial z} dz = - \int_0^{\bar{t}} dt \int_0^l f(x,t) \vartheta^*(x,0,t) dx$$

$$+ \int_0^{\bar{t}} dt \int_0^l f^*(x,t) \vartheta(x,0,t) dx + \int_0^{\bar{t}} dt \int_0^l dx \int_0^h \vartheta \frac{\partial}{\partial z} \nu \frac{\partial \vartheta^*}{\partial z} dz$$

Represent expression (7.1.16), with regard to the last relationship and transformations (7.1.17), (7.1.21), and (7.1.23), as

$$\int_0^{\bar{t}} dt \int_0^l dx \int_0^h \vartheta \left(-\frac{\partial \vartheta^*}{\partial t} - u \frac{\partial \vartheta^*}{\partial x} - w \frac{\partial \vartheta^*}{\partial z} - \mu \frac{\partial^2 \vartheta^*}{\partial x^2} - \frac{\partial}{\partial z} \nu \frac{\partial \vartheta^*}{\partial z} - p^* \right) dz$$

$$+ \int_0^l dx \int_0^h g^*(x,z) \vartheta(x,z,\bar{t}) dz - \int_0^l dx \int_0^h g(x,z) \vartheta^*(x,z,0) dz$$

$$+ \int_0^{\bar{t}} dt \int_0^l f(x,t) \vartheta^*(x,0,t) dx - \int_0^{\bar{t}} dt \int_0^l f^*(x,t) \vartheta(x,0,t) dx$$

$$- \int_0^{\bar{t}} dt \int_0^l dx \int_0^h q\vartheta^* dz + \int_0^{\bar{t}} dt \int_0^l dx \int_0^h p^* \vartheta dz = 0 \quad (7.1.27)$$

Here, we subtracted and then added the term $\int\limits_0^{\bar{t}} dt \int\limits_0^l dx \int\limits_0^h p^*\vartheta dz$.

Assuming

$$\int\limits_0^{\bar{t}} dt \int\limits_0^l dx \int\limits_0^h \vartheta \left(-\frac{\partial \vartheta^*}{\partial t} - u\frac{\partial \vartheta^*}{\partial x} \right.$$

$$\left. -w\frac{\partial \vartheta^*}{\partial z} - \mu\frac{\partial^2 \vartheta^*}{\partial x^2} - \frac{\partial}{\partial z}\nu\frac{\partial \vartheta^*}{\partial z} - p^* \right) dz = 0 \quad (7.1.28)$$

with regard to the notation for J as in (7.1.15), obtain from (7.1.27) the formula of reciprocity for J:

$$J = \int\limits_0^l dx \int\limits_0^h g(x,z)\vartheta^*(x,z,0)dz - \int\limits_0^{\bar{t}} dt \int\limits_0^l f(x,t)\vartheta^*(x,0,t)dx$$

$$+ \int\limits_0^{\bar{t}} dt \int\limits_0^l dx \int\limits_0^h q\vartheta^* dz \quad (7.1.29)$$

Formulas (7.1.15) and (7.1.29) are thus equivalent, but computing the functional J with (7.1.15) requires the knowledge of the temperature field ϑ, while using (7.1.29) implies the knowledge of the adjoint function ϑ^*. Relationship (7.1.28) will be satisfied if

$$L^*\vartheta^* \equiv -\frac{\partial \vartheta^*}{\partial t} - u\frac{\partial \vartheta^*}{\partial x} - w\frac{\partial \vartheta^*}{\partial z} - \mu\frac{\partial^2 \vartheta^*}{\partial x^2} - \frac{\partial}{\partial z}\nu\frac{\partial \vartheta^*}{\partial z} = p^* \quad (7.1.30)$$

Add the boundary conditions introduced in the previous transformations. These are, first of all, the boundary conditions

$$\vartheta^*(0,z,t) = \vartheta^*(l,z,t)$$

$$\nu\frac{\partial \vartheta^*}{\partial z} = f^*(x,t) \quad \text{at} \quad z = 0$$

$$\nu\frac{\partial \vartheta^*}{\partial z} = 0 \quad \text{at} \quad z = h \quad (7.1.31)$$

and the "initial" data

$$\vartheta^* = g^* \quad \text{at} \quad t = \bar{t} \quad (7.1.32)$$

As concerns the set of functions $D(L^*)$, it is possible to prove, just like in the main problem, that the adjoint problem solves uniquely within a class of continuous functions quadratically summable together with their first and second derivatives in spatial variables. Additional conditions for these functions are determined by the boundary conditions including the conditions of periodicity.

Formulate now the perturbation theory. Assume that along unperturbed problem (7.1.1)–(7.1.4) there is a perturbed problem

$$\frac{\partial \vartheta'}{\partial t} + u'\frac{\partial \vartheta'}{\partial x} + w'\frac{\partial \vartheta'}{\partial z} - \mu\frac{\partial^2 \vartheta'}{\partial x^2} - \frac{\partial}{\partial z}\nu\frac{\partial \vartheta'}{\partial z} = q' \qquad (7.1.33)$$

with boundary conditions

$$\vartheta'(0, z, t) = \vartheta'(l, z, t)$$

$$\nu\frac{\partial \vartheta'}{\partial z} = f'(x, t) \quad \text{at} \quad z = 0$$
$$\nu\frac{\partial \vartheta'}{\partial z} = 0 \quad \text{at} \quad z = h \qquad (7.1.34)$$

and initial condition

$$\vartheta'(x, z, 0) = g'(x, z) \qquad (7.1.35)$$

Assume that components of the perturbed velocity vector also satisfy the continuity equation

$$\frac{\partial u'}{\partial x} + \frac{\partial w'}{\partial z} = 0 \qquad (7.1.36)$$

Examine problem (7.1.33)–(7.1.36) together with the unperturbed problem for the adjoint equation

$$L^*\vartheta^* \equiv -\frac{\partial \vartheta^*}{\partial t} - u\frac{\partial \vartheta^*}{\partial x} - w\frac{\partial \vartheta^*}{\partial z} - \mu\frac{\partial^2 \vartheta^*}{\partial x^2} - \frac{\partial}{\partial z}\nu\frac{\partial \vartheta^*}{\partial z} = p^* \qquad (7.1.37)$$

with the boundary conditions

$$\vartheta^*(0, z, t) = \vartheta^*(l, z, t)$$

$$\nu\frac{\partial \vartheta^*}{\partial z} = f^*(x, t) \quad \text{at} \quad z = 0$$
$$\nu\frac{\partial \vartheta^*}{\partial z} = 0 \quad \text{at} \quad z = h \qquad (7.1.38)$$

and initial data

$$\vartheta^*(x, z, \bar{t}) = g^*(x, z) \qquad (7.1.39)$$

Let $\vartheta \in D(L)$, $\vartheta' \in D(L)$ and $\vartheta^* \in D(L^*)$. Multiply equation (7.1.33) by ϑ^* and equation (7.1.37) by ϑ', subtract the results from each other and integrate all over the definition domain of the solution to obtain

$$\int_0^{\bar{t}} dt \int_0^l dx \int_0^h \vartheta^* \left(\frac{\partial \vartheta'}{\partial t} + u' \frac{\partial \vartheta'}{\partial x} + w' \frac{\partial \vartheta'}{\partial z} - \mu \frac{\partial^2 \vartheta'}{\partial x^2} - \frac{\partial}{\partial z} \nu \frac{\partial \vartheta'}{\partial z} - q' \right) dz$$

$$+ \int_0^{\bar{t}} dt \int_0^l dx \int_0^h \vartheta' \left(\frac{\partial \vartheta^*}{\partial t} + u \frac{\partial \vartheta^*}{\partial x} + w \frac{\partial \vartheta^*}{\partial z} \right.$$

$$\left. + \mu \frac{\partial^2 \vartheta}{\partial x^2} + \frac{\partial}{\partial z} \nu \frac{\partial \vartheta^*}{\partial z} + p^* \right) dz = 0 \quad (7.1.40)$$

Transform the first integral in (7.1.40) just like in (7.1.16), where instead of ϑ, u, w, f, g, and q there are ϑ', u', w', f', g', and q'. As a result,

$$\int_0^{\bar{t}} dt \int_0^l dx \int_0^h \vartheta^* \left(\frac{\partial \vartheta'}{\partial t} + u' \frac{\partial \vartheta'}{\partial x} + w' \frac{\partial \vartheta'}{\partial z} - \mu \frac{\partial^2 \vartheta'}{\partial x^2} - \frac{\partial}{\partial z} \nu \frac{\partial \vartheta'}{\partial z} - q' \right) dz$$

$$= \int_0^{\bar{t}} dt \int_0^l dx \int_0^h \vartheta' \left(-\frac{\partial \vartheta^*}{\partial t} - u' \frac{\partial \vartheta^*}{\partial x} - w' \frac{\partial \vartheta^*}{\partial z} - \mu \frac{\partial^2 \vartheta^*}{\partial x^2} - \frac{\partial}{\partial z} \nu \frac{\partial \vartheta^*}{\partial z} - p^* \right) dz$$

$$+ \int_0^l dx \int_0^h g^*(x, z) \vartheta'(x, z, \bar{t}) dz - \int_0^l dx \int_0^{\bar{t}} f^*(x, t) \vartheta'(x, 0, t) dt$$

$$- \int_0^l dx \int_0^h g'(x, z) \vartheta^*(x, z, 0) dz + \int_0^l dx \int_0^{\bar{t}} f'(x, t) \vartheta^*(x, 0, t) dt$$

$$+ \int_0^{\bar{t}} dt \int_0^l dx \int_0^h p^* \vartheta' dz - \int_0^{\bar{t}} dt \int_0^l dx \int_0^h q' \vartheta^* dz \quad (7.1.41)$$

From (7.1.40) and (7.1.41),

$$\int_0^{\bar{t}} dt \int_0^l dx \int_0^h \vartheta' \left(-(u'-u)\frac{\partial \vartheta^*}{\partial x} - (w'-w)\frac{\partial \vartheta^*}{\partial z} \right) dz$$

$$+ \int_0^l dx \int_0^h g^*(x,z)\vartheta'(x,z,\bar{t})dz - \int_0^l dx \int_0^{\bar{t}} f^*(x,t)\vartheta'(x,0,t)dt$$

$$- \int_0^l dx \int_0^h g'(x,z)\vartheta^*(x,z,0)dz + \int_0^l dx \int_0^{\bar{t}} f'(x,t)\vartheta^*(x,0,t)dt$$

$$+ \int_0^{\bar{t}} dt \int_0^l dx \int_0^h p^*\vartheta'dz - \int_0^{\bar{t}} dt \int_0^l dx \int_0^h q'\vartheta^*dz = 0 \quad (7.1.42)$$

Assume that

$$J' = \int_0^l dx \int_0^h g^*(x,z)\vartheta'(x,z,\bar{t})dz - \int_0^l dx \int_0^{\bar{t}} f^*(x,z)\vartheta'(x,0,t)dt$$

$$+ \int_0^{\bar{t}} dt \int_0^l dx \int_0^h p^*\vartheta'dz \quad (7.1.43)$$

With regard to expression for J in (7.1.15), obtain

$$J' = J + \delta J, \quad (7.1.44)$$

where

$$\delta J = \int_0^l dx \int_0^h g^*(x,z)\delta\vartheta(x,z,\bar{t})dz$$

$$- \int_0^l dx \int_0^{\bar{t}} f^*(x,t)\delta\vartheta(x,0,t)dt + \int_0^{\bar{t}} dt \int_0^l dx \int_0^h p^*\delta\vartheta dz, \quad \delta\vartheta = \vartheta' - \vartheta$$

Next, use dual formula (7.1.29) to obtain

$$\int_0^l dx \int_0^h g'(x,z)\vartheta^*(x,z,0)dz - \int_0^l dx \int_0^t f'(x,t)\vartheta^*(x,0,t)dt$$

$$+ \int_0^{\bar{t}} dt \int_0^l dx \int_0^h q'\vartheta^*dz = J + \int_0^l dx \int_0^h \delta g\vartheta^*(x,z,0)dz$$

$$- \int_0^l dx \int_0^{\bar{t}} \delta f\vartheta^*(x,0,t)dt + \int_0^{\bar{t}} dt \int_0^l dx \int_0^h \delta q\vartheta^*dz, \quad (7.1.45)$$

where $\delta g = g' - g$, $\delta f = f' - f$, and $\delta q = q' - q$. By (7.1.42)–(7.1.45) and assuming that $u' = u + \delta u$, $w' = w + \delta w$, simple transformations produce the perturbation theory formula

$$\delta J = \int\limits_0^{\bar{t}} dt \int\limits_0^l dx \int\limits_0^h \left(\delta u \vartheta' \frac{\partial \vartheta^*}{\partial x} + \delta w \vartheta' \frac{\partial \vartheta^*}{\partial z} \right) dz$$

$$+ \int\limits_0^l dx \int\limits_0^h \delta g \vartheta^*(x,z,0) dz - \int\limits_0^l dx \int\limits_0^{\bar{t}} \delta f \vartheta^*(x,0,t) dt$$

$$+ \int\limits_0^{\bar{t}} dt \int\limits_0^l dx \int\limits_0^h \delta q \vartheta^* dz \quad (7.1.46)$$

If the perturbations $\delta \vartheta$ in (7.1.46) are small, ϑ' may be replaced everywhere with ϑ, which leads to the formula of small perturbation theory

$$\delta J = \int\limits_0^{\bar{t}} dt \int\limits_0^l dx \int\limits_0^h \left(\delta u \vartheta \frac{\partial \vartheta^*}{\partial x} + \delta w \vartheta \frac{\partial \vartheta^*}{\partial z} \right) dz + \int\limits_0^l dx \int\limits_0^h \delta g \vartheta^*(x,z,0) dz$$

$$- \int\limits_0^l dx \int\limits_0^{\bar{t}} \delta f \vartheta^*(x,0,t) dt + \int\limits_0^{\bar{t}} dt \int\limits_0^l dx \int\limits_0^h \delta q \vartheta^* dz \quad (7.1.47)$$

We have not yet fixed a concrete form of the functional J and assumed simply that

$$J = \int\limits_0^l dx \int\limits_0^h g^*(x,z) \vartheta(x,z,\bar{t}) dz$$

$$- \int\limits_0^l dx \int\limits_0^{\bar{t}} f^*(x,t) \vartheta(x,0,t) dt + \int\limits_0^{\bar{t}} dt \int\limits_0^l dx \int\limits_0^h p^* \vartheta dz \quad (7.1.48)$$

Assume now that we want the average temperature over the area $x_1 \leq x \leq x_2$ at the surface $z = 0$ for time interval $\bar{t} - \tau \leq t \leq \bar{t}$. Then the functional is

$$J = \frac{1}{\tau \Delta x} \int\limits_{\bar{t}-\tau}^{\bar{t}} dt \int\limits_{x_1}^{x_2} \vartheta(x,0,t) dx \quad (7.1.49)$$

where $\Delta x = x_2 - x_1$. This means that we have to select the "measurement functions" g^*, f^* and p^* as

$$g^*(x, z) = 0, \quad p^* = 0$$

$$f^*(x, t) = \begin{cases} \dfrac{1}{\tau \Delta x}, & \text{at} \quad \bar{t} - \tau \le t \le \bar{t} \quad x_1 \le x \le x_2 \\ 0, & \text{otherwise} \end{cases} \qquad (7.1.50)$$

Solving adjoint problem (7.1.37)–(7.1.39) with a "source" f^* as in (7.1.50) produces the function ϑ^* which reflects the sensitivity of the variation of the functional δJ assigned by formula (7.1.46), to variations in the input data of the main problem.

7.2. Temperature Anomalies in the Atmosphere and Ocean

We examined in § 7.1 a simple model and the main and adjoint problems in the atmosphere assuming the heat fluxes assigned at the bottom boundary of the atmosphere at $z = 0$. A statement like this does not help in determining the most important regions in the ocean where heat accumulates to be transported by currents to zones of intensive heat fluxes into the atmosphere. We just obtained an integral function of the heat flux f as a generalized result in the problem of interaction between the atmosphere and the ocean.

The initial mathematical model has to be therefore generalized, once the atmosphere and ocean are considered as a coupled system. Examine thus the following model problem:

$$L\vartheta \equiv \alpha \frac{\partial \vartheta}{\partial t} + \alpha u \frac{\partial \vartheta}{\partial x} + \alpha w \frac{\partial \vartheta}{\partial z} - \bar{\mu} \frac{\partial^2 \vartheta}{\partial x^2} - \frac{\partial}{\partial z} \bar{\nu} \frac{\partial \vartheta}{\partial z} = q(x, z, t) \qquad (7.2.1)$$
$$x \in (0, l), \quad z \in (-h_0, h), \quad t \in (0, \bar{t})$$

with conditions

$$\vartheta(0, z, t) = \vartheta(l, z, t)$$
$$\bar{\nu} \frac{\partial \vartheta}{\partial z} = 0 \quad \text{at} \quad z = h, \quad z = -h_0 \qquad (7.2.2)$$

and initial data

$$\vartheta(x, z, 0) = g(x, z) \quad \text{at} \quad t = 0 \qquad (7.2.3)$$

Here $\alpha = c_p \rho$, c_p, specific heat of a madium, $\rho = \rho(z)$, standard density, are different for atmosphere and ocean; $\bar{\mu} = \alpha \mu$, $\bar{\nu} = \alpha \nu$.

As before, assume that the vector of the velocity of currents in the atmosphere at $z \geq 0$ and in the ocean at $z < 0$ satisfies the continuity equation

$$\frac{\partial \alpha u}{\partial x} + \frac{\partial \alpha w}{\partial z} = 0 \quad \text{at any} \quad (x, z, t) \tag{7.2.4}$$

and, besides,

$$w = 0 \quad \text{at} \quad z = h \quad \text{and} \quad z = -h_0 \tag{7.2.5}$$

To make it simpler, assume that the functions u, w, μ and ν are continuous, though they change rapidly at the crossing of the surface of the ocean.

Since the function α is discontinuous in z at $z = 0$, the solution ϑ and its derivatives are not differentiable functions at the dividing surface "atmosphere–ocean"; therefore, it is necessary to require that the following additional boundary conditions in the neighborhood of $z = 0$ are to be satisfied:

$$\vartheta|_{z=-0} = \vartheta|_{z=+0}, \quad \bar{\nu} \frac{\partial \vartheta}{\partial z}\bigg|_{z=-0} = \bar{\nu} \frac{\partial \vartheta}{\partial z}\bigg|_{z=+0}$$

This model problem supposes that all the bottom surface of the atmosphere contacts the ocean of an equal depth of h_0.

Examine now the adjoint problem. Assume that the solution of the main problem belongs to a set $D(L)$ of functions periodic in x and that the properties of the set in the assigned conditions related to the equation coefficients and other input parameters imply a unique solution. This set $D(L)$ may be determined as discussed in § 7.1.

Multiply equation (7.2.1) by the function ϑ^* in the set $D(L^*)$ (properties of this one will be defined later), and integrate all over the definition domain to obtain

$$\int_0^{\bar{t}} dt \int_0^l dx \int_{-h_0}^h \vartheta^* \left(\alpha \frac{\partial \vartheta}{\partial t} + \alpha u \frac{\partial \vartheta}{\partial x} + \alpha w \frac{\partial \vartheta}{\partial z} - \bar{\mu} \frac{\partial^2 \vartheta}{\partial x^2} - \frac{\partial}{\partial z} \bar{\nu} \frac{\partial \vartheta}{\partial z} \right) dz$$

$$- \int_0^{\bar{t}} dt \int_0^l dx \int_{-h_0}^h q(x, z, t) \vartheta^*(x, z, t) dz = 0 \tag{7.2.6}$$

Integrate (7.2.6) by parts with regards to above assumptions on the solution ϑ, boundary conditions (7.2.2) and initial data (7.2.3), into

$$\int_{\iota}^{\bar{\iota}} dt \int_{0}^{l} dx \int_{-h_0}^{h} \vartheta \left(-\alpha \frac{\partial \vartheta^*}{\partial t} - \alpha u \frac{\partial \vartheta^*}{\partial x} - \alpha w \frac{\partial \vartheta^*}{\partial z} - \bar{\mu} \frac{\partial^2 \vartheta^*}{\partial x^2} - \frac{\partial}{\partial z} \bar{\nu} \frac{\partial \vartheta^*}{\partial z} \right) dz$$

$$- \int_{0}^{\bar{\iota}} dt \int_{0}^{l} dx \int_{-h_0}^{h} q(x,z,t)\vartheta^*(x,z,t)dz + \int_{0}^{l} dx \int_{-h_0}^{h} \alpha g^*(x,z)\vartheta(x,z,\bar{t})dz$$

$$- \int_{0}^{l} dx \int_{-h_0}^{h} \alpha g(x,z)\vartheta^*(x,z,0)dz = 0 \quad (7.2.7)$$

We used here the necessary smoothness and periodicity of ϑ^* and introduced the notation

$$\vartheta^* = g^*(x,z) \quad \text{at} \quad t = \bar{t} \quad (7.2.8)$$

Besides, we assumed, as we derived (7.2.7) the conditions

$$\bar{\nu} \frac{\partial \vartheta^*}{\partial z} = 0 \quad \text{at} \quad z = h$$

$$\bar{\nu} \frac{\partial \vartheta^*}{\partial z} = 0 \quad \text{at} \quad z = -h_0$$

are satisfied.

Introduce the main functional

$$J = \int_{0}^{l} dx \int_{-h_0}^{h} \alpha g^*(x,z)\vartheta(x,z,\bar{t})dz + \int_{0}^{\bar{\iota}} dt \int_{0}^{l} dx \int_{-h_0}^{h} p^*\vartheta dz \quad (7.2.9)$$

and assume

$$L^*\vartheta^* \equiv -\alpha \frac{\partial \vartheta^*}{\partial t} - \alpha u \frac{\partial \vartheta^*}{\partial x} - \alpha w \frac{\partial \vartheta^*}{\partial z} - \bar{\mu} \frac{\partial^2 \vartheta^*}{\partial x^2} - \frac{\partial}{\partial z} \bar{\nu} \frac{\partial \vartheta^*}{\partial z} = p^* \quad (7.2.10)$$

Then rewrite (7.2.7) as

$$\int\limits_0^{\bar{t}} dt \int\limits_0^l dx \int\limits_{-h_0}^h p^* \vartheta \, dz + \int\limits_0^l dx \int\limits_{-h_0}^h \alpha g^*(x,z)\vartheta(x,z,\bar{t}) dz$$

$$= \int\limits_0^l dx \int\limits_{-h_0}^h \alpha g \vartheta^*(x,z,0) dz + \int\limits_0^{\bar{t}} dt \int\limits_0^l dx \int\limits_{-h_0}^h q(x,z,t)\vartheta^*(x,z,t) dz \quad (7.2.11)$$

With regard to (7.2.9), the dual formula for the functional J is:

$$J = \int\limits_0^l dx \int\limits_{-h_0}^h \alpha g(x,z)\vartheta^*(x,z,0) dz$$

$$+ \int\limits_0^{\bar{t}} dt \int\limits_0^l dx \int\limits_{-h_0}^h q(x,z,t)\vartheta^*(x,z,t) dz \quad (7.2.12)$$

The functional J assigned by formula (7.2.12) is thus expressed by the initial fields of the function ϑ i. e., $g(x,z)$, and the radiation flux at the ocean surface $q(x,t)$, both taken with respective weights, which are the values of adjoint functions $\vartheta^*(x,z,0)$ and $\vartheta^*(x,0,t)$ respectively.

Summing up this section, we arrive at the adjoint problem:

$$L^*\vartheta^* = -\alpha\frac{\partial\vartheta^*}{\partial t} - \alpha u\frac{\partial\vartheta^*}{\partial x} - \alpha w\frac{\partial\vartheta^*}{\partial z} - \bar{\mu}\frac{\partial^2\vartheta^*}{\partial x^2} - \frac{\partial}{\partial z}\bar{\nu}\frac{\partial\vartheta^*}{\partial z} = p^* \quad (7.2.13)$$

with the boundary conditions

$$\vartheta^*(0,z,t) = \vartheta^*(l,z,t)$$

$$\bar{\nu}\frac{\partial\vartheta^*}{\partial z} = 0 \quad \text{at} \quad z = h, \quad \text{at} \quad z = -h_0 \quad (7.2.14)$$

and initial data

$$\vartheta^*(x,z,\bar{t}) = g^*(x,z) \quad (7.2.15)$$

It is necessary to set additional boundary conditions at dividing surface "atmosphere–ocean" ($z = 0$):

$$\vartheta^*|_{z=-0} = \vartheta^*|_{z=+0}, \quad \bar{\nu}\frac{\partial\vartheta^*}{\partial z}\bigg|_{z=-0} = \bar{\nu}\frac{\partial\vartheta^*}{\partial z}\bigg|_{z=+0}$$

Select, for example, the functions $g^*(x,z) = 0$ and

$$p^*(x,z,t) = \begin{cases} \dfrac{1}{\tau h \Delta x}, & \text{at } \bar{t}-\tau \leq t \leq \bar{t}, \ x_1 \leq x \leq x_2, \ 0 \leq z \leq h \\ 0, & \text{otherwise} \end{cases} \qquad (7.2.16)$$

Then

$$J = \frac{1}{\tau h \Delta x} \int\limits_{\bar{t}-\tau}^{\bar{t}} dt \int\limits_{x_1}^{x_2} dx \int\limits_{0}^{h} \vartheta(x,z,t)dz \qquad (7.2.17)$$

and functional (7.2.17) describes the average temperature in the region $-\tau \leq t \leq \bar{t}, \quad x_1 \leq x \leq x_2, \quad 0 \leq z \leq h.$

Formula (7.2.12) is applicable after the solution ϑ^* of adjoint problem (7.2.13)–(7.2.15) is found, where the function p^* in (7.2.16) will serve as the "source" in the right-hand part of equation (7.2.13). Numerical solution of such problem is exemplified in Appendix II.

7.3. Adjoint Functions of Atmosphere Dynamics

Examine a system of equations governing the dynamics of atmospheric processes in adiabatic approximation and the structure of the operator of the problem, e. g., a simple case of a barotropic atmosphere:

$$\frac{\partial u}{\partial t} + \bar{u}\frac{\partial u}{\partial x} + \bar{v}\frac{\partial u}{\partial y} - lv + R\overline{T}\frac{\partial \varphi}{\partial x} = 0$$

$$\frac{\partial v}{\partial t} + \bar{u}\frac{\partial v}{\partial x} + \bar{v}\frac{\partial v}{\partial y} + lu + R\overline{T}\frac{\partial \varphi}{\partial y} = 0 \qquad (7.3.1)$$

$$\frac{\partial u}{\partial x} + \frac{\partial v}{\partial y} = 0$$

Assume that $\bar{u}$ and $\bar{v}$ are known due to linearization and satisfy the continuity equation $\partial\bar{u}/\partial x + \partial\bar{v}/\partial y = 0$, $R\overline{T} = const$, while $\varphi(x,y,t)$ is a deviation from the standard pressure. Assume also that a square Ω is the definition domain of the solution and periodicity conditions are set at the boundary $\partial\Omega$. Introduce a solution vector and a matrix

$$\varphi = \begin{bmatrix} u \\ v \\ R\overline{T}\varphi \end{bmatrix}, \quad \mathbf{A} = \begin{bmatrix} \Lambda & -l & \partial/\partial x \\ l & \Lambda & \partial/\partial y \\ \partial/\partial x & \partial/\partial y & 0 \end{bmatrix}$$

where l is Coriolis' parameter,

$$\Lambda = \frac{\partial}{\partial x}\bar{u} \cdot + \frac{\partial}{\partial y}\bar{v}\cdot$$

Use this notation to obtain

$$\Lambda u = \mathrm{div}\,\bar{\mathbf{u}}u, \quad \Lambda v = \mathrm{div}\,\bar{\mathbf{u}}v, \quad \bar{\mathbf{u}} = \left(\begin{array}{c} \bar{u} \\ \bar{v} \end{array}\right)$$

Rewrite system (7.4.1) in the operator form:

$$\mathbf{L}\,\boldsymbol{\varphi} \equiv \mathbf{B}\frac{\partial\boldsymbol{\varphi}}{\partial t} + \mathbf{A}\boldsymbol{\varphi} = 0 \qquad (7.3.2)$$

where $\mathbf{B}$ is the matrix

$$\mathbf{B} = \left[\begin{array}{ccc} 1 & 0 & 0 \\ 0 & 1 & 0 \\ 0 & 0 & 0 \end{array}\right]$$

The solution $\boldsymbol{\varphi}$ is assumed to belong to a set of functions continuously differentiable in all arguments and satisfying the periodicity conditions.

Introduce the inner product in a real Hilbert space $\mathbf{H}$ at an arbitrary $t \in [0, \bar{t}]$ as

$$(\mathbf{g}, \mathbf{h})_\Omega = \sum_{i=1}^{3} \int_\Omega g_i h_i d\Omega$$

where g_i and h_i are respectively components of the vector-functions $\mathbf{g}$ and $\mathbf{h}$ from $\mathbf{H}$.

Find now an operator adjoint to $\mathbf{A}$ by the Lagrange identity

$$(\mathbf{A}\mathbf{h}, \mathbf{g})_\Omega = (\mathbf{h}, \mathbf{A}^*\mathbf{g})_\Omega$$

or

$$(\mathbf{A}\mathbf{h}, \mathbf{g})_\Omega = \int_\Omega \left[\left(\Lambda u + lv + R\bar{T}\frac{\partial\varphi}{\partial x}\right) u^* + \left(lu + \Lambda v + R\bar{T}\frac{\partial\varphi}{\partial y}\right) v^* \right.$$

$$\left. + \left(\frac{\partial u}{\partial x} + \frac{\partial v}{\partial y}\right) R\bar{T}\varphi^*\right] d\Omega \quad (7.3.3)$$

where

$$
\mathbf{h} = \begin{bmatrix} u \\ v \\ R\overline{T}\varphi \end{bmatrix}, \quad \mathbf{g} = \begin{bmatrix} u^* \\ v^* \\ R\overline{T}\varphi^* \end{bmatrix}
$$

It is assumed in (7.3.3) that $\mathbf{h}$ belongs to $D(\mathbf{A})$, which is the definition domain of the operator $\mathbf{A}$, at any $t \in [0, \overline{t}]$. Assume that the functions belonging to the set $D(\mathbf{A})$ are smooth as required so that all the following transformations in the domain Ω are valid. For example, we may take a set of vector-functions continuously differentiable in all the spatial variables and satisfying periodicity conditions for $D(\mathbf{A})$.

Integration by parts with periodicity of solutions assumed brings integral in the right-hand part of (7.3.3) to

$$
(\mathbf{Ah}, \mathbf{g})_\Omega = \int_\Omega \left[\left(\Lambda^* u^* + l v^* - R\overline{T}\frac{\partial \varphi^*}{\partial x} \right) u + \left(-l u^* + \Lambda^* v^* - R\overline{T}\frac{\partial \varphi^*}{\partial y} \right) v \right.
$$

$$
\left. - \left(\frac{\partial u^*}{\partial x} + \frac{\partial v^*}{\partial y} \right) R\overline{T}\varphi \right] d\Omega = (\mathbf{h}, \mathbf{A}^*\mathbf{g})_\Omega \quad (7.3.4)
$$

where

$$
\Lambda^* = - \left(\frac{\partial}{\partial x}\overline{u} \cdot + \frac{\partial}{\partial y}\overline{v} \cdot \right) = - \left(\overline{u}\frac{\partial \cdot}{\partial x} + \overline{v}\frac{\partial \cdot}{\partial y} \right) = -\Lambda \quad (7.3.5)
$$

Notice that $\overline{u}$ and $\overline{v}$ satisfy continuity equation $\partial \overline{u}/\partial x + \partial \overline{v}/\partial y = 0$. With regard to (7.3.4) and (7.3.5), obtain

$$
\varphi^* = \begin{bmatrix} u^* \\ v^* \\ R\overline{T}\varphi^* \end{bmatrix}, \quad \mathbf{A}^* = \begin{bmatrix} -\Lambda & l & -\partial/\partial x \\ -l & -\Lambda & -\partial/\partial y \\ -\partial/\partial x & -\partial/\partial y & 0 \end{bmatrix} = -\mathbf{A}
$$

The function φ^* belongs to the set $D(\mathbf{A}^*)$, which is the definition domain of the operator $\mathbf{A}^*$, at any $t \in [0, \overline{t}]$. Properties of the elements of $D(\mathbf{A}^*)$ are easily obtained by analysis of conditions necessary for transformations producing the operator $\mathbf{A}^*$ from $\mathbf{A}$. We may assume, in particular, that $D(\mathbf{A}) = D(\mathbf{A}^*)$.

We have so far assumed that $\overline{u}$ and $\overline{v}$ are assigned functions of x, y, and time. This assumption seriously restricted constructing the theory of adjoint problems. This author constructed the adjoint equations for hydrodynamics problems in the monograph [70] in 1974. The essence of the method is this: assume that we are dealing with a quasi-linear system

$$\frac{\partial u}{\partial t} + u\frac{\partial u}{\partial x} + v\frac{\partial u}{\partial y} - lv + R\overline{T}\frac{\partial \varphi}{\partial x} = 0$$

$$\frac{\partial v}{\partial t} + u\frac{\partial v}{\partial x} + v\frac{\partial v}{\partial y} + lu + R\overline{T}\frac{\partial \varphi}{\partial y} = 0 \qquad (7.3.6)$$

$$\frac{\partial u}{\partial x} + \frac{\partial v}{\partial y} = 0$$

and have found the solution to this system with periodicity conditions at the boundary, while the initial data are

$$u = u_0, \quad v = v_0, \quad \text{at} \quad t = 0 \qquad (7.3.7)$$

Consider the functions u and v obtained in the process of solution as coefficients in nonlinear operators $\mathbf{A}$ and $\mathbf{A}^*$ to obtain

$$\mathbf{A} = \begin{bmatrix} \Lambda & -l & \partial/\partial x \\ l & \Lambda & \partial/\partial y \\ \partial/\partial x & \partial/\partial y & 0 \end{bmatrix}, \quad \mathbf{A}^* = -\mathbf{A}$$

where Λ is now the operator

$$\Lambda = \frac{\partial}{\partial x}u \cdot + \frac{\partial}{\partial y}v \cdot$$

Add to problem (7.4.6) the adjoint problem

$$-\frac{\partial u^*}{\partial t} - u\frac{\partial u^*}{\partial x} - v\frac{\partial u^*}{\partial y} + lv^* - R\overline{T}\frac{\partial \varphi^*}{\partial x} = 0$$

$$-\frac{\partial v^*}{\partial t} - u\frac{\partial v^*}{\partial x} - v\frac{\partial v^*}{\partial y} - lu^* - R\overline{T}\frac{\partial \varphi^*}{\partial y} = 0 \qquad (7.3.8)$$

$$-\frac{\partial u^*}{\partial x} - \frac{\partial v^*}{\partial y} = 0$$

while

$$u^* = u_{\bar{t}}^*, \quad v^* = v_{\bar{t}}^* \quad \text{at} \quad t = \bar{t} \qquad (7.3.9)$$

Rewrite problems (7.3.6) and (7.3.7), and (7.3.8) and (7.3.9) in operator form:

$$\mathbf{L}\varphi \equiv \mathbf{B}\frac{\partial\varphi}{\partial t} + \mathbf{A}\varphi = 0, \quad \mathbf{B}\varphi = \mathbf{B}\varphi_0 \quad \text{at} \quad t = 0 \qquad (7.3.10)$$

$$\mathbf{L}^*\varphi^* \equiv -\mathbf{B}\frac{\partial\varphi^*}{\partial t} - \mathbf{A}\varphi^* = 0, \quad \mathbf{B}\varphi^* = \mathbf{B}\varphi_{\bar{t}}^* \quad \text{at} \quad t = \bar{t} \qquad (7.3.11)$$

Subtract from the inner product of (7.3.10) and φ^*, that of (7.3.11) and φ to obtain

$$\frac{d}{dt}(\mathbf{B}\varphi, \varphi^*) = 0 \qquad (7.3.12)$$

Integrate this at the assigned conditions $t = 0$ and $t = \bar{t}$ to obtain

$$(\mathbf{B}\varphi_{\bar{t}}, \varphi_{\bar{t}}^*)_\Omega = (\mathbf{B}\varphi_0, \varphi_0^*)_\Omega \qquad (7.3.13)$$

which we will use later, and now rewrite it component-wise:

$$\int_\Omega (u_{\bar{t}}u_{\bar{t}}^* + v_{\bar{t}}v_{\bar{t}}^*)d\Omega = \int_\Omega (u_0 u_0^* + v_0 v_0^*)d\Omega \qquad (7.3.14)$$

Notice that if we choose $u_{\bar{t}}$ and $v_{\bar{t}}$ for $u_{\bar{t}}^*$ and $v_{\bar{t}}^*$, we arrive at the kinetic energy conservation law:

$$\int_\Omega E_{\bar{t}}d\Omega = \int_\Omega E_0 d\Omega$$

The solution is then totally reversible. It means that once problem 7.3.6) and (7.3.7) is solved and assumed that $u_{\bar{t}}^* = u_{\bar{t}}$, $v_{\bar{t}}^* = v_{\bar{t}}$, problem 7.3.8) and (7.3.9) may be solved backward in time. This results in the same solutions of the main system as obtained by solving the main problem.

To conclude, prove that it is preferable sometimes to use a more common real phase space Φ with an inner product

$$(\mathbf{g}, \mathbf{h})_{\Omega \times \Omega_t} = \sum_{i=1}^{3} \int_\Omega d\Omega \int_0^{\bar{t}} g_i h_i dt$$

Introduce operators

$$\mathbf{L} = \mathbf{B}\frac{\partial}{\partial t} + \mathbf{A}, \quad \mathbf{L}^* = -\mathbf{B}\frac{\partial}{\partial t} - \mathbf{A}$$

which operate in the Hilbert space Φ with the definition domains $D(\mathbf{L})$, $D(\mathbf{L}^*)$. It is possible to assume, for example, $D(\mathbf{L}) = D(\mathbf{L}^*)$ as a set of functions continuously differentiable in all the variables and satisfying the periodicity conditions. Verify that

$$(\mathbf{L}\boldsymbol{\varphi}, \boldsymbol{\varphi}^*)_{\Omega \times \Omega_t} = (\boldsymbol{\varphi}, \mathbf{L}^*\boldsymbol{\varphi}^*)_{\Omega \times \Omega_t} - (\mathbf{B}\boldsymbol{\varphi}_{\bar{t}}, \boldsymbol{\varphi}_{\bar{t}}^*)_\Omega + (\mathbf{B}\boldsymbol{\varphi}_0, \boldsymbol{\varphi}_0^*)_\Omega \quad (7.3.15)$$

holds. Given (7.3.13), finally obtain

$$(\mathbf{L}\boldsymbol{\varphi}, \boldsymbol{\varphi}^*)_{\Omega \times \Omega_t} = (\boldsymbol{\varphi}, \mathbf{L}^*\boldsymbol{\varphi}^*)_{\Omega \times \Omega_t} \qquad (7.3.16)$$

where

$$\mathbf{L}^* = -\mathbf{L}$$

Examine now a system of the main equations with viscosity, i. e., let

$$\frac{\partial u}{\partial t} + u\frac{\partial u}{\partial x} + v\frac{\partial u}{\partial y} - lv + R\overline{T}\frac{\partial \varphi}{\partial x} - \mu\Delta u = 0$$

$$\frac{\partial v}{\partial t} + u\frac{\partial v}{\partial x} + v\frac{\partial v}{\partial y} + lu + R\overline{T}\frac{\partial \varphi}{\partial y} - \mu\Delta v = 0 \qquad (7.3.17)$$

$$\frac{\partial u}{\partial x} + \frac{\partial v}{\partial y} = 0$$

while

$$u = u_0, \quad v = v_0 \quad \text{at} \quad t = 0 \qquad (7.3.18)$$

where the solutions are assumed as periodic. Then, using the method discussed above and assuming that

$$\Lambda = u\frac{\partial}{\partial x} + v\frac{\partial}{\partial y} - \mu\Delta, \quad \Lambda^* = -u\frac{\partial}{\partial x} - v\frac{\partial}{\partial y} - \mu\Delta$$

obtain a system of adjoint equations

$$-\frac{\partial u^*}{\partial t} - u\frac{\partial u^*}{\partial x} - v\frac{\partial u^*}{\partial y} + lv^* - R\overline{T}\frac{\partial \varphi^*}{\partial x} - \mu\Delta u^* = 0$$

$$-\frac{\partial v^*}{\partial t} - u\frac{\partial v^*}{\partial x} - v\frac{\partial v^*}{\partial y} - lu^* - R\overline{T}\frac{\partial \varphi^*}{\partial y} - \mu\Delta v^* = 0 \qquad (7.3.19)$$

$$-\frac{\partial u^*}{\partial x} - \frac{\partial v^*}{\partial y} = 0$$

while
$$u^* = u_{\bar{t}}^*, \quad v^* = v_{\bar{t}}^* \quad \text{at} \quad t = \bar{t} \qquad (7.3.20)$$

The analysis of problems (7.3.17), (7.3.18) and (7.3.19), (7.3.20) proves that the main problem must be solved down the time within the interval $0 \leq t \leq \bar{t}$ and the adjoint problem up the time within the interval $\bar{t} \geq t \geq 0$. This way of calculation will only be correct for each problem because of dissipation terms present in the equations. The meaning of introduction of adjoint problems will be clear in the analysis of perturbation theory formulas.

7.4. Adjoint Functions for Baroclinic Atmosphere

Examine a model of the baroclinic atmosphere in adiabatic approximation:

$$\frac{\partial \bar{\rho} u}{\partial t} + \Lambda u - l\bar{\rho}v + \bar{p}\frac{\partial \varphi}{\partial x} = 0$$

$$\frac{\partial \bar{\rho} v}{\partial t} + \Lambda v + l\bar{\rho}u + \bar{p}\frac{\partial \varphi}{\partial y} = 0$$

$$-g\bar{\rho}\vartheta + \bar{p}\frac{\partial \varphi}{\partial z} = 0 \qquad (7.4.1)$$

$$\frac{\partial \bar{\rho} u}{\partial x} + \frac{\partial \bar{\rho} v}{\partial y} + \frac{\partial \bar{\rho} w}{\partial z} = 0$$

$$\frac{\partial \bar{\rho} \vartheta}{\partial t} + \Lambda \vartheta + \frac{\gamma_a - \gamma}{T}\bar{\rho}w = 0$$

while
$$\bar{\rho}w = 0 \quad \text{at} \quad z = 0$$
$$\bar{\rho}w = 0 \quad \text{at} \quad z = H \qquad (7.4.2)$$

Assume the solution to be periodic in a plane (x, y) and satisfying the initial conditions

$$u = u_0, \quad v = v_0, \quad \vartheta = \vartheta_0 \quad \text{at} \quad t = 0 \qquad (7.4.3)$$

Assume that $R\bar{T} = const$, $(\gamma_a - \gamma)/\bar{T} = const$, $\bar{p} = R\bar{\rho}\bar{T}$. Determine the operator Λ with the formula

$$\Lambda = \frac{\partial}{\partial x}\bar{\rho}u + \frac{\partial}{\partial y}\bar{\rho}v + \frac{\partial}{\partial z}\bar{\rho}w$$

Therefore,

$$\Lambda u = \mathrm{div}\bar{\rho}\mathbf{u}u, \quad \Lambda v = \mathrm{div}\bar{\rho}\mathbf{u}v, \quad \Lambda\vartheta = \mathrm{div}\bar{\rho}\mathbf{u}\vartheta$$

Introduce vector-functions and matrices

$$\varphi = \begin{bmatrix} u \\ v \\ w \\ \varphi \\ \vartheta \end{bmatrix}, \quad \mathbf{A} = \begin{bmatrix} \Lambda & -\bar{\rho}l & 0 & \bar{p}(\partial/\partial x) & 0 \\ \bar{\rho}l & \Lambda & 0 & \bar{p}(\partial/\partial y) & 0 \\ 0 & 0 & 0 & \bar{p}\dfrac{\partial}{\partial z} & -g\bar{\rho} \\ \dfrac{\partial}{\partial x}\bar{p} & \dfrac{\partial}{\partial y}\bar{p} & \dfrac{\partial}{\partial z}\bar{p} & 0 & 0 \\ 0 & 0 & g\bar{\rho} & 0 & \dfrac{\overline{T}g}{\gamma_a - \gamma}\Lambda \end{bmatrix}$$

$$\mathbf{B}\bar{\varphi}_0 = \begin{bmatrix} \bar{\rho}u_0 \\ \bar{\rho}v_0 \\ 0 \\ 0 \\ \dfrac{\overline{T}g}{\gamma_a - \gamma}\bar{\rho}\vartheta_0 \end{bmatrix}, \quad \mathbf{B} = \begin{bmatrix} \bar{\rho} & 0 & 0 & 0 & 0 \\ 0 & \bar{\rho} & 0 & 0 & 0 \\ 0 & 0 & 0 & 0 & 0 \\ 0 & 0 & 0 & 0 & 0 \\ 0 & 0 & 0 & 0 & 0 \\ 0 & 0 & 0 & 0 & \dfrac{\overline{T}g}{\gamma_a - \gamma}\bar{\rho} \end{bmatrix}$$

Rewrite problem (7.4.1) and (7.4.2) as

$$\mathbf{L}\varphi \equiv \mathbf{B}\frac{\partial\varphi}{\partial t} + \mathbf{A}\varphi = 0, \quad \mathbf{B}\varphi = \mathbf{B}\varphi_0 \quad \text{at} \quad t = 0 \qquad (7.4.4)$$

Assume that the solution belongs to a set of continuous differentiable functions satisfying boundary conditions (7.4.2) and the periodicity conditions. Introduce a real Hilbert space $\mathbf{H}$ of vector-functions with components squarely summable on Ω. The inner product is determined in $\mathbf{H}$ by

$$(\mathbf{g}, \mathbf{h})_\Omega = \sum_{i=1}^{5} \int_\Omega g_i h_i d\Omega$$

The operator $\mathbf{A}$ operates in the Hilbert space $\mathbf{H}$ with a definition domain $D(\mathbf{A})$. Take the set of functions continuously differentiable in spatial variables satisfying conditions (7.4.2) and periodicity conditions as $D(\mathbf{A})$.

Transformations like those made in Section 7.3 produce an operator adjoint to $\mathbf{A}$:

$$\boldsymbol{\varphi}^* = \begin{bmatrix} u^* \\ v^* \\ w^* \\ \varphi^* \\ \vartheta^* \end{bmatrix}, \quad \mathbf{A}^* = \begin{bmatrix} -\Lambda & \bar{\rho}l & 0 & -\bar{p}(\partial/\partial x) & 0 \\ -\bar{\rho}l & -\Lambda & 0 & -\bar{p}(\partial/\partial y) & 0 \\ 0 & 0 & 0 & -\bar{p}(\partial/\partial z) & g\bar{\rho} \\ -\dfrac{\partial}{\partial x}\bar{p} & -\dfrac{\partial}{\partial y}\bar{p} & -\dfrac{\partial}{\partial z}\bar{p} & 0 & 0 \\ 0 & 0 & -g\bar{\rho} & 0 & \dfrac{-\overline{T}g}{\gamma_a - \gamma}\Lambda \end{bmatrix}$$

As we constructed the adjoint operator, we used the verifiable fact that

$$\int_\Omega (u^*\mathrm{div}\bar{\rho}\mathbf{u}u + v^*\mathrm{div}\bar{\rho}\mathbf{u}v + w^*\mathrm{div}\bar{\rho}\mathbf{u}w)d\Omega$$

$$= -\int_\Omega (u\,\mathrm{div}\bar{\rho}\mathbf{u}u^* + v\,\mathrm{div}\bar{\rho}\mathbf{u}v^* + w\,\mathrm{div}\bar{\rho}\mathbf{u}w^*)d\Omega$$

This relationship holds when a number of conditions are met: that

$$\mathrm{div}\bar{\rho}\mathbf{u} = 0$$

$$\mathrm{div}\bar{\rho}\mathbf{u}^* = 0$$

that the components of the solution $\boldsymbol{\varphi}^*$ satisfy the smoothness conditions and the boundary conditions

$$\begin{aligned} \bar{\rho}w^* &= 0 \quad \text{at} \quad z = 0 \\ \bar{\rho}w^* &= 0 \quad \text{at} \quad z = H \end{aligned} \tag{7.4.5}$$

and finally, the periodicity conditions of the solutions in the plane (x, y). Then

$$\mathbf{A}^* = -\mathbf{A} \tag{7.4.6}$$

holds. The operator $\mathbf{A}^*$ operates in the Hilbert space $\mathbf{H}$ with the definition domain $D(\mathbf{A}^*) = D(\mathbf{A})$. The operator $\mathbf{A}$ is thus skew-symmetric.

Our goal is to construct equations adjoint to the evolutionary problems. Add then to (7.4.4) the adjoint problem

$$\mathbf{L}^*\boldsymbol{\varphi}^* \equiv -\mathbf{B}\frac{\partial \boldsymbol{\varphi}^*}{\partial t} - \mathbf{A}\boldsymbol{\varphi}^* = 0 \tag{7.4.7}$$

$$\boldsymbol{\varphi}^* = \boldsymbol{\varphi}_i^* \quad \text{at} \quad t = \bar{t} \tag{7.4.8}$$

There is obviously an identity for this problem like (7.3.13), but for a new, five-dimensional, phase space:

$$(\mathbf{B}\boldsymbol{\varphi}_{\bar{t}}, \boldsymbol{\varphi}_{\bar{t}}^*)_\Omega = (\mathbf{B}\boldsymbol{\varphi}_0, \boldsymbol{\varphi}_0^*)_\Omega \qquad (7.4.9)$$

Expand (7.4.9) into

$$\int_\Omega \left(\bar{\rho} u_{\bar{t}} u_{\bar{t}}^* + \bar{\rho} v_{\bar{t}} v_{\bar{t}}^* + \frac{g\overline{T}}{\gamma_a - \gamma} \bar{\rho} \vartheta_{\bar{t}} \vartheta_{\bar{t}}^* \right) d\Omega$$

$$= \int_\Omega \left(\bar{\rho} u_0 u_0^* + \bar{\rho} v_0 v_0^* + \frac{g\overline{T}}{\gamma_a - \gamma} \bar{\rho} \vartheta_0 \vartheta_0^* \right) d\Omega \quad (7.4.10)$$

If $u_{\bar{t}}^* = u_{\bar{t}}$, $v_{\bar{t}}^* = v_{\bar{t}}$, $\vartheta_{\bar{t}}^* = \vartheta_{\bar{t}}$, we arrive at the total energy conservation law:

$$\int_\Omega \bar{\rho} \pi_{\bar{t}} d\Omega = \int_\Omega \bar{\rho} \pi_0 d\Omega$$

where

$$\pi = u^2 + v^2 + \frac{g\overline{T}}{\gamma_a - \gamma} \bar{\rho} \vartheta^2$$

Assume that

$$u_{\bar{t}}^* = 0, \quad v_{\bar{t}}^* = 0, \quad \vartheta_{\bar{t}}^* = \frac{\gamma_a - \gamma}{g\overline{T}} \delta(x - x_0, y - y_0, z - z_0) \qquad (7.4.11)$$

Then, by virtue of (7.4.10), obtain

$$\bar{\rho} \vartheta_{\bar{t}}(x_0, y_0, z_0) = \int_\Omega \left(\bar{\rho} u_0 u_0^* + \bar{\rho} v_0 v_0^* + \frac{g\overline{T}}{\gamma_a - \gamma} \bar{\rho} \vartheta_0 \vartheta_0^* \right) d\Omega \qquad (7.4.12)$$

This formula indicates a link between the temperature in an assigned point of space at a moment of time $t = \bar{t}$ and the initial (at $t = 0$) state of atmosphere. Remember that u_0, v_0, and ϑ_0 are assigned at the initial moment of time and u_0^*, v_0^*, and ϑ_0^* are the solutions of the adjoint equations given condition (7.4.11).

7.5. Baroclinic Model of the Atmosphere with a Heat Flux from the Ocean

Examine a more complete system of equations governing the dynamics of the atmosphere with assigned heat sources with regard to the turbulent exchange:

$$\frac{\partial \bar{p}u}{\partial t} + \Lambda u - l\bar{p}v + \bar{p}\frac{\partial \varphi}{\partial x} - \mu\bar{p}\Delta u = 0$$

$$\frac{\partial \bar{p}v}{\partial t} + \Lambda v + l\bar{p}u + \bar{p}\frac{\partial \varphi}{\partial y} - \mu\bar{p}\Delta v = 0$$

$$-g\bar{p}\vartheta + \bar{p}\frac{\partial \varphi}{\partial z} = 0 \qquad (7.5.1)$$

$$\frac{\partial \bar{p}u}{\partial x} + \frac{\partial \bar{p}v}{\partial y} + \frac{\partial \bar{p}w}{\partial z} = 0$$

$$\frac{\partial \bar{p}\vartheta}{\partial t} + \Lambda\vartheta + \frac{\gamma_a - \gamma}{T}\bar{p}w - \frac{\partial}{\partial z}\nu_1\bar{p}\frac{\partial \vartheta}{\partial z} - \mu_1\bar{p}\Delta\vartheta = 0$$

Assume, as the boundary conditions,

$$\frac{\partial \vartheta}{\partial z} = \alpha_s(\vartheta - \bar{\vartheta}), \quad \bar{p}w = 0 \quad \text{at} \quad z = 0$$

$$\frac{\partial \vartheta}{\partial z} = 0, \quad \bar{p}w = 0 \quad \text{at} \quad z = H \qquad (7.5.2)$$

where α_s is the heat conductivity coefficient assigned, for the time being, equal to zero on-shore land and on the sea ice, $\bar{\vartheta}$ is the temperature of the surface layer of ocean, assumed to be known in this model. The motion equations do not contain the turbulent momentum exchange terms, since problem (7.5.1)–(7.5.2) is stated for free atmosphere.

A periodicity condition is assigned at the boundary of the region in plane (x, y). Assume the initial conditions as

$$u = u_0, \quad v = v_0, \quad \vartheta = \vartheta_0, \quad \text{at} \quad t = 0 \qquad (7.5.3)$$

Assume also that the solution has continuous derivatives of the first order for u, v, and ϑ in time, and of the second order in all the spatial variables. Introduce a matrix operator

$$
\mathbf{A} = \begin{bmatrix}
\Lambda - \mu\bar{\rho}\Delta & -l\bar{\rho} & 0 & \bar{p}(\partial/\partial x) & 0 \\
l\bar{\rho} & \Lambda - \mu\bar{\rho}\Delta & 0 & \bar{p}(\partial/\partial y) & 0 \\
0 & 0 & 0 & \bar{p}(\partial/\partial z) & -g\bar{\rho} \\
\dfrac{\partial}{\partial x}\bar{p} & \dfrac{\partial}{\partial y}\bar{p} & \dfrac{\partial}{\partial z}\bar{p} & 0 & 0 \\
0 & 0 & g\bar{\rho} & 0 & A_{55}
\end{bmatrix}
$$

where

$$
A_{55} = \frac{Tg}{\gamma_a - \gamma}\left(\Lambda - \frac{\partial}{\partial z}\nu_1\bar{\rho}\frac{\partial}{\partial z} - \mu_1\bar{\rho}\Delta\right)
$$

and a vector

$$
\boldsymbol{\varphi} = \begin{bmatrix} u \\ v \\ w \\ \varphi \\ \vartheta \end{bmatrix}
$$

The operator $\mathbf{A}$ operates in the Hilbert space $\mathbf{H}$ as introduced in Section 7.4, with the definition domain $D(\mathbf{A})$. Consider a set of vector-functions double continuously differentiable in the spatial variables and satisfying homogeneous conditions (7.5.2) and the periodicity conditions, as $D(\mathbf{A})$. Then the problem is

$$
\mathbf{L}\boldsymbol{\varphi} \equiv \mathbf{B}\frac{\partial\boldsymbol{\varphi}}{\partial t} + \mathbf{A}\boldsymbol{\varphi} = 0 \tag{7.5.4}
$$
$$
\boldsymbol{\varphi} = \boldsymbol{\varphi}_0 \quad \text{at} \quad t = 0
$$

Introduce the adjoint operator

$$
\mathbf{A}^* = \begin{bmatrix}
-\Lambda - \mu\bar{\rho}\Delta & l\bar{\rho} & 0 & -\bar{p}(\partial/\partial x) & 0 \\
-l\bar{\rho} & -\Lambda - \mu\bar{\rho}\Delta & 0 & -\bar{p}(\partial/\partial y) & 0 \\
0 & 0 & 0 & -\bar{p}(\partial/\partial z) & g\bar{\rho} \\
-\dfrac{\partial}{\partial x}\bar{p} & -\dfrac{\partial}{\partial y}\bar{p} & -\dfrac{\partial}{\partial z}\bar{p} & 0 & 0 \\
0 & 0 & -g\bar{\rho} & 0 & A_{55}^*
\end{bmatrix}
$$

where

$$
A_{55}^* = \frac{Tg}{\gamma_a - \gamma}\left(-\Lambda - \frac{\partial}{\partial z}\nu_1\bar{\rho}\frac{\partial}{\partial z} - \mu_1\bar{\rho}\Delta\right)
$$

and formulate the problem

$$\mathbf{L}^*\boldsymbol{\varphi}^* \equiv -\mathbf{B}\frac{\partial \boldsymbol{\varphi}^*}{\partial t} + \mathbf{A}^*\boldsymbol{\varphi}^* = 0 \tag{7.5.5}$$
$$\boldsymbol{\varphi}^* = \boldsymbol{\varphi}^*_{\bar{t}} \quad \text{at} \quad t = \bar{t}$$

The components of the vector-function $\boldsymbol{\varphi}^*$ meet the conditions of smoothness, periodicity conditions at the region's boundary, and the boundary conditions

$$\frac{\partial \vartheta^*}{\partial z} = \alpha_s \vartheta^*, \quad \bar{\rho} w^* = 0 \quad \text{at} \quad z = 0$$
$$\frac{\partial \vartheta^*}{\partial z} = 0, \quad \bar{\rho} w^* = 0 \quad \text{at} \quad z = H \tag{7.5.6}$$

The operator $\mathbf{A}^*$ operates in a Hilbert space $\mathbf{H}$ with a definition domain $D(\mathbf{A}^*)$. Consider a set of vector-functions continuously double differentiable in spatial variables and satisfying conditions (7.5.6) and periodicity conditions, as $D(\mathbf{A}^*)$. Assume now that the solution $\boldsymbol{\varphi}^*$ of problem (7.5.5) is continuously differentiable in t and for each $t \in [0, \bar{t}]$ belongs to the set $D(\mathbf{A}^*)$. The component-wise form of problem (7.5.5) is

$$-\frac{\partial \bar{\rho} u^*}{\partial t} - \Lambda u^* + l\bar{\rho} v^* - \bar{p}\frac{\partial \varphi^*}{\partial x} - \mu \bar{\rho} \Delta u^* = 0$$

$$-\frac{\partial \bar{\rho} v^*}{\partial t} - \Lambda v^* + l\bar{\rho} u^* - \bar{p}\frac{\partial \varphi^*}{\partial y} - \mu \bar{\rho} \Delta v^* = 0$$

$$g\bar{\rho}\vartheta^* - \bar{p}\frac{\partial \varphi^*}{\partial z} = 0 \tag{7.5.7}$$

$$\frac{\partial \bar{\rho} u^*}{\partial x} + \frac{\partial \bar{\rho} v^*}{\partial y} + \frac{\partial \bar{\rho} w^*}{\partial z} = 0$$

$$-\frac{\partial \bar{\rho}\vartheta^*}{\partial t} - \Lambda\vartheta^* - \frac{\gamma_a - \gamma}{T}\bar{\rho} w^* - \frac{\partial}{\partial z}\nu_1\bar{\rho}\frac{\partial \vartheta^*}{\partial z} - \mu_1\bar{\rho}\Delta\vartheta^* = 0$$

with boundary conditions (7.5.6) and initial data

$$u^* = u_{\bar{t}}, \quad v^* = v_{\bar{t}}, \quad \vartheta^* = \vartheta_{\bar{t}} \quad \text{at} \quad t = \bar{t} \tag{7.5.8}$$

Since the operators $\mathbf{A}$ and $\mathbf{A}^*$ are adjoint, the condition

$$(\mathbf{A}\boldsymbol{\varphi}, \boldsymbol{\varphi}^*)_\Omega = (\boldsymbol{\varphi}, \mathbf{A}^*\boldsymbol{\varphi}^*)_\Omega$$

holds at $\bar{\vartheta} = 0$.

Multiply equation (7.5.4) by φ^* and equation (7.5.5) by φ, integrate over the time from 0 to $\bar{t}$ and subtract the results from each other to obtain

$$\int_\Omega \left(u_{\bar{t}} u_{\bar{t}}^* + v_{\bar{t}} v_{\bar{t}}^* + \frac{g\overline{T}}{\gamma_a - \gamma} \vartheta_{\bar{t}} \vartheta_{\bar{t}}^* \right) \bar{\rho} d\Omega$$

$$- \int_\Omega \left(u_0 u_0^* + v_0 v_0^* + \frac{g\overline{T}}{\gamma_a - \gamma} \vartheta_0 \vartheta_0^* \right) \bar{\rho} d\Omega$$

$$- q \int_0^{\bar{t}} dt \int_S \alpha_s \bar{\vartheta} \vartheta^* dS = 0 \quad (7.5.10)$$

where $q = \nu_1 g \bar{\rho} \overline{T} / (\gamma_a - \gamma)$, S is the ocean surface with the prescribed surface layer temperature $\bar{\vartheta}(x, y, 0, t)$.

If we are interested in the anomalies of the average field of temperature over the domain $G = \{(x, y, z) : \ x, y \in \sigma, \ 0 \leq z \leq h\}$, select the "initial" data for the adjoint equations as

$$u_{\bar{t}}^* = 0, \quad v_{\bar{t}}^* = 0$$

$$\frac{g\overline{T}}{\gamma_a - \gamma} \vartheta_{\bar{t}}^* = \frac{1}{\mathrm{mes}\, G} \quad \text{at} \quad (x, y, z) \in G$$

$$v_{\bar{t}}^* = 0 \qquad\qquad\qquad \text{outside} \quad G$$

(7.5.11)

Introduce a notation for an anomaly of temperature average over a region G at a moment of time $t = \bar{t}$:

$$J \equiv \frac{1}{\mathrm{mes}\, G} \int_G \bar{\rho} \vartheta_{\bar{t}} d\Omega = \overline{\bar{\rho} \vartheta}_{\bar{t}}^G$$

Rewrite formula (7.6.10) as

$$J = \int_\Omega \left(u_0 u_0^* + v_0 v_0^* + \frac{g\overline{T}}{\gamma_a - \gamma} \vartheta_0 \vartheta_0^* \right) \bar{\rho} d\Omega + q \int_0^{\bar{t}} dt \int_S \alpha_s \bar{\vartheta} \vartheta^* dS \quad (7.5.12)$$

The expression $\overline{\bar{\rho} \vartheta}_{\bar{t}}^G$ means that the average temperature anomaly is computed with the initial data at the interval $0 \leq t \leq \bar{t}$. The problem of average temperature anomaly forecasting was thus reduced to solving adjoint problem (7.5.7) and (7.5.8) on condition (7.5.11).

The perturbation theory was constructed in this section at specially assigned initial conditions in a system of adjoint equations and homogeneous boundary conditions. Prove that other problem statements are possible in the system of adjoint equations which lead to conveniently applicable formulas of the perturbation theory. Examine system of equations (7.5.1) at boundary conditions (7.5.2) and initial conditions (7.5.3), and adjoint system of equations (7.5.7).

Determine boundary conditions for system (7.5.7) as

$$\frac{\partial \vartheta^*}{\partial z} = \alpha_s \vartheta^* + f^*, \quad \bar{\rho} w^* = 0 \quad \text{at} \quad z = 0,$$
$$\frac{\partial \vartheta^*}{\partial z} = 0, \quad \bar{\rho} w^* = 0 \quad \text{at} \quad z = H, \tag{7.5.13}$$

where $f^*(x, y, t)$ is assigned as

$$f^* = \frac{\bar{\rho}}{q \operatorname{mes} G_0} \delta(t - \bar{t}) \quad \text{at} \quad (x, y) \in G_0$$
$$f^* = 0 \qquad \text{outside this region}$$

Here G_0 is a region of the Earth's surface for which an average anomaly of temperature is to be forecast.

Assume

$$u^* = 0, \quad v^* = 0, \quad \vartheta^* = 0 \quad \text{at any} \quad t > \bar{t} \tag{7.5.14}$$

as the initial data for the adjoint problem. Use the main and adjoint problems and the technique discussed above to obtain the functional

$$\overline{\bar{\rho}\vartheta_{\bar{t}}}^{G_0} = \int\limits_{\Omega} \left(u_0 u_0^* + v_0 v_0^* + \frac{g\overline{T}}{\gamma_a - \gamma} \vartheta_0 \vartheta_0^* \right) \bar{\rho} d\Omega + q \int\limits_0^{\bar{t}} dt \int\limits_S \alpha_s \bar{\vartheta} \vartheta^* dS \tag{7.5.15}$$

where

$$\overline{\bar{\rho}\vartheta_{\bar{t}}}^{G_0} = \frac{1}{\operatorname{mes} G_0} \int\limits_{G_0} \bar{\rho}\vartheta_{\bar{t}} d\Omega \tag{7.5.16}$$

Simple models of the general circulation in the atmosphere have been examined above as examples. More complex models are build in a similar way. The reader can do it easily on the basis of principles discussed in this chapter. A general atmosphere-ocean circulation adjoint problem accounting for the continents may be formulated in spherical coordinates, for example. Details like this, however, are the subject of quite another research.

7.6. Problems of Climate Change Sensitivity in Various Regions of the World

While modelling mathematically the climate change for separate regions of the Earth's continents, the sensitivity theory of chosen functionals is important as related to the continents, World Ocean, initial data, external sources and internal parameters of the problem. The problem of climate sensitivity allows on the basis of real data evaluate retrospectively a quality of the models and find new mechanisms responsible for formation of climate.

We will discuss in this section general approaches to the estimation of sensitivity on the basis of simple theoretical models. These results may be generalized then for the most complex statements of problems.

Let us consider the thermal interaction of atmosphere with the World Ocean and continents. Consider three-dimensional model domain Ω in spherical system of coordinates (λ, ψ, z), where λ is the longitude, ψ is the latitude and z is the height measured from the Earth surface, which is assumed to be spherical, as well as that of ocean. Let h_1 be the height of atmosphere layer, h_2 – thickness of active ocean layer, h_3 – thickness of soil layer, S – the Earth's surface, $S = S_1 \cup S_2 \cup S_3$, S_1 – part of the Earth covered with ice and/or snow, S_2 – oceanic surface, S_3 – continental surface free of ice and snow.

The domain $\Omega = \Omega_1 \cup \Omega_2 \cup \Omega_3$ consists of:
- atmosphere layer $\Omega_1 = \{(\lambda, \psi, z) : (\lambda, \psi) \in S_1, \ 0 < z < h_1\}$;
- active ocean layer $\Omega_2 = \{(\lambda, \psi, z) : (\lambda, \psi) \in S_2, \ -h_2 < z < 0\}$;
- upper layer of the soil $\Omega_3 = \{(\lambda, \psi, z) : (\lambda, \psi) \in S_3, \ -h_3 < z < 0\}$.

Examine the following problem for the temperature field $T(\lambda, \psi, z, t)$ at the interval $\Omega_t = (0, T)$:

$$\alpha \frac{\partial T}{\partial t} + \text{div}(\alpha \mathbf{u} T) - \frac{\partial}{\partial z}(\bar{\nu}\frac{\partial T}{\partial z}) - \bar{\mu}\Delta T = \varepsilon \qquad (7.6.1)$$

with initial condition

$$T = T_0(\mathbf{r}) \quad \text{at} \quad t = 0 \qquad (7.6.2)$$

where $\mathbf{r} = (\lambda, \psi, z)^T$, $\bar{\nu} = \alpha\nu$, $\bar{\mu} = \alpha\mu$, and ν, μ are coefficients of vertical and horizontal turbulent exchange respectively assumed to be dependent on the height z since they are different for atmosphere and ocean. The function ν depends also on horizontal coordinates and time. Meaning of other variables in equation (7.6.1): ε is the source of radiation energy, the function which differs from zero just in the atmosphere, that is, in the domain Ω_1; $\alpha = c_p\rho(z)$ where c_p is the specific heat of a medium, $\rho(z)$ is standard density of atmosphere (at $z > 0$) and of ocean (at $z < 0$); $\mathbf{u}(\mathbf{r}, t)$ is a velocity vector of wind in the atmosphere Ω_1 and of ocean currents in Ω_2.

Assume that the values $\mathbf{u}$, ν and μ are assigned in the definition domain of the problem with $\mathbf{u}(\mathbf{r}, t) = 0$, $\mu = 0$ in Ω_3. Assume also that the vector function $\mathbf{u}(\mathbf{r}, t)$ satisfies in atmosphere and ocean the simplest continuity equation

$$\operatorname{div}(\rho\mathbf{u}) = 0 \qquad (7.6.3)$$

and its normal component equals zero at lateral surface of the World Ocean surface as well as at the surfaces $z = 0$, $z = h_1$, $z = -h_2$. Naturally, for the atmosphere, if needed, the continuity equation can be considered with regard to changing with time density of air masses.

Equation (7.6.1) is considered in the domain $\Omega = \Omega_1 \cup \Omega_2 \cup \Omega_3$ at $t \in \Omega_t = (0, \bar{t})$. Boundary conditions must be formulated at the boundary of the domain Ω at $z = 0$, $z = h_1$, $z = -h_2$, $z = -h_3$ and at the atmosphere-continent contact surfaces, which will provide for the existence of unique solution to the initial boundary value problem.

Conditions of temperature equality and jump of heat fluxes are assigned at the surfaces dividing atmosphere and ocean, atmosphere and continent $(z = 0)$:

$$[T] = 0, \quad \left[\nu\frac{\partial T}{\partial z}\right] = F(\lambda, \psi, z) \quad \text{at} \quad z = 0, \quad (\lambda, \psi) \in S_2 \cup S_3 \quad (7.6.4)$$

where $[f] = f|_{z=-0} - f|_{z=+0}$ is a jamp of the function f in z at the point $z = 0$, and $F(\lambda, \psi, z)$ is assumed to be a function known from observations.

The following condition is assigned at the boundary between the atmosphere and the part of the Earth's surface covered with ice and/or snow:

$$\nu\frac{\partial T}{\partial z} + F = 0 \quad \text{at} \quad z = 0, \ (\lambda, \psi) \in S_1 \qquad (7.6.5)$$

The condition of no heat flux is assigned at the top boundary of atmosphere (at $z = h_1$) and at lower boundary of active oceanic and soil layers:

$$\bar{\nu}\frac{\partial T}{\partial z} = 0 \quad \text{at} \quad z = h_1, \ z = -h_2, \ z = -h_3 \qquad (7.6.6)$$

Finally, for simplicity, assume the boundaries between the continents and the World Ocean as cylindrical and

$$\bar{\mu}\frac{\partial T}{\partial n} = 0 \quad \text{at} \quad \partial\Omega_2 \qquad (7.6.7)$$

where $\partial\Omega_2$ is the lateral surface of Ω_2 and n is a normal to $\partial\Omega_2$.

So, problem (7.6.1)–(7.6.7) is posed completely and we will suggest that initial data of the problem determine its unique solution which belongs to the Hilbert space $L_2(\Omega \times \Omega_t)$. Assume that this solution is sufficiently smooth function so that all the further transformations are justified. The questions of existence and uniqueness of solution to problem (7.6.1)–(7.6.7) are discussed in [115].

Turn to the sensitivity of problem's (7.6.1)–(7.6.7) solution to various functionals.

The following functional is the most interesting for us:

$$J = \int_0^{\bar{t}} dt \int_S F^*(\lambda, \psi, t) T(\lambda, \psi, 0, t) dS \qquad (7.6.8)$$

where $F^*(\lambda, \psi, t)$ is some weight function connected with a temperature field at the surface $z = 0$. For example, if we want to determine an average temperature in some selected region of a continent ω at $z = 0$ in the interval $\bar{t} - \tau \le t \le \bar{t}$, then we choose for F^* the function

$$F^*(\lambda, \psi, t) = \begin{cases} 1/(\tau \mathrm{mes}\,\omega), & \text{if} \ (\lambda, \psi) \in \omega, \ \bar{t} - \tau \le t \le \bar{t} \\ 0, & \text{otherwise} \end{cases} \qquad (7.6.9)$$

where $\mathrm{mes}\,\omega$ means, as usual, an area of the region ω. Then functional (7.6.8) will be rewritten in the form:

$$J = \frac{1}{\tau} \int_{\bar{t}-\tau}^{\bar{t}} dt \left(\frac{1}{\tau \mathrm{mes}\,\omega} \int_\omega T(\lambda, \psi, 0, t) dS \right) \qquad (7.6.10)$$

Expression (7.6.10) represents an average temperature over the interval $\bar{t} - \tau \leq t \leq \bar{t}$ for selected region ω. These types of functionals are the most interesting in the theory of climate change.

Since the functional of the problem is defined in the form (7.6.8) or (7.6.10), we may formulate an adjoint problem to (7.6.1)–(7.6.7) using the methods we discussed in preceding chapters. It will have the form:

$$-\alpha\frac{\partial T^*}{\partial t} - \operatorname{div}(\alpha \mathbf{u} T^*) - \frac{\partial}{\partial z}(\bar{\nu}\frac{\partial T^*}{\partial z}) - \bar{\mu}\Delta T^* = 0 \qquad (7.6.11)$$

with "initial" condition

$$T^*(\mathbf{r}, t) = 0 \quad \text{at} \quad t = \bar{t} \qquad (7.6.12)$$

Equation (7.6.11) is considered in the domain $\Omega = \Omega_1 \cup \Omega_2 \cup \Omega_3$ at $t \in \Omega_t$. As for the surface $z = 0$ as well as for another "special" surfaces, where the necessary condition of differentiability of the solution T^* and its derivatives is not met, one must determine additional boundary conditions:

$$[T^*] = 0, \quad \left[\bar{\nu}\frac{\partial T^*}{\partial z}\right] = F^* \quad \text{at} \quad z = 0, \quad (\lambda, \psi) \in S_2 \cup S_3 \qquad (7.6.13)$$

where F^* is defined by functional (7.6.8) or, in particular, (7.6.9).

The following condition is assigned at the boundary between the atmosphere and the part of the Earth's surface covered with ice and/or snow:

$$\bar{\nu}\frac{\partial T^*}{\partial z} + F^* = 0 \quad \text{at} \quad z = 0, \ (\lambda, \psi) \in S_1 \qquad (7.6.14)$$

The conditions at the top boundary of atmosphere (at $z = h_1$) and ıt lower boundary of active oceanic and soil layers are as follows:

$$\bar{\nu}\frac{\partial T^*}{\partial z} = 0 \quad \text{at} \quad z = h_1, \ z = -h_2, \ z = -h_3 \qquad (7.6.15)$$

We assume at the boundaries between the continents and the World)cean

$$\bar{\mu}\frac{\partial T^*}{\partial n} = 0 \quad \text{at} \quad \partial\Omega_2 \qquad (7.6.16)$$

ʾhere n is an external normal to the boundary cylindrical surface ocean–land".

The theorem of existence and uniqueness of generalized solution of the main problem (7.6.1)–(7.6.7) and adjoint problem (7.6.11)–(7.6.16) is proved in [115].

The following relationships are valid, by equation (7.6.3) and since the solutions of the main and adjoint problems belong to $L_2(\Omega \times \Omega_t)$, possess the necessary smoothness and satisfy formulated boundary conditions:

$$\frac{d}{dt}\int_\Omega \alpha T \, d\Omega = \int_S F \, dS - \int_{\Omega_1} \varepsilon \, d\Omega_1$$

$$\text{(7.6.17)}$$

$$-\frac{d}{dt}\int_\Omega \alpha T^* \, d\Omega = \int_S F^* \, dS$$

$$\frac{d}{dt}\int_\Omega \alpha T^2 \, d\Omega \le \int_S FT \, dS - \int_{\Omega_1} \varepsilon T \, d\Omega_1$$

$$\text{(7.6.18)}$$

$$-\frac{d}{dt}\int_\Omega \alpha T^{*2} \, d\Omega = \int_S F^* T^* \, dS$$

Normal component of a velocity vector turns to zero at the boundary of continents with oceans; this fact is regarded to in (7.6.17), (7.6.18).

It follows from (7.6.18) that the solution of the main problem is stable, as well as that of adjoint problem which is to be solved in the direction from $t = \bar{t}$ to $t = 0$.

Next, examine another form of the functional (7.6.8), expressed through the solution T^* of the adjoint problem. Multiply equation (7.6.1) by T^* and equation (7.6.11) by T, integrate in time at Ω_t and over all solution's definition domain Ω with regard to initial data (7.6.2), (7.6.12) and boundary conditions (7.6.4)–(7.6.7), (7.6.13)–(7.6.16) and subtract the results from each other. Integrate by parts to obtain

$$J = \int_\Omega \alpha T^*(\mathbf{r}, 0) T_0(\mathbf{r}) \, d\Omega + \int_0^t dt \int_S F(\lambda, \psi, t) T^*(\lambda, \psi, 0, t) \, dS$$

$$+ \int_0^t dt \int_{\Omega_1} \varepsilon T^*(\lambda, \psi, z, t) \, d\Omega_1 \quad \text{(7.6.19)}$$

Notice, that the left-hand part of this relationship coincides exactly with the functional J defined by formula (7.6.8).

Relationship (7.6.19) describes a sensitivity of the functional J to the location of various regions and input parameters of the problem, since the weight functions T^* represent an influence of initial data $T_0(\mathbf{r})$, heat fluxes $F(\lambda, \psi, t)$ at the boundary $z = 0$ and radiation sources in the atmosphere ε.

Representing the input functions T, $\mathbf{u}$, ε, F of original problem (7.6.1)–(7.6.7) as a sum of standard climatic values and deviations

$$T = \overline{T} + \delta T, \quad \mathbf{u} = \overline{\mathbf{u}} + \delta \mathbf{u}, \quad \varepsilon = \overline{\varepsilon} + \delta \varepsilon, \quad F = \overline{F} + \delta F \quad (7.6.20)$$

we arrive at more complex perturbation theory formula for this functional.

Assume that climatic components satisfy the problem

$$\alpha \frac{\partial \overline{T}}{\partial t} + \operatorname{div}(\alpha \overline{\mathbf{u}} \overline{T}) - \frac{\partial}{\partial z}(\overline{\nu} \frac{\partial \overline{T}}{\partial z}) - \overline{\mu} \Delta \overline{T} = \overline{\varepsilon} \quad (7.6.21)$$

with initial data

$$\overline{T} = \overline{T}_0 \quad \text{at} \quad t = 0 \quad (7.6.22)$$

and boundary conditions

$$[\overline{T}] = 0, \quad \left[\overline{\nu} \frac{\partial \overline{T}}{\partial z} \right] = \overline{F}(\lambda, \psi, z) \quad \text{at} \quad z = 0, \quad (\lambda, \psi) \in S_2 \cup S_3$$

$$\overline{\nu} \frac{\partial \overline{T}}{\partial z} + \overline{F} = 0 \quad \text{at} \quad z = 0, \ (\lambda, \psi) \in S_1 \quad (7.6.23)$$

$$\overline{\nu} \frac{\partial \overline{T}}{\partial z} = 0 \quad \text{at} \quad z = h_1, \ z = -h_2, \ z = -h_3$$

$$\overline{\mu} \frac{\partial \overline{T}}{\partial n} = 0 \quad \text{at} \quad \partial \Omega_2 \quad (7.6.24)$$

Then the problem for deviations δT will be formulated as follows

$$\alpha \frac{\partial \delta T}{\partial t} + \operatorname{div}(\alpha \mathbf{u} \delta T) - \frac{\partial}{\partial z}(\overline{\nu} \frac{\partial \delta T}{\partial z}) - \overline{\mu} \Delta \delta T = \varepsilon_1 \quad (7.6.25)$$

where

$$\varepsilon_1 = \delta \varepsilon - \operatorname{div}(\alpha \overline{T} \delta \mathbf{u}) \quad (7.6.26)$$

Join to (7.6.25) corresponding initial data

$$\delta T = \delta T_0 \quad \text{at} \quad t = 0 \tag{7.6.27}$$

and boundary conditions:

$$[\delta T] = 0, \quad \left[\bar{\nu}\frac{\partial \delta T}{\partial z}\right] = \delta F \quad \text{at} \quad z = 0, \quad (\lambda, \psi) \in S_2 \cup S_3$$

$$\bar{\nu}\frac{\partial \delta T}{\partial z} + \delta F = 0 \quad \text{at} \quad z = 0, \ (\lambda, \psi) \in S_1 \tag{7.6.28}$$

$$\bar{\nu}\frac{\partial \delta T}{\partial z} = 0 \quad \text{at} \quad z = h_1, \ z = -h_2, \ z = -h_3$$

$$\bar{\mu}\frac{\partial \delta T}{\partial n} = 0 \quad \text{at} \quad \partial\Omega_2 \tag{7.6.29}$$

Note the function $\mathbf{u} = \bar{\mathbf{u}} + \delta\mathbf{u}$ standing in the second term of equation (7.6.25) under the div sign. As for the functional J, write its variation $\delta J = J - \bar{J}$ as

$$\delta J = \int\limits_0^{\bar{t}} dt \int\limits_S F^*(\lambda, \psi, t)\delta T(\lambda, \psi, 0, t)\, dS \tag{7.6.30}$$

or, using the above transformations

$$\delta J = \int\limits_\Omega \alpha T^*(\mathbf{r}, 0)\delta T_0(\mathbf{r})\, d\Omega + \int\limits_0^{\bar{t}} dt \int\limits_S \delta F(\lambda, \psi, t) T^*(\lambda, \psi, 0, t)\, dS$$

$$+ \int\limits_0^{\bar{t}} dt \int\limits_{\Omega_1} \delta\varepsilon T^*(\lambda, \psi, z, t)\, d\Omega_1 - \int\limits_0^{\bar{t}} dt \int\limits_\Omega \mathrm{div}(\alpha\overline{T}\delta\mathbf{u}) T^*(\lambda, \psi, z, t)\, d\Omega \tag{7.6.31}$$

If variations $\delta\mathbf{u}$ of a velocity vector are small, so that $\mathbf{u} \approx \bar{\mathbf{u}}$, then the last term in (3.7.31) may be neglected and we obtain the approximate formula:

$$\delta J = \int\limits_\Omega \alpha T^*(\mathbf{r}, 0)\delta T_0(\mathbf{r})\, d\Omega + \int\limits_0^{\bar{t}} dt \int\limits_S \delta F(\lambda, \psi, t) T^*(\lambda, \psi, 0, t)\, dS$$

$$+ \int\limits_0^{\bar{t}} dt \int\limits_{\Omega_1} \delta\varepsilon T^*(\lambda, \psi, z, t)\, d\Omega_1 \tag{7.6.32}$$

We will use this simplified formula further to illustrate the results on simple models, since variations $\delta\mathbf{u}$ in the ocean are unknown. Therefore, both for the ocean and for the atmosphere we will use correlated climatic data and will assume that $\mathbf{u} = \bar{\mathbf{u}}$.

Representing the time interval in equation (7.6.32) as a sum of subintervals $I_k = (t_k, t_{k-1})$ we rewrite formula (7.6.32) as

$$\delta J = \int_\Omega \alpha \delta T_0(\mathbf{r}, 0) T^*(\mathbf{r}, 0)\, d\Omega + \sum_{k=1}^{N} \int_{t_k}^{t_{k-1}} dt \int_S \delta F T^*\, dS$$

$$+ \sum_{k=1}^{N} \int_{t_k}^{t_{k-1}} dt \int_{\Omega_1} \delta \varepsilon T^*\, d\Omega \quad (7.6.33)$$

where $t_0 = \bar{t}$, $t_N = 0$, $t_k = t_{k-1} - \tau_k$, $k = \overline{1, N}$.

Now, consider a problem of vertical turbulent exchange in the system atmosphere–ocean which plays the most important role in the formation of interaction between these two mediums.

We used, in numerical experiments with the model of thermal interaction between the atmosphere and underlying surface, the fields of vertical turbulent diffusion coefficients in the ocean prepared with the method discussed in [107]. Describe this method for the upper active layer of the ocean.

Introduce an operator of running average for arbitrary integrable function $b = b(t)$ on a characteristic temporal scale τ:

$$\bar{b} = \frac{1}{\tau} \int_{t-\tau/2}^{t+\tau/2} b(t')dt' \quad (7.6.34)$$

and representation $b = \bar{b} + b'$, where average values are marked by bar and deviations from average by a prime.

Let us assume that an integral coefficient of vertical mixing in the ocean is determined on seasonal time scale ($\tau \sim 1\text{--}3$ months) by vertical sheer of currents' velocity Σ, vertical temperature gradient γ, its seasonal variability σ, and buoyancy parameter λ that is, by the following set of characteristics:

$$\Sigma = \left[\left(\frac{\partial \bar{u}}{\partial z}\right)^2 + \left(\frac{\partial \bar{v}}{\partial z}\right)^2\right]^{1/2}, \quad \gamma = \frac{\partial \overline{T}}{\partial z}$$
$$\sigma = \left(\overline{T'^2}\right)^{1/2}, \quad \lambda = g\alpha_T$$

(7.6.35)

where g is the gravity acceleration, α_T is the thermal expansion coefficient for the water, $\lambda = 0.2 \text{ cm}/(\text{c}^2\text{grad})$, $\bar{u}$ and $\bar{v}$ are average horizontal components of currents' velocity vectors.

The only one dimensionless parameter may be deduced from these parameters: Richardson's number

$$R_i = \frac{\lambda\gamma}{\Sigma^2}$$

(7.6.36)

and the only one dimension combination for mixing coefficients (cm^2/c):

$$\nu = C(R_i)\lambda^{\frac{1}{2}}|\gamma|^{-\frac{3}{2}}\sigma^2, \quad \bar{\nu} = \alpha\nu$$

(7.6.37)

where C is a dimensionless value depending on R_i.

There is an estimate of the coefficient $C(R_i)$ in (7.6.37) obtained in the work [107] in assumption that while using large-scale characteristics σ and γ variations of this coefficient are small and using the information from some registration points in the ocean; the estimate happend to be $C = 2 \cdot 10^{-5}$.

Use now formula (7.6.37) to obtain the coefficient of vertical heat mixing (for the seasonal time scale) on the basis of observations of large-scale characteristics of upper boundary layer of the ocean with a diagnostic method. The data on spatial distribution of seasonal variability of sea surface temperature σ will be used in the nodes of a grid $10° \times 10°$ [130]. We will calculate the temperature gradient γ in an upper layer of the ocean from climatic fields of temperature for winter and summer seasons, using unfiltered Hellerman's data (a version of filtered data is published in [207]).

Fig. 52 shows the resulting spatial distribution of the coefficient ν for a winter season with isolines of a function $\ln(1+\nu)$ in the upper layer of the ocean (contours starting with 0 are spaced by 0.7). A maximum of a field $\nu(\lambda, \vartheta)$ is located in a winter period in the Atlantic to the south from the Newfoundland island. One can see, that the positions of local coefficient ν maxima coincide with known climatic energetically active zones of the World Ocean – Newfoundland, Norwayan, Aleutian and energetically active zones of the Gulf Stream and Kuroshio, where the heat exchange between the atmosphere and ocean is the most active in the Northern Hemisphere.

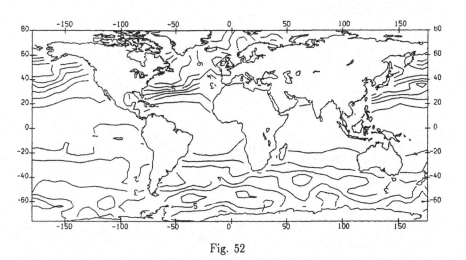

Fig. 52

In order to compute the thermal influence functions a numerical model was constructed based on the splitting method (see Appendix I). Original three-dimensional problem (7.6.1)–(7.6.7) was reduced to a set of simple one-dimensional problems of transport–diffusion along the directions λ, ψ, z solved easily by numerical methods with no iteration procedures. Two-cyclic splitting method using Crank-Nicolson scheme in each stage was used while solving the problems in time, which provided for a stable algorithm with the second order of accuracy in time.

While solving one-dimensional transport–diffusion problems along λ and ϑ, Richardson's extrapolation method [73] was used to increase an approximation order, where a linear combination of solutions on the grids with the steps h and $h/3$ ensured an accuracy of $O(h^4)$ order.

The model used latitude-longitude grid with $5° \times 4°$ resolution. Non-uniform grid was used in a vertical direction with calculation levels at geopotential heights 850, 500, 300 mb in the atmosphere, at the depths 25, 75, 150 m in the ocean, and at the depths 0.5, 1.0 and 2.0 m in the soil.

The main parameters of the model:

$$h_1 = 2 \cdot 10^6 \text{ cm}, \quad h_2 = 2 \cdot 10^4 \text{ cm}, \quad h_3 = 300 \text{ cm}$$

in the atmosphere:

$$c_p = 10^7 \frac{\text{erg}}{\text{g} \cdot \text{grad}}, \quad 0 \le \rho \le 1.2 \cdot 10^{-3} \frac{\text{g}}{\text{cm}^3}, \quad \mu = {}^9\frac{\text{cm}^2}{\text{c}}, \quad 10^4 \le \nu \le 10^5 \frac{\text{cm}^2}{\text{c}};$$

in the ocean:

$$c_p = 4.2 \cdot 10^7 \frac{\text{erg}}{\text{g} \cdot \text{grad}}, \quad \rho = 1.0 \frac{\text{g}}{\text{cm}^3}, \quad \mu = 10^8 \frac{\text{cm}^2}{\text{c}}, \quad \nu = \nu(\lambda, \psi, t);$$

in the soil:

$$c_p = 10^7 \frac{\text{erg}}{\text{g} \cdot \text{grad}}, \quad \rho = 2.7 \frac{\text{g}}{\text{cm}^3}, \quad \mu = 0, \quad \nu = 3 \cdot 10^{-3} \frac{\text{cm}^2}{\text{c}}.$$

The field of oceanic currents used in these calculations, was computed from climatic data on temperature and salinity [250], the monthly mean fields of u and v for the atmosphere for observed in 1961–1970 were taken from aerological climate archive of the World Data Center (Obninsk, Russia).

Let us present the results of solutions of adjoint problems for climatic mean December anomaly of a surface temperature over the territories of Europe and Northern America. The functionals of (7.6.8) type were examined being determined by the characteristic (7.6.9). The interval $I_1 = (\bar{t} - \tau, \bar{t})$ equal to one month corresponds to the December. The whole time interval $(0, \bar{t})$ corresponds to 12 months, that is, $N = 12$, $\tau_k = 30$ days in formula (7.6.33).

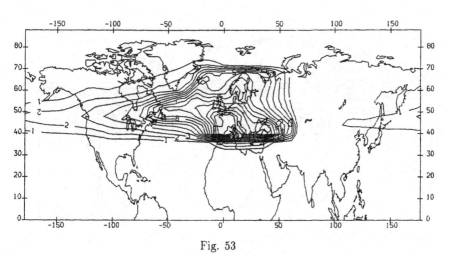

Fig. 53

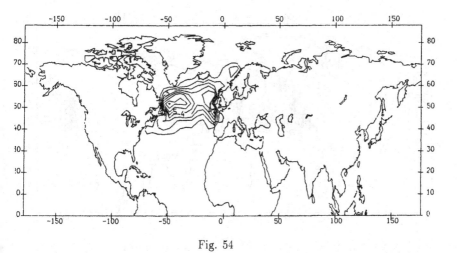

Fig. 54

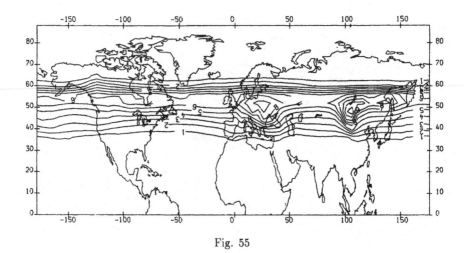

Fig. 55

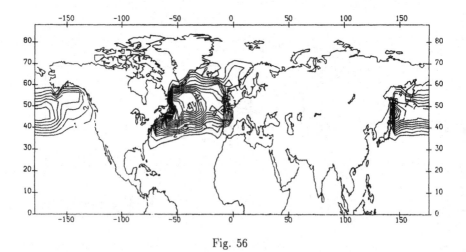

Fig. 56

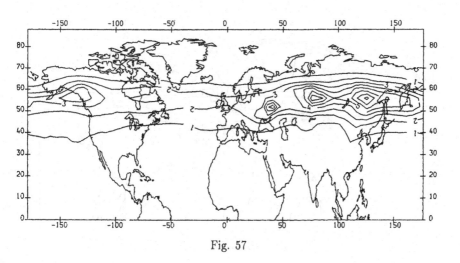

Fig. 57

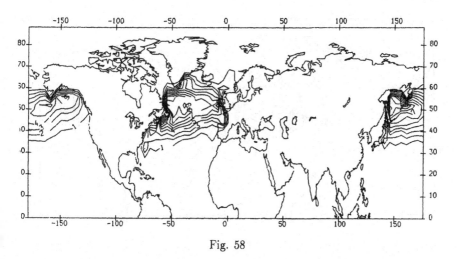

Fig. 58

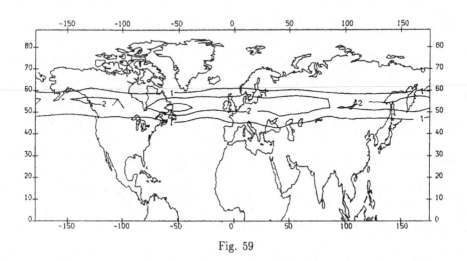

Fig. 59

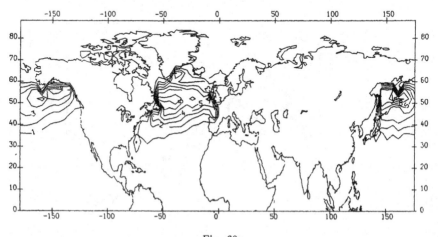

Fig. 60

In order to illustrate the qualitative picture we will limit ourselves
with averaged characteristics for all the atmosphere layer $(0 < z < h_1)$,

$$T_k^* = \frac{1}{\tau_k} \frac{1}{h_1} \int_{t_k}^{t_{k-1}} \int_0^{h_1} T^*(\lambda, \psi, z, t) dz dt \qquad (7.6.38)$$

and for the oceanic layer $(-h_2 < z < 0)$,

$$T_k^* = \frac{1}{\tau_k} \frac{1}{h_2} \int_{t_k}^{t_{k-1}} \int_{-h_2}^0 T^*(\lambda, \psi, z, t) dz dt \qquad (7.6.39)$$

since all the functions T^* at all levels of the system atmosphere–ocean
are too numerous. Of course, while computing the functionals in applications all the values of T^* and other parameters are needed at all
levels in atmosphere and ocean.

For simplicity we will refer to the functions (7.6.38), (7.6.39) on the
diagrams as influence functions for the atmosphere and ocean.

Consider two experiments. The source $F^*(\lambda, \psi, t)$ (see (7.6.9)) in the
first experiment differed from zero in December over the Europe, in the
second experiment – over the Northern America. Figures 53, 55, 57,
59 show the isolines of the influence function for the atmosphere in the
first experiment for $k = 1$, 3, 6, 9 respectively. Figures 54, 56, 58, 60
show the development of integral influence function in the upper 200-m
.ayer of the ocean in the same experiment (contours starting with 0 are
;paced by 0.2).

Since the adjoint problem is solved in the direction opposite to the
·eal time direction, the local maxima of the function T_k^* (and, respec-
ively the influence zones) evolve backwards in time (with increasing k),
ιp the flows of air and water masses.

By (7.6.9) the source $F^*(\lambda, \psi, t)$ is localized in the region ω and
liffers from zero just on the interval $I_1 = (\bar{t} - \tau, \bar{t})$. The atmospheric
ιfluence function in the first experiment for this time propagates over
ll the Europe, part of Atlantic and Northern America (see Fig. 53).
·he atmospheric influence function T_3^* (see Fig. 55) propagates due
) the west-east transport in the atmosphere over all the latitude belt
etween 35° and 65° N with small local maxima in Europe (in the
eighborhood of 52° N and 25° E) and in the region of Mongolia (48° N

and 105° E). The first maximum is caused by the residual influence of
the source $F^*(\lambda, \psi, t)$ and the second by local convergence of climatic
wind in this region. Substantial dissipation of the atmospheric signal
is observed to the sixth time interval with an amplitude decreased by
the factor of 5 as compared to the amplitude of T_1^* to the nineth time
interval. As we see from the Fig. 59, the integral influence function for
the atmosphere dissipates nearly completely during 9 time intervals.

The temporal-spatial structure of the function T_k^* in the upper 200-
meter layer of the ocean (Fig. 54, 56, 58, 60) is determined with increasing k by the processes of advection, turbulent mixing and vertical
exchange at the separation surfaces of interacting media. One must
mark the coincidence of positions of local maxima of adjoint function
and those of spatial distribution of vertical turbulent mixing coefficient
ν in the ocean (see Fig. 52). So, during the first time interval gradual
"penetration" begins of influence function into the upper ocean layers
of Northern Atlantic (see Fig. 54). In three time intervals local maxima
of T_3^* in the ocean (see Fig. 52) begin to appear distinctly in the main
energetically active zones (Newfoundland, Gulf Stream and Kuroshio).
Thus, energetically active zones affect decisively the position and development of local maxima of the functions T_k^* and, therefore, the formation of monthly mean anomalies of air temperature over the Europe.

Fig. 54, 56, 58, 60 show propagating influence function in the ocean
up the main streams and especially along the eastern coast of Northern
America in the Gulf Stream region. As the amplitude of T_k^* decreases
gradually, its intensity in the Northern Atlantic and Western part of
the Pacific become comparable (at $k = 6, 9$, see Fig. 58, 60).

The analogous processes are observed in the second experiment carried out for the territory of Northern America. The influence function
for the atmosphere evolve during the first time interval (Fig. 61) in
correspondence with a structure of climatic wind over the Northern
American continent. Later it fills the latitude belt $20° - 65°$ N (k=3,
Fig. 63) and dissipates gradually at $k = 6, 9$ (Figs. 65, 67). The Figs.
62, 64, 66, 68 show the dynamics of the influence function in the second
experiment in the upper layer of the ocean for $k = 1, 3, 6, 9$ respectively
(contours staring with 0 are spaced by 0.2). It increases during the first

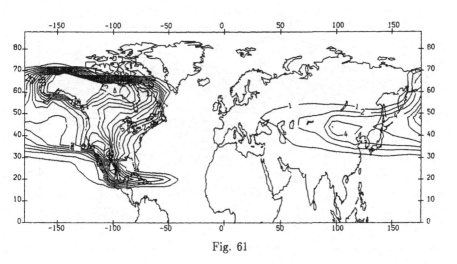

Fig. 61

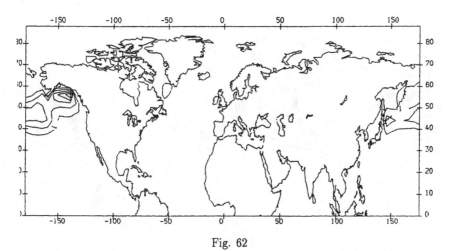

Fig. 62

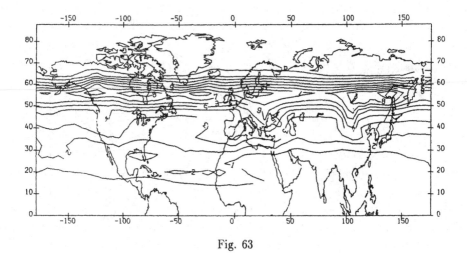

Fig. 63

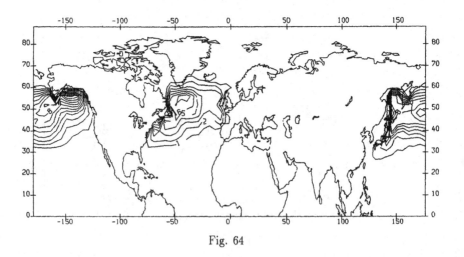

Fig. 64

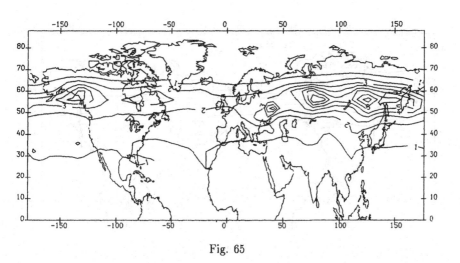

Fig. 65

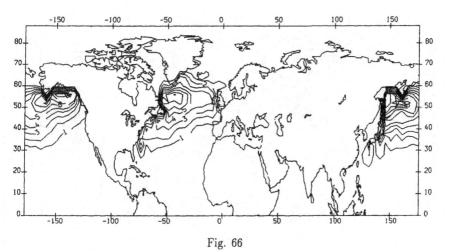

Fig. 66

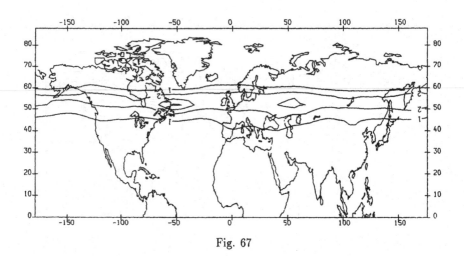

Fig. 67

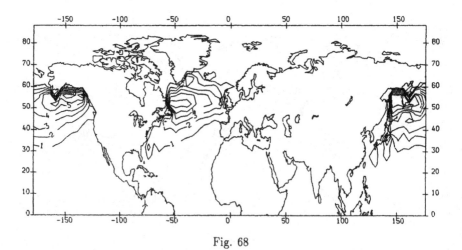

Fig. 68

time interval in the Pacific in the region of Kuroshio and Aleutian islands. At $k = 3$, 6, 9 the influence function, propagating along the latitude belt in the atmosphere, begins to penetrate into the Northern Atlantic though its amplitude stays maximum in Pacific energetically active zones.

The analysis of these diagrams suggests that during the first time intervals atmospheric processes dominate in formation of anomalies over Europe and Northern America and then a role of the ocean increases continuously, and in six time intervals of numerical simulation begins to weaken.

It is possible to obtain more precise information on sensitivity regions on the basis of calculations of T^* at all the levels in the atmosphere, ocean and soil of the continents. Naturally, calculations will require the knowledge of actual fields of wind velocity and oceanic currents' vector, radiation fluxes at $z = 0$ and the values of coefficients ν and μ.

The analysis of numerical experiments shows also, that spatial-temporary structure of solutions of adjoint problems reveals what energetically active zones of the World Ocean, and in what time may affect the formation of monthly mean anomalies of temperature over the selected limited region. This opens the possibility to determine the spatial-temporary characteristics of the processes of formation of linear response of the interacting system atmosphere–ocean–soil to perturbations of initial fields of temperature and external heat sources. This is the basis for the development of optimal network of observations over the World Ocean and adaptation of its data to the processes in the atmosphere.

'.7. Application of the Sensitivity Theory to the Comparison of Mathematical Models

Range of investigation of various mathematical models describing those r another processes in science and technology is rather wide, therefore he researchers use for the same problems various models, parameterizations and numerical algorithms. For example, there are more than

30 different models designed for investigating the general circulation of atmosphere which, naturally, produce different results, though many of them reflect general features of the atmosphere dynamics. W. L. Gates [193] suggested the idea to compare the results of different models and formulated some general principles of realizations, convenient for the comparison of the results. Numerical experiments with different models and comparison of the results were carried out under his guidance in the National Livermore Laboratory, USA, yielding enormous material for suggestions on quality and features of those models. Conservation laws allowed to refine the conclusions and estimate objectively the models. So, hopefully, the start was done to the comparison of the results of solutions of complex problems.

So, let we have two different linear models varying a bit with their operators. Let them be the evolutionary problems

$$\frac{\partial \varphi_1}{\partial t} + A_1 \varphi_1 = f, \quad t \in (0, \bar{t})$$
$$\varphi_1 = g \quad \text{at} \quad t = 0 \tag{7.7.1}$$

and

$$\frac{\partial \varphi_2}{\partial t} + A_2 \varphi_2 = f, \quad t \in (0, \bar{t})$$
$$\varphi_2 = g \quad \text{at} \quad t = 0 \tag{7.7.2}$$

We assume here that the right-hand parts of the equations – the sources (forcings) – coincide and are equal to f. Initial data g coincide too. Let $f = f(t)$ and g be assigned functions from some real Hilbert space H. Assume that A_1, A_2 are linear operators in the Hilbert space H with the definition domains $D(A_1)$ and $D(A_2)$ respectively. Each element of the sets $D(A_1)$ and $D(A_2)$ is assumed to satisfy corresponding smoothness conditions, additional conditions (boundary, for instance) and some other requirements according to the essence of problem. Assume that the solutions $\varphi_1(t)$ and $\varphi_2(t)$ of problems (7.7.1) and (7.7.2) exist, are unique and belong to $D(A_1)$ and $D(A_2)$ respectively at any $t \in [0, \bar{t}]$ and are sufficiently smooth.

Let us examine the following functional:

$$J = \int_0^{\bar{t}} (p, \varphi) dt \tag{7.7.3}$$

where φ is a solution of problem (7.7.1) or (7.7.2), p is a fixed element from H characterizing a measurement and $(\cdot, \cdot)$ denotes an inner product in H. Let

$$J_1 = \int_0^{\bar{t}} (p, \varphi_1) dt \qquad (7.7.4)$$

$$J_2 = \int_0^{\bar{t}} (p, \varphi_2) dt \qquad (7.7.5)$$

According to the general theory, construct the adjoint equations corresponding to problems (7.7.1) and (7.7.2):

$$-\frac{\partial \varphi_1^*}{\partial t} + A_1^* \varphi_1^* = p, \quad t \in (0, \bar{t}) \\ \varphi_1^* = 0 \quad \text{at} \quad t = \bar{t} \qquad (7.7.6)$$

and

$$-\frac{\partial \varphi_2^*}{\partial t} + A_2^* \varphi_2^* = p, \quad t \in (0, \bar{t}) \\ \varphi_2^* = 0 \quad \text{at} \quad t = \bar{t} \qquad (7.7.7)$$

where p is a function determining the functional (7.7.3), A_1^* and A_2^* are the operators adjoint to A_1 and A_2 respectively, satisfying the Lagrange identities:

$$(A_\alpha u, v) = (u, A_\alpha^* v), \quad u \in D(A_\alpha), \quad v \in D(A_\alpha^*), \quad \alpha = 1, 2$$

Assume that solutions $\varphi_1^*(t)$ and $\varphi_2^*(t)$ of adjoint problems (7.7.6) and (7.7.7) exist, are unique and belong to $D(A_1^*)$ and $D(A_2^*)$ respectively at any $t \in [0, \bar{t}]$ and are sufficiently smooth.

Perform some transformations. Multiply the equation from (7.7.1) by φ_1^* in H, the equation from (7.7.6) by φ_1 subtract the results from each other and integrate in t on the interval $0 \leq t \leq \bar{t}$ to obtain

$$\int_0^{\bar{t}} \frac{\partial}{\partial t}(\varphi_1, \varphi_1^*) dt + \int_0^{\bar{t}} [(A_1 \varphi_1, \varphi_1^*) - (\varphi_1, A_1^* \varphi_1^*)] dt$$

$$= \int_0^{\bar{t}} (f, \varphi_1^*) dt - \int_0^{\bar{t}} (\varphi_1, p) dt \quad (7.7.8)$$

By virtue of the Lagrange identity

$$(A_1\varphi_1, \varphi_1^*) - (\varphi_1, A_1^*\varphi_1^*) = 0$$

at all $t \in [0, \bar{t}]$. As concerns the first term in the left-hand part of
(7.7.8), if φ_1, φ_1^* are sufficiently smooth, it will be equal to

$$\int_0^{\bar{t}} \frac{\partial}{\partial t}(\varphi_1, \varphi_1^*)dt = (\varphi_1, \varphi_1^*)|_{t=\bar{t}} - (\varphi_1, \varphi_1^*)|_{t=0} \qquad (7.7.9)$$

Since $\varphi_1^* = 0$ at $t = \bar{t}$, the first term in the right-hand part of (7.7.9)
turns to zero, and the second term due to the condition $\varphi_1 = g$ at $t = 0$
equals

$$-(\varphi_1, \varphi_1^*)|_{t=0} = -(g, \varphi_1^*)|_{t=0}$$

As a result it follows from (7.7.8) that

$$J_1 = \int_0^{\bar{t}} (f, \varphi_1^*)dt - (g, \varphi_1^*|_{t=0}) \qquad (7.7.10)$$

The relationships of (7.7.10) type may be obtained for the second couple
of problems (7.7.2) and (7.7.7) in the same way. Here the functional
under study is J_2, so we have

$$J_2 = \int_0^{\bar{t}} (f, \varphi_2^*)dt - (g, \varphi_2^*|_{t=0}) \qquad (7.7.11)$$

Subtract relationship (7.7.11) from (7.7.10) to obtain the formula for
$\delta J = J_1 - J_2$

$$\delta J = \int_0^{\bar{t}} (f, \delta\varphi^*)dt - (g, \delta\varphi^*|_{t=0}) \qquad (7.7.12)$$

where $\delta\varphi^* = \varphi_1^* - \varphi_2^*$

The difference $\delta J = J_1 - J_2$ of the functionals of two models will be
thus determined by the function $\delta\varphi^* = \varphi_1^* - \varphi_2^*$. The function $\delta\varphi^*$ will
be an influence function for δJ or a weight determining a contribution
of the forcing f and initial value g to the functional δJ. Generally
speaking, the less is δJ, the smaller is a difference between two models
(as related to the functional J).

But, what is more essential, formula (7.7.12) allows to detect the regions in space and in time where the differences in the functions $\delta\varphi^*$ are the biggest, which is very important information for further analysis and refinement of the models.

We have considered the case when the source function f is the same for both models. If the sources are different, the problems (7.7.1) and (7.7.2) will be rewritten as

$$\frac{\partial\varphi_1}{\partial t} + A_1\varphi_1 = f_1, \quad t \in (0,\bar{t})$$
$$\varphi_1 = g \quad \text{at} \quad t = 0 \tag{7.7.13}$$

and

$$\frac{\partial\varphi_2}{\partial t} + A_2\varphi_2 = f_2, \quad t \in (0,\bar{t})$$
$$\varphi_2 = g \quad \text{at} \quad t = 0 \tag{7.7.14}$$

where $f_1 = f_1(t)$, $f_2 = f_2(t) \in H$. We assume here that initial data coincide in both models, i. e., $\varphi_1 = \varphi_2 = g$ at $t = 0$.

Consider, as before, the values of functionals

$$J_1 = \int_0^{\bar{t}} (p, \varphi_1)dt \tag{7.7.15}$$

$$J_2 = \int_0^{\bar{t}} (p, \varphi_2)dt \tag{7.7.16}$$

Assume again, that the function $p \in H$ which is a characteristic of the functional J, is the same for both models.

Consider the adjoint problems

$$-\frac{\partial\varphi_1^*}{\partial t} + A_1^*\varphi_1^* = p, \quad t \in (0,\bar{t})$$
$$\varphi_1^* = 0 \quad \text{at} \quad t = \bar{t} \tag{7.7.17}$$

and

$$-\frac{\partial\varphi_2^*}{\partial t} + A_2^*\varphi_2^* = p, \quad t \in (0,\bar{t})$$
$$\varphi_2^* = 0 \quad \text{at} \quad t = \bar{t} \tag{7.7.18}$$

Obtain, as before, the relationships of (7.7.10) and (7.7.11) type:

$$J_1 = \int_0^{\bar{t}} (f_1, \varphi_1^*)dt - (g, \varphi_1^*|_{t=0}) \tag{7.7.19}$$

$$J_2 = \int_0^{\bar{t}} (f_2, \varphi_2^*) dt - (g, \varphi_2^*|_{t=0}) \qquad (7.7.20)$$

Subtract from each other to obtain

$$\delta J = \int_0^{\bar{t}} [(f_1, \varphi_1^*) - (f_2, \varphi_2^*)] dt - (g, \delta \varphi^*|_{t=0}) \qquad (7.7.21)$$

where $\delta \varphi^* = \varphi_1^* - \varphi_2^*$. Let

$$\begin{aligned} f_1 &= f + \delta f_1 \\ f_2 &= f + \delta f_2 \end{aligned} \qquad (7.7.22)$$

where $f = f(t)$, $\delta f_1 = \delta f_1(t)$, $\delta f_2 = \delta f_2(t)$ are assigned functions from H. For example, $f = (f_1 + f_2)/2$ may be assumed as f, or one of the functions, say, $f = f_1$ (then $\delta f_1 = 0$), or climatic values of these functions. Then

$$\delta J = \int_0^{\bar{t}} (f, \delta \varphi^*) dt - (g, \delta \varphi^*|_{t=0}) + \int_0^{\bar{t}} [(\delta f_1, \varphi_1^*) - (\delta f_2, \varphi_2^*)] dt \quad (7.7.23)$$

The formula just obtained is somewhat more complicated than (7.7.12) but it is more informative since helps analyzing two more complex models.

If the functions p are taken as different, then the comparison of the models may be carried out with different functionals, which is important for obtaining more complete comparison.

Finally, a few words on nonlinear models. Examine the problems

$$\begin{aligned} \frac{\partial \varphi_1}{\partial t} + A_1(\varphi_1)\varphi_1 &= f, \quad t \in (0, \bar{t}) \\ \varphi_1 &= g \quad \text{at} \quad t = 0 \end{aligned} \qquad (7.7.24)$$

and

$$\begin{aligned} \frac{\partial \varphi_2}{\partial t} + A_2(\varphi_2)\varphi_2 &= f, \quad t \in (0, \bar{t}) \\ \varphi_2 &= g \quad \text{at} \quad t = 0 \end{aligned} \qquad (7.7.25)$$

where $A_1(\varphi_1)$, $A_2(\varphi_2)$ are linear operators in the Hilbert space H with the definition domains $D(A_1)$, $D(A_2)$ respectively, $f = f(t)$, g are assigned functions from H. The definition domains $D(A_1)$, $D(A_2)$ of the operators A_1 and A_2 are determined by the essence of the problem. Assume that the solutions $\varphi_1(t)$ and $\varphi_2(t)$ of problems (7.7.24) and (7.7.25) at each $t \in [0, \bar{t}]$ belong to $D(A_1)$ and $D(A_2)$ respectively and are sufficiently smooth functions.

Consider the functionals

$$J_1 = \int_0^{\bar{t}} (p, \varphi_1) dt \tag{7.7.26}$$

$$J_2 = \int_0^{\bar{t}} (p, \varphi_2) dt \tag{7.7.27}$$

where p is an assigned function from H.

Write the adjoint problems

$$-\frac{\partial \varphi_1^*}{\partial t} + A_1^*(\varphi_1)\varphi_1^* = p, \quad t \in (0, \bar{t}) \tag{7.7.28}$$
$$\varphi_1^* = 0 \quad \text{at} \quad t = \bar{t}$$

and

$$-\frac{\partial \varphi_2^*}{\partial t} + A_2^*(\varphi_2)\varphi_2^* = p, \quad t \in (0, \bar{t}) \tag{7.7.29}$$
$$\varphi_2^* = 0 \quad \text{at} \quad t = \bar{t}$$

where $A_1^*(\varphi_1)$ and $A_2^*(\varphi_2)$ are the operators adjoint to $A_1(\varphi_2)$ and $A_2(\varphi_2)$ respectively, satisfying the Lagrange identities (see Section 1.8 of Chapter 1)

$$(A_\alpha(u)v, w) = (v, A_\alpha^*(u)w), \quad u, v \in D(A_\alpha), \quad w \in D(A_\alpha^*), \quad \alpha = 1, 2 \tag{7.7.30}$$

Assume that solutions $\varphi_1^*(t)$ and $\varphi_2^*(t)$ of problems (7.7.6) and (7.7.7) exist at known $\varphi_1(t)$ and $\varphi_2(t)$, are unique, belong to $D(A_1^*)$ and $D(A_2^*)$ respectively at each $t \in [0, \bar{t}]$ and are sufficiently smooth functions.

Obtain new expressions for J_1 and J_2 through φ_1^* and φ_2^* with regard to the Lagrange identity (7.7.30):

$$J_1 = \int_0^{\bar{t}} (f, \varphi_1^*) dt - (g, \varphi_1^*|_{t=0}) \tag{7.7.31}$$

$$J_2 = \int_0^{\bar{t}} (f, \varphi_2^*) dt - (g, \varphi_2^*|_{t=0}) \tag{7.7.32}$$

and as a result arrive at the formula analogues to (7.7.12)

$$\delta J = \int_0^{\bar{t}} (f, \delta\varphi^*)dt - (g, \delta\varphi^*|_{t=0}) \qquad (7.7.33)$$

where $\delta J_1 = J_1 - J_2$, $\delta\varphi^* = \varphi_1^* - \varphi_2^*$.

Another generalizations of the formula for δJ are possible corresponding to assumptions for various models.

We discussed just differential statements of problems, though it is possible to compare the models based on various difference approximations of problems with regard to parameterizations used there.

Now, consider the most important question of modelling the processes. Let us, for definiteness, deal with estimation of a change of certain climatic characteristic determined by the functional of (7.7.3) type:

$$J = \int_0^{\bar{t}} (p, \varphi)dt \qquad (7.7.34)$$

Assume that the functional computed by formula (7.7.34) differs from the functional obtained by direct measurements. We need to find the reason of this difference.

Since numerous factors affect the functional J, it is practically impossible to get a definite answer using the main equations, except, may be, for some simple cases.

It is very convenient to solve an adjoint problem and obtain the functional in the above terms:

$$J = \int_0^{\bar{t}} (f, \varphi^*)dt - (g, \varphi^*|_{t=0}) \qquad (7.7.35)$$

The influence function φ^* computed from initial data allows to detect the most meaningful (sensitive) regions and other input data determining the functional J. This function helps understand the causes of those of another changes of J which are sometimes unpredictable. At the same time it provides the ground for refining the mathematical model which more or less adequately reflects the sensitivity regions of the functional on real data.

CHAPTER 8

Adjoint Equations in Data Processing Problems

The problem of obtaining and processing measurement data in various fields of knowledge grows more important nowadays. This problem may be modelled mathematically as multidimensional (space and time) data assimilation and data processing problem which is one of optimal control problems.

The problems of this type attracted an attention of many specialists applying optimal control methods to practical solving various problems, and were studied by R. Bellman [175], L. S. Pontryagin [139, 240], N. N. Krasovski [53, 54], J.-L. Lions [210–214], R. Glowinski [195, 196], A. Balakrishnan [170], G. I. Marchuk [74, 79, 220]. These methods were advanced most of all in nuclear power engineering, physics of atmosphere and ocean, environment protection (see G. I. Marchuk [70, 81, 88], V. V. Penenko and N. N. Obraztsov [238], G. R. Kontarev [200], J. Lewis and J. Derber [185, 186, 209], F. X. Le Dimet et. al. [204, 205, 263, 267], P. Courtier and O. Talagrand [184, 259], A. C. Lorenc [217], G. I. Marchuk and V. I. Agoshkov [6, 95, 227], G. I. Marchuk and V. B. Zalesny [233]), J. M. Wallace [262].

It was found, that the statements of these problems may be formulated with properly selected adjoint equations. In this chapter we shall consider the role of adjoint equations in data processing problems. We shall start with exemplifying the idea by simple evolutionary problem and then examine general numerical algorithm for solution of data assimilation problem. The material of this chapter is based on the works of this author together with V. I. Agoshkov, V. B. Zalesny and V. P. Shutyaev [227, 232, 233].

8.1. Data Assimilation problem for Evolutionary Equation

Examine the evolutionary problem

$$\frac{\partial \varphi}{\partial t} + A(t)\varphi = f \quad \text{in} \quad \Omega \times (0, \bar{t})$$
$$\varphi = u \quad \text{at} \quad t = 0 \tag{8.1.1}$$

where $A(t)$ is linear operator, $\varphi(x, t)$ for each t belongs to a set of functions $D(A)$ with a definition domain $\Omega \subset \mathbf{R}^n$, $n \geq 1$. The functions $f(x, t)$, $u(x)$ for each $t \in [0, \bar{t}]$ are assumed to be quadratically summable in x on Ω and to belong to real Hilbert space $H = L_2(\Omega)$ where conventional inner product $(\cdot, \cdot)^{1/2}$ and a norm $\|\cdot\| = (\cdot, \cdot)^{1/2}$ are assigned. The operator $A(t)$ operates in the Hilbert space H with the definition domain $D(A)$. Assume that $\varphi(x, t)$ is sufficiently smooth so that the functions φ, $\partial \varphi / \partial t$, and $A(t)\varphi$ are quadratically summable on $\Omega \times (0, \bar{t})$.

Examine the functional $S(\varphi)$ of the type:

$$S(\varphi) = \frac{\alpha}{2} \|\varphi|_{t=0}\|^2 + \frac{\beta}{2} \int_0^{\bar{t}} \|\varphi - \hat{\varphi}\|^2 dt \tag{8.1.2}$$

where α, $\beta = const$, $\hat{\varphi} = \hat{\varphi}(x, t)$ is an assigned function.

The function $\hat{\varphi}(x, t)$ is determined a priori, as a rule, by observation data. In some cases $\hat{\varphi}$ may be obtained by interpolating the measurement data assigned on discrete set of the points (x_i, t_j) [233, 267]. The numbers α, β are weight coefficients usually referred to as regularization parameters [148, 227].

Examine problem (8.1.1) with unknown function $u \in H$ (i. e., with unknown control u) in initial condition:

$$\frac{\partial \varphi}{\partial t} + A(t)\varphi = f \quad \text{in} \quad \Omega \times (0, \bar{t})$$
$$\varphi = u \quad \text{at} \quad t = 0 \tag{8.1.3}$$

The data assimilation problem may be formulated as follows: find the functions φ, u, satisfying problem (8.1.3) with the functional $S(\varphi)$ adopting the least value:

$$S(\varphi) = \inf_u S(\varphi)$$

Assume that this problem has a solution and formulate a system of equations for the functions φ, u, using an adjoint problem. Assume instead of u in problem (8.1.3) the function $u_1 = u + \varepsilon u_0$, $u_0 \in H$, $\varepsilon > 0$ with corresponding solution φ_1:

$$\frac{\partial \varphi_1}{\partial t} + A(t)\varphi_1 = f \quad \text{in} \quad \Omega \times (0, \bar{t})$$
$$\varphi_1 = u_1 \quad \text{at} \quad t = 0 \tag{8.1.4}$$

Since problem (8.1.4) is linear,

$$\varphi_1 = \varphi + \varepsilon \varphi_0$$

where φ is a solution of problem (8.1.3) and φ_0 is a solution of the problem

$$\frac{\partial \varphi_0}{\partial t} + A(t)\varphi_0 = 0 \quad \text{in} \quad \Omega \times (0, \bar{t})$$
$$\varphi_0 = u_0 \quad \text{at} \quad t = 0 \tag{8.1.5}$$

We obtain for the functional $S(\varphi_1)$

$$S(\varphi_1) = \frac{\alpha}{2}\|u + \varepsilon u_0\|^2 + \frac{\beta}{2}\int_0^{\bar{t}} \|\varphi + \varepsilon\varphi_0 - \hat{\varphi}\|^2 dt \tag{8.1.6}$$

Since φ, u represent solution to data assimilation problem, the functional $S(\varphi_1)$ is minimal at $\varphi_1 = \varphi$, i. e., when $\varepsilon = 0$. This means that

$$\frac{dS}{d\varepsilon}\bigg|_{\varepsilon=0} = 0 \tag{8.1.7}$$

Since

$$\|u + \varepsilon u_0\|^2 = (u + \varepsilon u_0, \, u + \varepsilon u_0) = (u, \, u) + 2\varepsilon(u_0, \, u) + \varepsilon^2(u_0, \, u_0)$$

$$\|\varphi + \varepsilon\varphi_0 - \hat{\varphi}\|^2 = (\varphi - \hat{\varphi}, \, \varphi - \hat{\varphi}) + 2\varepsilon(\varphi_0, \, \varphi - \hat{\varphi}) + \varepsilon^2(\varphi_0, \, \varphi_0)$$

then by (8.1.6) we have

$$\frac{dS}{d\varepsilon} = \alpha(u_0, \, u) + \alpha\varepsilon(u_0, \, u_0) + \beta\int_0^{\bar{t}}(\varphi_0, \, \varphi_0 - \hat{\varphi})dt + \beta\varepsilon\int_0^{\bar{t}}(\varphi_0, \, \varphi_0)dt$$

Hence by condition (8.1.7) we arrive at the equation

$$\alpha(u_0, \, u) - J(\varphi_0) = 0 \tag{8.1.8}$$

where the functional $J(\varphi_0)$ is defined by the equality

$$J(\varphi_0) = \int\limits_0^{\bar{t}} (\varphi_0,\, p)dt \qquad (8.1.9)$$

at $p = \beta(\hat{\varphi} - \varphi)$.

Equation (8.1.8) contains the function φ_0 which is a solution to problem (8.1.5). Let us employ the dual representation for the functional $J(\varphi_0)$ in the left-hand part of (8.1.8) through a solution of adjoint problem, considering problem (8.1.5) as the main problem. According to § 1.7 from Chapter 1, match the latter with an adjoint problem of the type:

$$\frac{\partial \varphi^*}{\partial t} + A^*\varphi^* = p \quad \text{in} \quad \Omega \times (0, \bar{t})$$
$$\varphi^* = 0 \quad \text{at} \quad t = \bar{t} \qquad (8.1.10)$$

where p is a function assigned by formula (8.1.9), A^* is an operator adjoint to $A(t)$ and satisfying for each t the Lagrange identity:

$$(A\varphi,\, \varphi^*) = (\varphi,\, A^*\varphi^*), \quad \varphi \in D(A), \quad \varphi^* \in D(A^*)$$

Assume, that solutions to problems (8.1.5) and (8.1.10) are sufficiently smooth so that the functions φ_0, $\partial\varphi_0/\partial t$, $A(t)\varphi_0$, φ^*, $\partial\varphi^*/\partial t$, $A^*\varphi^*$ are quadratically summable on $\Omega \times (0, \bar{t})$. As in § 1.7, perform some transformations. Multiply the equation from (8.1.5) by φ^* and the equation from (8.1.10) by φ_0, subtract the results from each other and integrate over the interval $0 \leq t \leq \bar{t}$. Then by Lagrange identity we obtain the adjointment relationship

$$\int\limits_0^{\bar{t}} (\varphi_0,\, p)dt = (u_0,\, \varphi^*|_{t=0}) \qquad (8.1.11)$$

By (8.1.11) we obtain another representation for the functional $J(\varphi_0)$ defined by formula (8.1.9):

$$J(\varphi_0) = (u_0,\, \varphi^*|_{t=0}) \qquad (8.1.12)$$

With regard to (8.1.12) rewrite condition (8.1.8) as

$$\alpha(u_0,\, u) - (u_0,\, \varphi^*|_{t=0}) = 0$$

or

$$(u_0, \alpha u - \varphi^*|_{t=0}) = 0 \qquad (8.1.13)$$

Since the function $u_0 \in H$ is arbitrary, rewrite condition (8.1.13) as

$$\alpha u - \varphi^*|_{t=0} = 0 \qquad (8.1.14)$$

Thus, if φ, u represent a solution to the problem formulated above, then they satisfy problem (8.1.3) and condition (8.1.14), where φ^* is a solution to adjoint problem (8.1.10). Then, with regard to the form of p (see (8.1.9)) we obtain for φ, u the system of equations:

$$\frac{\partial \varphi}{\partial t} + A(t)\varphi = f \quad \text{in} \quad \Omega \times (0, \bar{t})$$
$$\varphi = u \quad \text{at} \quad t = 0 \qquad (8.1.15)$$

$$\frac{\partial \varphi^*}{\partial t} + A^*\varphi^* = \beta(\hat{\varphi} - \varphi) \quad \text{in} \quad \Omega \times (0, \bar{t})$$
$$\varphi^* = 0 \quad \text{at} \quad t = \bar{t} \qquad (8.1.16)$$

$$\alpha u - \varphi^*|_{t=0} = 0 \qquad (8.1.17)$$

So, the assimilation problem was reduced to system (8.1.15)–(8.1.17) for the functions φ and u.

If the function u is already found, then φ is to be found by solving the main problem (8.1.15). An equation for the control u is easily obtained from system (8.1.15)–(8.1.17) excluding there the functions p and φ^*. A method for solving problem (8.1.15)–(8.1.17) is chosen depending on this control equation.

We have considered in this section simple data assimilation problem for evolutionary equation. More general statements may be found in the works by J.-L. Lions [210–214] and by G. I. Marchuk and V. I. Agoshkov 6, 227]. The latter examine data assimilation problem for evolutionary equation in the case when problem's operator is nonlinear, solvability of these problems is investigated in proper functional spaces, numerical algorithms are proposed and justified.

We will discuss in the next section a numerical algorithm for solving this data assimilation problem.

8.2. Numerical Algorithm for Solving the Data Assimilation Problem

Let us consider one of numerical algorithms for solving the data assimilation problem. As before, examine an evolutionary problem of the type (8.1.1) as an object of investigation. Assume for simplicity, that we have already approximated the original problem (8.1.1) in all variables (except for t) and obtained the system

$$\frac{d\varphi^h}{dt} + A^h(t)\varphi^h = f^h \quad \text{in} \quad \Omega^h \times (0, \bar{t})$$
$$\varphi^h = u^h \quad \text{at} \quad t = 0 \tag{8.2.1}$$

where Ω^h is a grid domain; $A^h(t)$ is a real square matrix of an order n; φ^h, f^h, u^h are grid vector-functions of a length n. Let H denote a space of real grid functions with a norm $\| \cdot \|_h = (\cdot, \cdot)_h^{1/2}$. A grid parameter h will be omitted hereafter to simplify the notations.

Consider the functional $S(\varphi)$, as before, in the form:

$$S(\varphi) = \frac{\alpha}{2}\|\varphi|_{t=0}\|^2 + \frac{\beta}{2} \int_0^{\bar{t}} \|\varphi - \hat{\varphi}\|^2 dt \tag{8.2.2}$$

The data assimilation problem is nothing more than a task to find the vector-function $\varphi(t)$ and control vector u so, that

$$\frac{d\varphi}{dt} + A(t)\varphi = f \quad \text{in} \quad \Omega \times (0, \bar{t})$$
$$\varphi = u \quad \text{at} \quad t = 0$$
$$S(\varphi) = \inf_u S(\varphi) \tag{8.2.3}$$

and the last is reduced, as follows from § 8.1, to the system

$$\frac{d\varphi}{dt} + A(t)\varphi = f \quad \text{in} \quad \Omega \times (0, \bar{t})$$
$$\varphi = u \quad \text{at} \quad t = 0 \tag{8.2.4}$$

$$-\frac{d\varphi^*}{dt} + A^*\varphi^* = \beta(\hat{\varphi} - \varphi) \quad \text{in} \quad \Omega \times (0, \bar{t})$$
$$\varphi^* = 0 \quad \text{at} \quad t = \bar{t} \tag{8.2.5}$$

$$\alpha u - \varphi^*|_{t=0} = 0 \tag{8.2.6}$$

vhere $A^* = A^T$ is the matrix transposed to A.

Assume, that the vector-functions $f(t)$ and $\hat{\varphi}(t)$ are continuous in t ιn the interval $[0, T]$ and the matrix $A(t)$ from (8.2.3) is positive definite ιnd does not depend on t i. e., $A(t) = A > 0$. It is easy to show [23] that ιn this simple case problems (8.2.3) and (8.2.4)–(8.2.6) are equivalent ιnd have unique solutions. Here the solutions of the main and adjoint ιroblems are understood in classic meaning.

Before we formulate an algorithm for numerical solution of problεm (8.2.4)–(8.2.6), let us obtain an equation for the unknown vectorιunction u. Assuming that $f(t)$ is continuous, write the solution to ιroblem (8.2.4) at $A(t) = A$ in explicit form [37]:

$$\varphi(t) = e^{-At}u + \varphi_1(t) \tag{8.2.7}$$

ιhere

$$\varphi_1(t) = \int_0^t e^{-A(t-s)} f(s)\, ds$$

ιere e^{-At} is a function of a matrix which may be determined [37] ιhrough the formula:

$$e^{-At} = \sum_{k=0}^{\infty} \frac{(-At)^k}{k!}, \quad (A_0 = E)$$

ιhere E is a unit matrix. Similarly we obtain from (8.2.5) that

$$\varphi^*(t) = \int_t^{\bar{t}} e^{-A(t-s)} \beta \left[\hat{\varphi}(s) - \varphi(s)\right] ds \tag{8.2.8}$$

ιubstitute (8.2.7) into (8.2.8) to obtain

$$\varphi^*(t) = \beta \int_t^{\bar{t}} e^{A^*(t-s)} \left[\hat{\varphi}(s) - \varphi_1(s)\right] ds - \beta \int_t^{\bar{t}} e^{-A^*(t-s)} e^{-As} u \, ds$$

ιence, assuming that A is normal matrix (i. e., $AA^* = A^*A$) we arrive ι the equality

$$\varphi^*(0) = \beta \int_0^{\bar{t}} e^{-A^*s} \left[\hat{\varphi}(s) - \varphi_1(s)\right] ds - \beta \int_0^{\bar{t}} e^{-(A+A^*)s} u \, ds \tag{8.2.9}$$

By (8.2.6) and (8.2.9) we obtain the equation for the control u:

$$Lu = F \qquad (8.2.10)$$

where

$$F = \beta \int_0^{\bar{t}} e^{-A^* s} \left[\hat{\varphi}(s) - \varphi_1(s) \right] ds$$

$$L = \alpha E + \beta \int_0^{\bar{t}} e^{-(A+A^*)s} ds = \alpha E + \beta (A + A^*)^{-1} (E - e^{-(A^*+A)\bar{t}})$$

Write the matrix L of system (8.2.10) in the form

$$L = \alpha E + \beta M^{-1}(E - e^{-M\bar{t}}) \qquad (8.2.11)$$

where

$$M = A + A^*$$

Obviously, at α, $\beta > 0$ the matrix L is symmetric and positive definite. Thus, system (8.2.10) has a unique solution even if one of regularization parameters (α or β) is equal to zero. We assume hereinafter, that $\beta > 0$.

If $\alpha = 0$ then the parameter β in original statement may be regarded as equal to the unit, and the equation for the control u takes the form

$$L_1 u = F_1 \qquad (8.2.12)$$

with symmetric positive definite matrix $L_1 = M^{-1}(E - e^{-M\bar{t}})$ and the right-hand part

$$F_1 = \int_0^{\bar{t}} e^{-A^* s} [\hat{\varphi}(s) - \varphi_1(s)] ds$$

Examine a problem on eigenvalues for the matrix M:

$$M u_k = \lambda_k u_k \qquad (8.2.13)$$

With these assumptions the matrix M has positive eigenvalues λ_k and corresponding eigenvectors compose a basis in H. It is easy to show that u_k are also eigenvectors of the matrix L i. e., they are solutions to the problem

$$L u_k = \mu_k u_k \qquad (8.2.14)$$

with

$$\mu_k = \alpha + \beta \frac{1 - e^{-\lambda_k \bar{t}}}{\lambda_k} \tag{8.2.15}$$

where λ_k are eigenvalues of the matrix M.

Since the matrix L is symmetric, the inequality

$$(Lu, u) \geq \mu_{\min}(u, u), \quad u \in H \tag{8.2.16}$$

holds where

$$\mu_{\min} = \alpha + \beta \frac{1 - e^{-\lambda_{\max} \bar{t}}}{\lambda_{\max}}$$

and $\lambda_{\max} = \lambda_{\max}(M)$ is a maximum eigenvalue of the matrix $M = A + A^*$.

The following estimate for the solution u of problem (8.2.10) is valid by (8.2.16):

$$\|u\| \leq \frac{1}{\mu_{\min}} \|F\| \tag{8.2.17}$$

Equation (8.2.10) transforms into equation (8.2.12) in the case of $\alpha = 0$. Let $u = \bar{u}$ be a solution of equation (8.2.12) for which the estimate holds

$$\|\bar{u}\| \leq \frac{1}{\bar{\mu}_{\min}} \|F_1\| \tag{8.2.18}$$

where

$$\bar{\mu}_{\min} = \frac{1 - e^{-\lambda_{\max} \bar{t}}}{\lambda_{\max}}$$

By (8.2.12) and (8.2.10) obtain an equation for the difference $\xi = u - \bar{u}$:

$$\beta L_1(u - \bar{u}) = -\alpha u \tag{8.2.19}$$

Hence, by the estimates (8.2.17) and (8.2.18) we arrive at the inequality

$$\|u - \bar{u}\| \leq \frac{\alpha}{\mu_{\min} \bar{\mu}_{\min}} \|F_1\|$$

or

$$\|u - \bar{u}\| \leq \frac{\alpha \|F_1\|}{\frac{1 - e^{-\lambda_{\max} \bar{t}}}{\lambda_{\max}} \left(\alpha + \beta \frac{1 - e^{-\lambda_{\max} \bar{t}}}{\lambda_{\max}} \right)} \tag{8.2.20}$$

The solution u of problem (8.2.10) thus converges to the solution $\bar{u}$ of problem (8.2.12) at $\alpha \to 0$.

Consider now iterative methods for solving the system (8.2.10). Most of iterative methods applied to solving linear systems may be united by single universal formula:

$$B_j \frac{u^{j+1} - u^j}{\tau_j} = -(Lu^j - F) \qquad (8.2.21)$$

where $\{B_j\}$ is a sequence of nonsingular matrices and $\{\tau_j\}$ is a sequence of real parameters. Some extra restrictions are applied, as a rule, to the matrices B_j. We will discuss them later. Iterative process (8.2.21) may be written in the form:

$$u^{j+1} = T_j u^j + F^j \qquad (8.2.22)$$

where

$$T_j = E - \tau_j B_j^{-1} L, \quad F^j = \tau_j B_j^{-1} F$$

Consider stationary iterative methods (when τ_j, B_j do not depend on iteration number j) of the type:

$$B \frac{u^{j+1} - u^j}{\tau} = -(Lu^j - F) \qquad (8.2.23)$$

with an initial vector $u_0 \in H$ and real parameter τ.

The following condition [73] is necessary and sufficient for convergence of the iterative method (8.2.23):

$$\max_k |\lambda_k(T_\tau)| < 1 \qquad (8.2.24)$$

where $T_\tau = E - \tau B^{-1} L$ and $\lambda_k(T_\tau)$ denote eigenvalues of the matrix T_τ. Let B hereafter be a symmetric positive definite matrix.

If eigenvectors of matrices L and B coincide, then (8.2.24) produces the following condition on the parameter τ:

$$0 < \tau < \frac{2\lambda_k(B)}{\mu_k} \qquad (8.2.25)$$

where $\lambda_k(B)$ are eigenvalues of the matrix B and μ_k are eigenvalues of the matrix L defined by formula (8.2.22). In this case a problem on optimal choice of the parameter τ is easily solved (see [73]).

By (8.2.24) and using an explicit form of the operator L write the iterative process for system (8.2.4)–(8.2.6):

$$\frac{d\varphi^j}{dt} + A\varphi^j = f$$
$$\varphi^j|_{t=0} = u^j \tag{8.2.26}$$

$$-\frac{d\varphi^{*j}}{dt} + A^*\varphi^{*j} = \beta(\hat{\varphi} - \varphi^j)$$
$$\varphi^{*j}|_{t=\bar{t}} = 0 \tag{8.2.27}$$

$$B\frac{u^{j+1} - u^j}{\tau} = \varphi^{*j}|_{t=0} - \alpha u^j. \tag{8.2.28}$$

If eigenvectors of matrices B and $M = A + A^*$ coincide, then condition (8.2.25) is necessary and sufficient condition for convergence of iterative process (8.2.26)–(8.2.28).

Examine in more detail the case when $B = E$. Then iterative process (8.2.26)–(8.2.28) may be written in the form:

$$\frac{d\varphi^j}{dt} + A\varphi^j = f$$
$$\varphi^j|_{t=0} = u^j \tag{8.2.29}$$

$$-\frac{d\varphi^{*j}}{dt} + A^*\varphi^{*j} = \beta(\hat{\varphi} - \varphi^j)$$
$$\varphi^{*j}|_{t=\bar{t}} = 0 \tag{8.2.30}$$

$$u^{j+1} = u^j + \tau(\varphi^{*j}|_{t=0} - \alpha u^j) \tag{8.2.31}$$

With regard to (8.2.25) and using the form of μ_k from (8.2.15) conclude, that following condition is necessary and sufficient condition for the convergence of this iterative process:

$$0 < \tau < \frac{2}{\alpha + \beta\frac{1-e^{-\lambda_{\min}\bar{t}}}{\lambda_{\min}}} \tag{8.2.32}$$

where $\lambda_{\min} = \lambda_{\min}(M)$ is minimum eigenvalue of the matrix $M = A + A^*$.

It is easy to show using standard technique [73], that for the error $\psi^j = u^j - u$ the inequality holds:

$$\|\psi^j\| \le q^j(\tau)\|\psi^0\| \tag{8.2.33}$$

where

$$q(\tau) = \max\left\{\left|1 - \tau\left(\alpha + \beta\frac{1 - e^{-\lambda_{min}\bar{t}}}{\lambda_{min}}\right)\right|, \left|1 - \tau\left(\alpha + \beta\frac{1 - e^{-\lambda_{max}\bar{t}}}{\lambda_{max}}\right)\right|\right\}$$

Optimal parameter τ is determined by the formula

$$\tau = \tau_{opt} = 1/\left\{\alpha + \frac{\beta}{2}\left(\frac{1 - e^{-\lambda_{min}\bar{t}}}{\lambda_{min}} + \frac{1 - e^{-\lambda_{max}\bar{t}}}{\lambda_{max}}\right)\right\}$$

It satisfies (8.2.32) and, therefore, ensures a convergence. In this case

$$q_{opt} = q(\tau_{opt}) = \frac{\beta\left(\frac{1-e^{-\lambda_{min}\bar{t}}}{\lambda_{min}} - \frac{1-e^{-\lambda_{max}\bar{t}}}{\lambda_{max}}\right)}{2\alpha + \beta\left(\frac{1-e^{-\lambda_{min}\bar{t}}}{\lambda_{min}} + \frac{1-e^{-\lambda_{max}\bar{t}}}{\lambda_{max}}\right)} \qquad (8.2.34)$$

The quantity $q(\tau)$ characterizes the convergence rate of iterative process (see (8.2.32)). The smaller is $q(\tau)$, the faster is a convergence of iterative process. It follows from (8.2.34), that if $\alpha = 0$ then q_{opt} increases, i. e., iterative process converges more slowly. With $\alpha = 0$ and large λ_{max} the quantity $q(\tau)$ is close to 1, which implies also slow convergence of an algorithm. So the role of the regularizator is obvious here: with $\alpha > 0$ an iterative process converges faster.

We have discussed here just the case when $B = E$. It is possible to select B in (8.2.26)–(8.2.28) in other ways, obtaining thus various iterative methods. Having selected proper iterative algorithm, we can solve the main and adjoint problems (8.2.26) and (8.2.27) at every iteration with known conventional methods, such as finite-difference schemes in t, splitting methods and so on.

8.3. Numerical Method of Solution of Geophysical Data Assimilation Problem Using the Splitting Scheme

Observation data assimilation problem in mathematical models describing geophysical processes is one of urgent problems in contemporary geophysical hydrodynamics. The problem of adjustment of observational data and the data obtained in numerical experiments is important and interesting due to following theoretical and practical questions:

- reconstruction (updating) of observational data arrays in space and time and four-dimensional analysis;
- construction of "initial" geophysical fields using archived information and short-range observational data for preceding periods;
- management of observational data and results of numerical experiments (data processing), accumulation of united observational–experimental data;

Data assimilation problem is especially important in dynamic oceanology, which is explained first of all by high cost of oceanic observations and complex character of real ocean variability.

Mathematical methods based on the theory of adjoint equations [70, 200, 229] and the optimal control theory [23, 139, 212] have recently become very popular as one of effective tools for the analysis and assimilation of observational data.

A number of problems of physical, mathematical and computational character arise while solving complex nonlinear data assimilation problems in the framework of the theory of adjoint equations and the optimal control theory:

- physical aspects of a problem are connected with physical statement of problem and with a choice of informative functionals to be optimized [254];
- mathematical aspects are connected with an equivalence of original variational problems and differential problems obtained after transformations and with their solvability [23, 227].
- numerical aspects are connected with a choice of good discrete approximations of high accuracy and efficient algoritms to solve the discrete problems [95].

This section discusses the construction of numerical methods for solving the data assimilation problem. We are mainly aiming at ocean dynamics problems, though the methods suggested here may be used in various fields of geophysical hydrodynamics as well. The peculiarities of large-scale ocean dynamics problems are due to following features of oceanic medium:

- the ocean is in many respects a forced system affected by the atmosphere;

– large-scale oceanic processes are of well manifested linear character;
– the ocean is poorly lit by synchronous data, but it is a system with a "long-lasting memory".

These peculiarities of the ocean determine the statements of problems and the data assimilation methods we use in this and the next section.

Examine the evolutionary problem

$$\frac{d\varphi}{dt} + A\varphi = f \quad \text{in} \quad \Omega \times (0, \bar{t})$$
$$\varphi = u \quad \text{at} \quad t = 0 \tag{8.3.1}$$

Assume, as in § 8.2, that the original problem is approximated in all variables (save t), Ω is a grid domain, A is real square matrix of an order n, obtained as a result of approximation of some linear operator, $\varphi = \varphi(t)$, $f = f(t)$, u are grid functions, vectors of a length n. Parameter index h of a grid is omitted to simplify the notations. Let H denote a real space of grid functions with a norm $\| \cdot \| = (\cdot, \cdot)^{1/2}$.

Practically total absence of information on oceanic fields at initial moment of time is a bright feature of data assimilation problem for the ocean. But, often instead of lacking initial data there is some information on a solution φ. Let we know the vector-function $\hat{\varphi} = \hat{\varphi}(t) \in H$ related to observational data.

Examine a functional S in the form

$$S(\varphi) = \frac{C_1}{2} \| \varphi|_{t=0} - \hat{\varphi}^0 \|^2 + \frac{C_2}{2} \int_0^{\bar{t}} \| \varphi - \hat{\varphi} \|^2 dt \tag{8.3.2}$$

where C_1, C_2 are some positive constants which are independent of t, $\hat{\varphi}^0 = \hat{\varphi}|_{t=0}$.

Let us formulate data assimilation problem as follows: find a couple of vector-functions φ, u satisfying problem (8.3.1) and providing minimum value to the functional $S(\varphi)$ on the set of solutions of (8.3.1)

$$S(\varphi) = \inf_u S(\varphi)$$

This problem is reduced to a system of equations for two vector-functions $\varphi = \varphi(t)$ and $\varphi^* = \varphi^*(t)$ and control vector u with the method discussed in § 8.1:

$$\frac{d\varphi}{dt} + A\varphi = f \quad \text{in} \quad \Omega \times (0, \bar{t})$$
$$\varphi = u \quad \text{at} \quad t = 0 \tag{8.3.3}$$

$$-\frac{d\varphi^*}{dt} + A^*\varphi^* = C_2(\hat{\varphi} - \varphi) \quad \text{in} \quad \Omega \times (0, \bar{t})$$
$$\varphi^* = 0 \quad \text{at} \quad t = \bar{t} \tag{8.3.4}$$

$$C_1(u - \hat{\varphi}^0) - \varphi^*|_{t=0} = 0 \tag{8.3.5}$$

where $A^* = A^T$.

Assume that the vector-functions $f(t)$ and $\hat{\varphi}(t)$ are continuous in t on the interval $[0, \bar{t}]$ and the matrix A from (8.3.1) is positive-definite and independent of t. It is easily shown [23] that in this case problem (8.3.3)–(8.3.5) has a unique solution. The solutions of the main and adjoint problems are understood in a classic sense.

Notice that system (8.3.3)–(8.3.5) represents a factorized form of a boundary-value problem. Actually, with sufficiently smooth φ^*, $\hat{\varphi}$ it is possible to rewrite (8.3.3)–(8.3.5) as

$$\left(\frac{d}{dt} + A\right) C_2^{-1} \left(\frac{d}{dt} - A^*\right) \varphi^* = f_1$$

$$C_1 C_2^{-1} \left(-\frac{d\varphi^*}{dt} + A^*\varphi^*\right) - \varphi^* = 0 \quad \text{at} \quad t = 0 \tag{8.3.6}$$

$$\varphi^* = 0 \quad \text{at} \quad t = \bar{t}$$

where $f_1 = f - \frac{d\hat{\varphi}}{dt} - A\hat{\varphi}$.

Hence it follows that an approximation of boundary-value problem (8.3.6) may serve as one of natural ways to check a numerical scheme for original equations (8.3.3)–(8.3.5).

Let us describe an algorithm for solving the problem (8.3.3)–(8.3.5). Introduce a uniform grid with a step τ on the interval $0 \le t \le \bar{t}$ and let $t_j = j\tau$, $j = \overline{0, J}$, $J = \bar{t}/\tau$. Approximate (8.3.3)–(8.3.5) with the Crank-Nicolson scheme:

$$\frac{\varphi^{j+1} - \varphi^j}{\tau} + A\frac{\varphi^{j+1} + \varphi^j}{2} = f^{j+1/2} \qquad (8.3.7)$$

$$-\frac{\psi^{j+1} - \psi^j}{\tau} + A^*\frac{\psi^{j+1} + \psi^j}{2} + C_2\frac{\varphi^{j+1} + \varphi^j}{2} = C_2\hat{\varphi}^{j+1/2} \qquad (8.3.8)$$
$$j = 0,\ldots,J-1$$

$$C_1(\varphi^0 - \hat{\varphi}^0) - \psi^0 = 0, \quad \psi^J = 0 \qquad (8.3.9)$$

where $\psi = \varphi^*$, $\varphi^0 = u$, $f^{j+1/2} = f(t_{j+1/2})$, $t_{j+1/2} = (j+1/2)\tau$.

It was shown in [233] that system (8.3.7)–(8.3.9) has a unique solution and approximates original problem (8.3.3)–(8.3.5) with an order of $O(\tau^2)$. System (8.3.7)–(8.3.9) approximates also boundary-value problem (8.3.6), at any rate at $C_1 = 0$ [233].

Let us modify approximation (8.3.7)–(8.3.9) using the splitting method. Assume that the following representations are valid:

$$A = A_1 + A_2, \quad A^* = A_1^* + A_2^* \qquad (8.3.10)$$

where A_i, A_i^* are positive semi-definite matrices, $i = 1, 2$. Let us use a two-cyclic splitting scheme (see Appendix I) preserving the second order of approximation in time of original problem (8.3.3)–(8.3.5) [233].

$$\begin{cases} \left(E + \frac{\tau}{4}A_1\right)\varphi_1 = \left(E - \frac{\tau}{4}A_1\right)\varphi^j \\ \left(E + \frac{\tau}{4}A_2\right)\left(\varphi_2 - \frac{\tau}{2}f^{j+1/2}\right) = \left(E - \frac{\tau}{4}A_2\right)\varphi_1 \\ \left(E + \frac{\tau}{4}A_2\right)\varphi_3 = \left(E - \frac{\tau}{4}A_2\right)\left(\varphi_2 + \frac{\tau}{2}f^{j+1/2}\right) \\ \left(E + \frac{\tau}{4}A_1\right)\varphi^{j+1} = \left(E - \frac{\tau}{4}A_1\right)\varphi_3, \quad j = 0,\ldots,J-1 \end{cases} \qquad (8.3.11)$$

$$\begin{aligned} &\left(E - \frac{\tau}{4}A_1^*\right)\psi_1 = \left(E + \frac{\tau}{4}A_1^*\right)\psi^j \\ &\left(E - \frac{\tau}{4}A_2^*\right)\left[\psi_2 - \frac{\tau C_2}{2}\left(\frac{\varphi^{j+1} + \varphi^j}{2} - \hat{\varphi}^{j+1/2}\right)\right] = \left(E + \frac{\tau}{4}A_2^*\right)\psi_1 \\ &\left(E - \frac{\tau}{4}A_2^*\right)\psi_3 = \left(E + \frac{\tau}{4}A_2^*\right)\left[\psi_2 + \frac{\tau C_2}{2}\left(\frac{\varphi^{j+1} + \varphi^j}{2} - \hat{\varphi}^{j+1/2}\right)\right] \\ &\left(E - \frac{\tau}{4}A_1^*\right)\psi^{j+1} = \left(E + \frac{\tau}{4}A_1^*\right)\psi_3, \quad j = 0,\ldots,J-1 \end{aligned}$$
$$(8.3.12)$$

$$C_1(\varphi^0 - \hat{\varphi}^0) - \psi^0 = 0, \quad \psi^J = 0 \tag{8.3.13}$$

Exclude intermediate values $\varphi_1, \ldots, \psi_3$ from (8.3.11) and (8.3.12) to obtain

$$\varphi^{j+1} = T_1 T_2 T_2 T_1 \varphi^j + \tau T_1 T_2 f^{j+1/2} \tag{8.3.14}$$

$$\psi^{j+1} = S_1 S_2 S_2 S_1 \psi^j + \tau C_2 S_1 S_2 \left(\frac{\varphi^{j+1} + \varphi^j}{2} - \hat{\varphi}^{j+1/2} \right) \tag{8.3.15}$$

$$j = 0, \ldots, J - 1$$

where

$$T_i = \left(E + \frac{\tau}{4} A_i \right)^{-1} \left(E - \frac{\tau}{4} A_i \right), \quad S_i = \left(E - \frac{\tau}{4} A_i^* \right)^{-1} \left(E + \frac{\tau}{4} A_i^* \right), \quad i = 1, 2$$

Let us present an iterative procedure for solving consequently the main and adjoint problems (8.3.14), (8.3.15). Suppose, we have obtained some value of the vector $\varphi_{(\nu-1)}$ at $(\nu-1)$th iteration including its value at initial moment $\varphi_{(\nu-1)}^j = \varphi^j$ for $j = 0$. Then the νth iteration consists of the following steps:

1. by (8.3.14) we have (solution of the main equation "forward" in time)

$$\varphi^{j+1} = T_1 T_2 \left(T_2 T_1 \varphi^j + \tau f^{j+1/2} \right), \quad j = 0, \ldots, J - 1$$

or

$$\varphi^{j+1} = T_\tau^{j+1} \varphi^0 + \sum_{i=0}^{j} T_\tau^i \left(\tau T_1 T_2 f^{j-i+1/2} \right), \quad j = 0, \ldots, J - 1 \tag{8.3.16}$$

where $T_\tau = T_1 T_2 T_2 T_1$.

2. by (8.3.15) we have (solution of adjoint equation "backward" in time)

$$\psi^J = 0$$

$$\psi^{j-1} = S_1^{-1} S_2^{-1} S_2^{-1} S_1^{-1} \psi^j - \tau C_2 S_1^{-1} S_2^{-1} \left(\frac{\varphi^{j-1} + \varphi^j}{2} - \hat{\varphi}^{j-1/2} \right)$$

$$j = J, \ldots, 2$$

or

$$\psi^{j-1} = -\tau C_2 \sum_{i=j}^{J} S^{i-j} S_1^{-1} S_2^{-1} \left(\frac{\varphi^i + \varphi^{i-1}}{2} - \hat{\varphi}^{i-1/2} \right) \tag{8.3.17}$$

$$j = J, \ldots, 2$$

where $S = S_1^{-1} S_2^{-1} S_2^{-1} S_1^{-1}$. Substitute the expression for φ from (8.3.15) into (8.3.16) to obtain

$$\psi^{j-1} = -\frac{\tau C_2}{2} \sum_{i=j}^{J} S^{i-j} S_1^{-1} S_2^{-1} \left[T_\tau^i \varphi^0 + T_\tau^{i-1} \varphi^0 + \tau \sum_{k=0}^{i-1} T_\tau^k \left(T_1 T_2 f^{i-1-k+1/2} \right) \right.$$

$$\left. + \tau \sum_{k=0}^{i-2} T_\tau^k \left(T_1 T_2 f^{i-2-k+1/2} \right) - 2\hat{\varphi}^{i-1/2} \right] \quad j = J, \dots, 2 \quad (8.3.18)$$

3. concluding step of νth iteration finds new value of $\varphi_{(\nu)}^0$ as a result of solution of the equations

$$\frac{\tau}{2} C_2 S_1^{-1} S_2^{-1} \varphi_{(\nu)}^0 + \psi_{(\nu)}^0 = S \psi_{(\nu-1)}^1 - \frac{\tau}{2} C_2 S_1^{-1} S_2^{-1} \left(\varphi_{(\nu-1)}^1 - 2\hat{\varphi}^{1/2} \right)$$

$$C_1 \varphi_{(\nu)}^0 - \psi_{(\nu)}^0 = C_1 \hat{\varphi}^0$$

Whence we obtain for $\varphi_{(\nu)}^0$

$$\left(\frac{\tau}{2} C_2 S_1^{-1} S_2^{-1} + C_1 E \right) \varphi_{(\nu)}^0$$

$$= C_1 \hat{\varphi}^0 + S \psi_{(\nu-1)}^1 - \frac{\tau}{2} C_2 S_1^{-1} S_2^{-1} \left(\varphi_{(\nu-1)}^1 - 2\hat{\varphi}^{1/2} \right) \quad (8.3.19)$$

Having found $\varphi_{(\nu)}^0$ repeat the cycle of computations (see item 1) until the convergence of iterations occurs.

Let us show that there is a convergence of above iteration algorithm provided some condition for τ is satisfied. To do so, estimate a norm of a transition operator from $(\nu-1)$th to νth iteration while computing φ^0.

By (8.3.19)

$$\tilde{S} \varphi_{(\nu)}^0 + \psi_{(\nu)}^0 = S \psi_{(\nu-1)}^1 - \tilde{S} \left(\varphi_{(\nu-1)}^1 - 2\hat{\varphi}^{1/2} \right)$$
$$\varphi_{(\nu)}^0 - C_1^{-1} \psi^0 = \hat{\varphi}_\nu^0 \quad (8.3.20)$$

where $\tilde{S} = \frac{\tau}{2} C_2 S_1^{-1} S_2^{-1}$, hence

$$\varphi_{(\nu)}^0 = (E + C_1^{-1} \tilde{S})^{-1} \left\{ \hat{\varphi}^0 - C_1^{-1} \left[S^{J-1} \tilde{S} \left(\varphi_{(\nu-1)}^J + \varphi_{(\nu-1)}^{J-1} - 2\hat{\varphi}^{J-1/2} \right) \right. \right.$$

$$\left. + \dots + S \tilde{S} \left(\varphi_{(\nu-1)}^2 + \varphi_{(\nu-1)}^1 - 2\hat{\varphi}^{3/2} \right) \right]$$

$$\left. - C_1^{-1} \tilde{S} \left(T_\tau \varphi_{(\nu-1)}^0 + \tau T_1 T_2 f^{1/2} - 2\hat{\varphi}^{1/2} \right) \right\} \quad (8.3.21)$$

Here we made use of (8.3.16) and (8.3.18) to represent respectively $o^1_{(\nu-1)}$ and $\psi^1_{(\nu-1)}$. Estimate $\|\varphi^0_{(\nu-1)}\|$ according to (8.3.21) to obtain

$$\|\varphi^0_{(\nu)}\| \leq \|(E + C_1^{-1}\tilde{S})^{-1}\| \{\|\hat{\varphi}^0\| + |C_1^{-1}| [\|S^{J-1}\| \cdot \|\tilde{S}\| (\|\varphi^J_{(\nu-1)}\|$$

$$+ \|\varphi^{J-1}_{(\nu-1)}\| + 2\|\hat{\varphi}^{J-1/2}\|) + \ldots + \|S\| \cdot \|\tilde{S}\| (\|\varphi^2_{(\nu-1)}\| + \|\varphi^1_{(\nu-1)}\| + 2\|\hat{\varphi}^{3/2}\|)]$$

$$+ |C_1^{-1}|\|\tilde{S}\| (\|T_\tau\| \cdot \|\varphi^0_{(\nu-1)}\| + \tau\|T_1 T_2\| \cdot \|f^{1/2}\| + 2\|\hat{\varphi}^{1/2}\|)\} \quad (8.3.22)$$

By the Kellogg lemma [73]

$$\|S\| = \|S_1^{-1} S_2^{-1} S_2^{-1} S_1^{-1}\| \leq \|\left(E + \frac{\tau}{4}A_1^*\right)^{-1}\left(E - \frac{\tau}{4}A_1^*\right)\|$$

$$\times \|\left(E + \frac{\tau}{4}A_2^*\right)^{-1}\left(E - \frac{\tau}{4}A_2^*\right)\| \cdot \|\left(E + \frac{\tau}{4}A_2^*\right)^{-1}\left(E - \frac{\tau}{4}A_2^*\right)\|$$

$$\cdot \|\left(E + \frac{\tau}{4}A_1^*\right)^{-1}\left(E - \frac{\tau}{4}A_1^*\right)\| \leq 1, \quad \|T_\tau\| \leq 1$$

Let us continue inequality (8.3.22) using the estimates $\|\varphi^j_{(\nu-1)}\|$, $i = 1, \ldots, J$, equality (8.3.16) and with regard to (8.3.20):

$$\|\varphi^0_{(\nu)}\| \leq \frac{\tau}{2} C_2 C_1^{-1}(2J - 1)\|\left(E + \frac{\tau}{2}C_2 C_1^{-1} S_1^{-1} S_2^{-1}\right)^{-1}\| \cdot \|\varphi^0_{(\nu-1)}\|$$

$$+ \|G\| \quad (8.3.23)$$

where G is a vector-function depending just on assigned vectors $f^{j+1/2}$, $g^{j+1/2}$. It is sufficient for the iteration process to converge if a norm of a transition operator is less than unit:

$$\frac{\tau}{2} C_2 C_1^{-1}(2J - 1)\|\left(E + \frac{\tau}{2}C_2 C_1^{-1} S_1^{-1} S_2^{-1}\right)^{-1}\| < 1 \quad (8.3.24)$$

Since

$$\|\left(E + \frac{\tau}{2}C_2 C_1^{-1} S_1^{-1} S_2^{-1}\right)^{-1}\| \leq 1$$

then it is sufficient to require that

$$\frac{\tau}{2} C_2 C_1^{-1}(2J - 1) \leq 1$$

The iteration process thus converges under the condition

$$\tau < \frac{2}{2J - 1} C_1 C_2^{-1} \qquad (8.3.25)$$

Since $T = \tau J$, inequality (8.3.25) may be strengthened and rewritten in the terms of T which is an interval of assimilation. Then one may suggest that the iteration process converges under the condition

$$T < C_2^{-1} C_1$$

being independent of the number of points at an assimilation interval.

8.4. Numerical Solution of Oceanic Data Initialization Problem

Let us exemplify this problem by the algorithm of observation data assimilation presented in Section 8.3. From the physical point of view the problem may be stated as follows.

Suppose that

– we have some numerical model simulating large-scale thermodynamic characteristics (temperature and currents fields) in some oceanic aquatory;

– we possess an exact information on the sources acting at the ocean surface, such as a wind stress, and a heat flux during assigned time interval $0 \leq t \leq \bar{t}$;

– we have no exact information on three-dimensional oceanic fields at initial moment of time, which is typical situation in reality, but we know some general regularities in these fields;

– we got the observation data $\hat{T}$ on the interval $[0, \bar{t}]$ after the time of a real forecast range.

Our task is to reconstruct the temperature field T at the moment of time $t = 0$ which could serve as an initial data for our numerical model. In other words, we must construct such a temperature field at the moment of time $t = 0$ which would ensure the reproduction by our model the temporary evolution of T at the interval $[0, \bar{t}]$ with minimum deviation from the observed data $\hat{T}$.

The initialization problem is exceptionally difficult because of two causes: a model simulating the characteristics of ocean thermodynamics is very complex and nonlinear; the measurements in the deep ocean is extremely expansive and time consuming. In order to compute a concrete example we assume a number of simplifications.

Let us exemplify the above considerations by computation of initial temperature field for the aquatory of Arabian Sea. We will try to reconstruct large-scale structure of the temperature field with the numerical model of Arabian Sea [10] with resolution $1° \times 1°$ in latitude and longitude and 15 levels in vertical.

The equations of ocean thermodynamics in spherical coordinates λ, ψ, z) have the form:

$$\frac{du}{dt} - (l - mu\cos\psi)v = -\frac{m}{\rho_0}\frac{\partial p}{\partial \lambda} + \frac{\partial}{\partial z}\nu\frac{\partial u}{\partial z} + F^u \qquad (8.4.1)$$

$$\frac{dv}{dt} + (l - mu\cos\psi)u = -\frac{n}{\rho_0}\frac{\partial p}{\partial \psi} + \frac{\partial}{\partial z}\nu\frac{\partial v}{\partial z} + F^v \qquad (8.4.2)$$

$$\frac{\partial p}{\partial z} = \rho g \qquad (8.4.3)$$

$$m\left(\frac{\partial u}{\partial \lambda} + n\frac{\partial}{\partial \psi}\frac{v}{m}\right) + \frac{\partial w}{\partial z} = 0 \qquad (8.4.4)$$

$$\frac{dT}{dt} = \frac{\partial}{\partial z}\nu_T\frac{\partial T}{\partial z} + F^T - \gamma_T w \qquad (8.4.5)$$

$$\frac{dS}{dt} = \frac{\partial}{\partial z}\nu_s\frac{\partial S}{\partial z} + F^s - \gamma_s w \qquad (8.4.6)$$

$$\rho = f(T + \overline{T}, S + \overline{S}) - \bar{p} - \rho_0 \qquad (8.4.7)$$

here

$$F^\varphi = m^2\frac{\partial}{\partial \lambda}\mu_\varphi\frac{\partial \varphi}{\partial \lambda} + mn\frac{\partial}{\partial \psi}\mu_\varphi\frac{n}{m}\frac{\partial \varphi}{\partial \psi}, \quad \varphi = \{u, v, T, S\} \qquad (8.4.8)$$

$$\frac{d}{dt} = \frac{\partial}{\partial t} + mu\frac{\partial}{\partial \lambda} + nv\frac{\partial}{\partial \psi} + w\frac{\partial}{\partial z}$$

here ψ is a latitude counted from the South Pole; λ is a longitude; is a depth; $u = u(\lambda, \psi, z, t)$, $v = v(\lambda, \psi, z, t)$, $w = w(\lambda, \psi, z, t)$ are the components of a velocity vector $\mathbf{u}$; p – pressure, T – temperature, S – salinity, ρ – density, g – gravity acceleration, l – Coriolis' parameter, $m = 1/(a\sin\psi)$, $n = 1/a$, a – the Earth's radius, ρ_0 – standard mean density, ν, ν_T, ν_s are the coefficients of vertical turbulent diffusion; $\overline{T}(z)$, $\overline{S}(z)$, $\bar{\rho}(z)$ are standard characteristic distributions of temperature, salinity and density in vertical; $\gamma_T = \dfrac{d\overline{T}}{dz}$, $\gamma_s = \dfrac{d\overline{S}}{dz}$; F^u, F^v, F^T, F^s are the terms, describing the horizontal turbulent friction; μ_u, μ_v, μ_T, μ_s are the coefficients of horizontal turbulent diffusion; τ^λ, τ^ψ are the components of a wind stress at the sea surface; α_T, β_T, α_s, β_s are assigned coefficients characterizing an external forcing.

System of equations (8.4.1)–(8.4.7) is considered on a time interval $(0, \bar{t}]$ in a cylindrical domain D bounded by a piecewise smooth boundary ∂D composed by unperturbed ocean surface Ω, lateral vertical surface Σ, and the bottom relief $H = H(\lambda, \psi)$. The following boundary (if any) and initial conditions are set for problem (8.4.1)–(8.4.7):

at $z = 0$

$$\nu\frac{\partial u}{\partial z} = -\frac{\tau^\lambda}{\rho_0}, \quad \nu\frac{\partial v}{\partial z} = -\frac{\tau^\psi}{\rho_0}, \quad w = 0$$

$$\nu_T\frac{\partial T}{\partial z} = \alpha_T T + \beta_T Q_T, \quad \nu_s\frac{\partial S}{\partial z} = \alpha_s S + \beta_s Q_s \tag{8.4.9}$$

at $z = H(\lambda, \psi)$

$$(\boldsymbol{u}, \boldsymbol{n}_H) = 0, \quad (\boldsymbol{Du}, \boldsymbol{n}_H) = 0, \quad (\boldsymbol{Dv}, \boldsymbol{n}_H) = 0$$
$$(\boldsymbol{DT}, \boldsymbol{n}_H) = 0, \quad (\boldsymbol{DS}, \boldsymbol{n}_H) = 0 \tag{8.4.10}$$

where

$$\boldsymbol{n}_H = \frac{\boldsymbol{i}_z - \boldsymbol{i}_\lambda m\frac{\partial H}{\partial \lambda} - \boldsymbol{i}_\psi n\frac{\partial H}{\partial \psi}}{\sqrt{1 + \left(m\frac{\partial H}{\partial \lambda}\right)^2 + \left(n\frac{\partial H}{\partial \psi}\right)^2}}$$

$$\boldsymbol{D}\varphi = \mu_\varphi\left(\boldsymbol{i}_\lambda m\frac{\partial\varphi}{\partial\lambda} + \boldsymbol{i}_\psi n\frac{\partial\varphi}{\partial\psi}\right) + \boldsymbol{i}_z\nu_\varphi\frac{\partial\varphi}{\partial z}$$

$\boldsymbol{i}_\lambda$, $\boldsymbol{i}_\psi$, $\boldsymbol{i}_z$ are the unit orts of a vector normal to the surface $H(\lambda, \psi)$;

on the lateral surface Σ

$$u = 0, \quad v = 0, \quad (\nabla T, \boldsymbol{n}_\Sigma) = 0, \quad (\nabla S, \boldsymbol{n}_\Sigma) = 0 \qquad (8.4.11)$$

where

$$\boldsymbol{n}_\Sigma = \boldsymbol{i}_\lambda \cos(\boldsymbol{n}_\Sigma, \boldsymbol{i}_\lambda) + \boldsymbol{i}_\psi \cos(\boldsymbol{n}_\Sigma, \boldsymbol{i}_\psi)$$

$$\nabla \xi = \boldsymbol{i}_\lambda m \frac{\partial \xi}{\partial \lambda} + \boldsymbol{i}_\psi n \frac{\partial \xi}{\partial \psi}, \quad \xi = T, S$$

at $t = 0$

$$u = u^0, \quad v = v^0, \quad T = T^0, \quad S = S^0 \qquad (8.4.12)$$

where $\boldsymbol{n}_\Sigma$ is a normal to Σ, u^0, v^0, T^0, S^0 are prescribed functions.

System of equations of ocean thermodynamics (8.4.1)–(8.4.12) is nonlinear and complex. It is extremely difficult to construct a computation process to solve it as a data assimilation problem and, in particular, as an initialization problem. Let us construct the solution process of initialization problem in a following way.

– Assume that we know an exact solution of problem (8.4.1)–(8.4.12) on the interval $(0, \bar{t}]$ together with exact initial condition for the temperature $\hat{T}_0$, which is a result of solution of complete nonlinear problem (8.4.1)–(8.4.12) in our numerical example. Let us refer to the fields of currents and temperature thus obtained as the "measurement data".

– Assume that the evolution of temperature on the interval $(0, \bar{t}]$ is determined mainly by locally one-dimensional process of transport-diffusion in vertical at each point (λ, ψ) of a grid domain. Averaged in time "measurement data" i. e., averaged in time field of w computed with nonlinear model (8.4.1)–(8.4.12) is used in numerical solution of initialization problem.

In this case the initialization problem for oceanic fields is simplified and reduced to the restoration of a vertical structure of initial temperature field. This very problem will exemplify the numerical algorithm for data assimilation problem presented in Section 8.3.

The numerical experiment includes the following computations.

Step 1. Solution of complete nonlinear problem (8.4.1)–(8.4.12) for one month.

Adjusted fields of currents (Fig. 69) and temperature (Fig. 70, isolines start with 27°C, spaced by 0.5°) corresponding to the spring period (middle of April) were used as initial data. The arrow of maximum
length in logarithmic scale corresponds to velocity of 0.84 m/c. Just
the first computational 10 m level is presented at these and following
figures. The wind stress observed in 1988 (year of well manifested In-
dian monsoon) was set at the sea surface for the computation for a
month (till the middle of May). Solution of the problem resulted in
averaged in time fields of the velocity $\hat{u}$ and temperature $\hat{T}$:

$$\hat{u} = \frac{1}{\bar{t}} \int_0^{\bar{t}} u(\lambda, \psi, z, t)dt, \quad u = (u, v, w)^T \tag{8.4.13}$$

$$\hat{T} = \frac{1}{\bar{t}} \int_0^{\bar{t}} T(\lambda, \psi, z, t)dt \tag{8.4.14}$$

These fields were used later as "measurement data". Fig. 71 shows
monthly mean velocity field. The arrow of maximum length in logarithmic scale corresponds to velocity of 0.72 m/c. Fig. 72 represents
monthly mean vertical velocity (isolines start with $-5 \cdot 10^{-6}$ m/s, spaced
by 10^{-6} m/s). Fig. 73 demonstrates monthly mean temperature field
(isolines start with 27°C, spaced by 0.5°).

Step 2. Solution of three-dimensional transport–diffusion problem
for one month with initial temperature field obtained on Step 1, but
with fixed velocities $\hat{u}$ computed with (8.4.13). This computation allows to estimate what will be the loss in accuracy if we neglect the
nonlinear and nonstationary effects of currents (monthly variability)
while restoring the temperature fields on the computation interval.
Fig. 74 presents mean square difference between two solutions obtained
on Steps 1 and 2 (isolines start with 0, spaced by 0.1):

$$\delta_{1,2} = \frac{1}{\bar{t}} \int_0^{\bar{t}} \int_0^H (T_1 - T_2)^2 dz dt$$

One may see that it is possible to neglect the effects of adaptation of
the fields of currents and temperature and non-stationary state of a
velocity field over all the aquatory of the Arabian Sea in computations
of this kind preserving satisfactory accuracy.

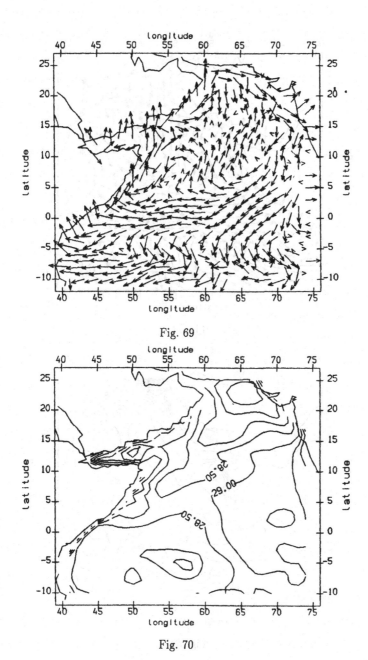

Fig. 69

Fig. 70

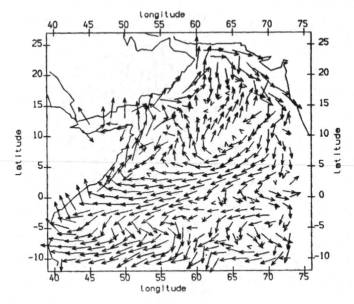

Fig. 71

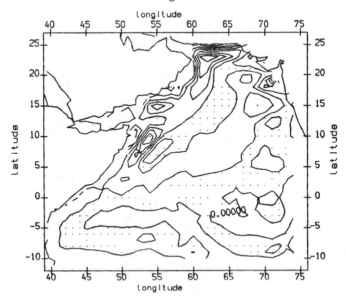

Fig. 72

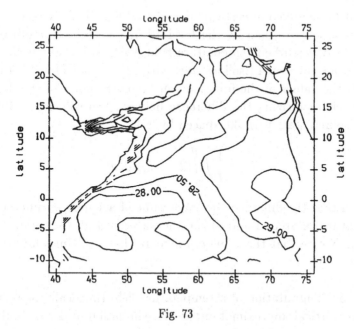

Fig. 73

Step 3. Solution of locally one-dimensional in vertical transport–diffusion problem.

Initial data are the same as in Steps 1 and 2, the field of $\hat{w}$ is taken from the computation of Step 1 by (8.4.13). Fig. 75 presents mean square difference $\delta_{2,3}$ (isolines start with 0, spaced by 2.0):

$$\delta_{2,3} = \frac{1}{\bar{t}} \int\limits_{\bar{t}}^{\bar{t}} \int\limits_0^H (T_2 - T_3)^2 dz dt$$

One may say considering the functionals $\delta_{1,2}$ and $\delta_{2,3}$, that while describing the temperature structure it is possible to use the locally one-dimensional approximation in the open sea except for a narrow coastal stripe.

Step 4. Computation of a temperature evolution in the locally one-dimensional approximation for a month starting with a perturbed initial condition.

The initial condition was perturbed as follows. The top layer of the ocean (from $z = 0$ to $z = z_1/H(\lambda, \psi)$, where $z_1 = 100$ m) was assumed to be mixed in vertical. Fig. 70 and Fig. 76 show respectively initial temperature field at the level 10 m (the initial condition on Step 3) and perturbed temperature field which is initial condition in computations of Step 4 (isolines start with 22°C, spaced by 1°). Fig. 78 presents the values of the functional $\delta_{1,4}$ in every point of the domain in λ, ψ characterizing the difference between two solutions (Steps 1 and 4, isolines start with 30, spaced by 20):

$$\delta_{1,4} = \frac{1}{\bar{t}} \int\limits_0^{\bar{t}} \int\limits_0^H (T_1 - T_4)^2 dz dt$$

It is seen that the difference (average value of a square of difference of temperature on Steps 1 and 4) is rather large over all the computation domain. Notice that the value $\delta_{1,4}$ is more than ten times larger than $\delta_{2,3}$ (Fig. 75).

Step 5. Computation of a temperature field (in locally one-dimensional in vertical approximation) with assimilation of data simulating the measurements (Step 1).

We assumed in this case that an initial condition is "spoiled" in a top layer (from the surface to $z = z_1/H(\lambda, \psi)$, where $z_1 = 100$ m) and the temperature is homogeneous in vertical. The computations were aimed at restoration of the temperature field i. e., the result of Step 1, including the initial state on the interval of one month, using the above algorithm of data assimilation. Here the data to be assimilated were represented by the averaged in time temperature field obtained on Step 1. In other words, averaged in time values of exact solution of the problem were assimilated.

Fig. 77 presents the result of assimilation as the functional $\delta_{1,5}$ (isolines start with 0, spaced by 2). The temperature field obtained here approximates efficiently the "measurement data" in the open sea. The errors in coastal stripe are due to locally one-dimensional in vertical approximation of the temperature evolution. Fig. 77 shows that this approximation is valid only outside of the zones of coastal currents.

Fig. 79 (isolines start with 30, spaced by 50) and Fig. 80 (isolines start with 0, spaced by 0.1) show respectively the differences between

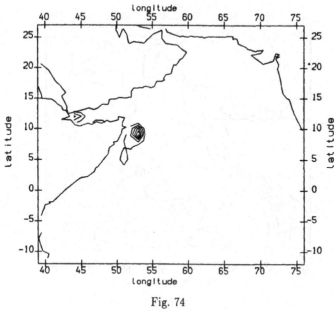

Fig. 74

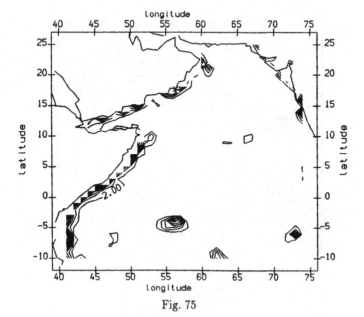

Fig. 75

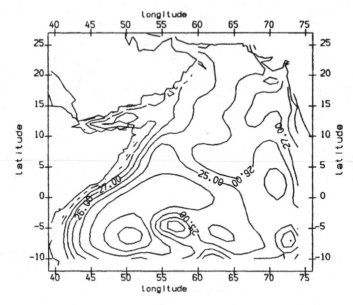

Fig. 76

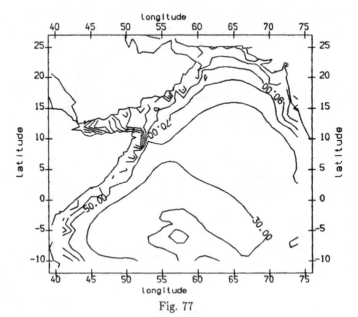

Fig. 77

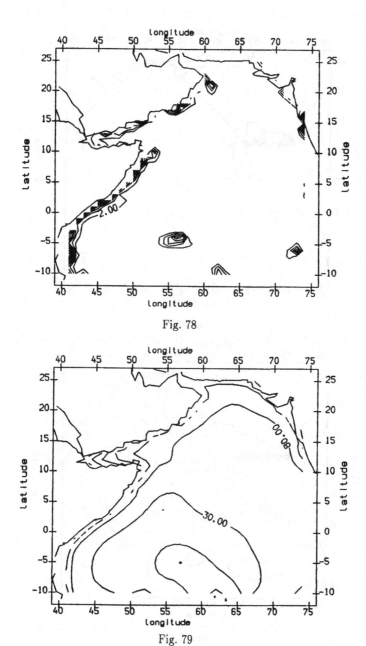

Fig. 78

Fig. 79

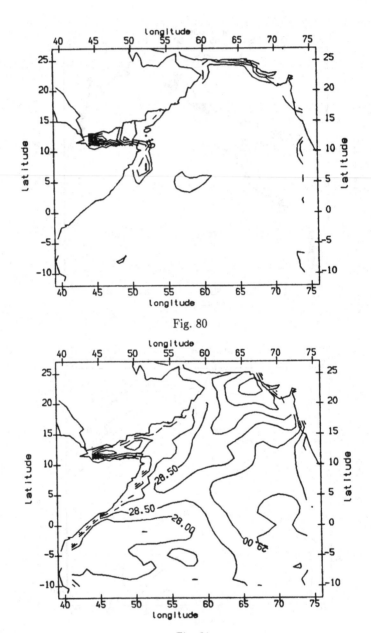

Fig. 80

Fig. 81

initial and perturbed and between initial and computed in the assimilation experiment distributions of temperature at initial moment of
time:

$$\delta_0 = \int_0^H (T_0 - \hat{T}_0)^2 dz$$

Solution of the data assimilation problem resulted in restoration of
initial temperature field (Fig. 81, isolines start with 27°C, spaced by
0.5°) with a satisfactory accuracy (compare Fig. 76 and Fig. 81). Recall
that we assume as the "data" the solution obtained on Step 1.

The results of numerical experiments demonstrate the applicabi-
lity of the method of data assimilation for the restoration of a priori
unknown or poorly lit by observations distributions of temperature
at some initial moment of time. This restoration is performed as a
synthesis of computed (model) data and observational data by solving
the joint system of the main and adjoint equations.

APPENDIX I

SPLITTING METHODS IN THE SOLUTION

OF GLOBAL PROBLEMS

1. Splitting Methods for Solution of Non-Stationary Problems

In many cases, when complicated problem of mathematical physics is to be solved, there is a possibility to reduce this problem to a sequence of simpler problems which are solved effectively by computers. This reduction of complex problems to more simple is possible when original positive semi-definite operator of the problem may be represented as a sum of positive semi-definite simple operators. We will refer to these methods as *splitting methods*.

Initially splitting methods were formulated and theoretically justified for simple problems with commutative positive definite operators. It is clear now that splitting methods introduced for such problems by different authors proved to be as a matter of fact either equivalent, differing only with schemes of realization, or close.

Later a circle of nontrivial problems solved by splitting methods grew wider and to this day the splitting methods became powerful tool to solve very complicated problems in mathematical physics. Since the theory of splitting methods was most thoroughly developed for the case when original operator may be presented as a sum of two more simple ones, we will begin with considering this very case. The componentwise splitting method investigated below is, to our opinion, the most universal for applications.

So, examine the evolutionary equation

$$\frac{d\varphi}{dt} + A\varphi = f \quad \text{in} \quad \Omega \times \Omega_t, \quad \text{at} \quad t = 0 \qquad (1.1)$$

Assume for simplicity that the original problem is approximated over all variables (except for t), Ω is a grid domain, $\Omega_t = (0, T)$, a real square matrix $A = A(t)$ of an order of n results from approximation of a linear operator, $\varphi = \varphi(t)$, $f = f(t)$, g are grid vector-functions of the length of n. Index of a grid parameter h is omitted for simplicity. Let H denote a space of real grid functions with a norm $\| \cdot \| = (\cdot, \cdot)^{1/2}$. Assume the matrix A as positive semi-definite, i. e., $(A\varphi, \varphi) \geq 0$, which will be written hereafter as $A \geq 0$.

We assume that a solution of the problem (3.1) possesses all the necessary smoothness.

Introduce the problem adjoint to (1.1):

$$-\frac{d\varphi^*}{dt} + A^*\varphi^* = f^* \quad \text{in} \quad \Omega \times \Omega_t, \quad \varphi^* = g^* \quad \text{at} \quad t = T \qquad (1.2)$$

where A^* is a matrix transposed to A, f^*, g^* are assigned functions. By definition, $A^* \geq 0$, i. e., $(A^*\varphi^*, \varphi^*) \geq 0$.

The solution of problem (1.2) is performed "from the right to the left" i. e., beginning with $t = T$ and then for $t < T$ till $t = 0$.

Introduce new independent variable $t' = T - t$ to write problem (1.2) in usual equivalent form

$$-\frac{d\psi^*}{dt} + A^*\psi^* = F^* \quad \text{in} \quad \Omega \times \Omega_t, \quad \psi^* = G^* \quad \text{at} \quad t' = 0 \qquad (1.3)$$

where $\psi^*(t') = \varphi^*(T - t')$, $F^*(t') = f^*(T - t')$, $G^* = g^*$.

Comparing (1.1) and (1.3) we see that the problems are formally equivalent but differ in their statements. Just transformation $t = T - t'$ is needed in adjoint problem.

This means that solution algorithm for problems (1.1) and (1.3) is the same, therefore, we will consider just the algorithm for the main problem (1.1) which is valid for adjoint problem (1.3) too.

Actually there is no need to pass to new variable t', it is sufficient to solve problem (1.2) "from the right to the left" and just renumber discrete indices, assuming that $j' = 0$ corresponds to the moment of time $t = T$ and $j' = 1$ to the moment $t = T - \tau$ and so on.

Assume now (see [218], [257]) that the matrix A from (1.1) is represented as

$$A = A_1 + A_2 \qquad (1.4)$$

where $A_1 \geq 0$, $A_2 \geq 0$.

The componentwise splitting method does not require that the operators A_α must be independent of time. So, the stability analysis may be carried out when

$$A_1(t) \geq 0, \quad A_2(t) \geq 0$$

Consider the approximations of these matrices on the time interval $t_j \leq t \leq t_{j+1}$ in the form

$$\Lambda_\alpha^j = A_\alpha(t_{j+1/2}) \qquad (1.5)$$

assuming that the elements are sufficiently smooth. Let us construct a system of difference equations consisting of the sequence of simple Crank-Nicolson schemes:

$$\frac{\varphi^{j+1/2} - \varphi^j}{\tau} + \Lambda_1^j \frac{\varphi^{j+1/2} + \varphi^j}{2} = 0$$
$$\frac{\varphi^{j+1} - \varphi^{j+1/2}}{\tau} + \Lambda_2^j \frac{\varphi^{j+1} + \varphi^{j+1/2}}{2} = 0 \qquad (1.6)$$

The system of difference equations (1.6) may be reduced to one equation excluding auxiliary functions $\varphi^{j+1/2}$:

$$\varphi^{j+1} = T^j \varphi^j \qquad (1.7)$$

where

$$T^j = (E + \frac{\tau}{2}\Lambda_2^j)^{-1}(E - \frac{\tau}{2}\Lambda_2^j)(E + \frac{\tau}{2}\Lambda_1^j)^{-1}(E - \frac{\tau}{2}\Lambda_1^j) \qquad (1.8)$$

First of all we investigate the problem of approximation. To this end expand the operator T^j into a power series in τ, assuming that

$$\frac{\tau}{2}\|\Lambda_\alpha^j\| < 1$$

As a result we have

$$T^j = E - \tau \Lambda^j + \frac{\tau^2}{2}[(\Lambda_1^j)^2 + 2\Lambda_2^j \Lambda_1^j + (\Lambda_2^j)^2] - \ldots \qquad (1.9)$$

where $\Lambda^j = \Lambda_1^j + \Lambda_2^j$.

If the operators Λ_α^j commute, i. e., $\Lambda_1^j \Lambda_2^j = \Lambda_2^j \Lambda_1^j$, then expansion (1.9) may be written in the form

$$T^j = E - \tau \Lambda^j + \frac{\tau^2}{2}(\Lambda^j)^2 - \ldots \qquad (1.10)$$

So, if $A_1(t) \geq 0$, $A_2(t) \geq 0$, then the difference scheme (1.6) is absolutely stable (it follows immediately from the inequality $\|T^j\| \leq 1$ which is valid according to Kellogg lemma) and, with an assumption that the elements of matrices and the solution of problem (1.1) are sufficiently smooth, the scheme (1.6) approximates the original equation (1.1) at $f = 0$ with the second order in τ in the case when operators Λ_1^j and Λ_2^j commute, and with the first order if they do not.

Now, let us approximate the operators $A_1(t)$ and $A_2(t)$ not on the interval $t_j \leq t \leq t_{j+1}$ as in (1.6), but on the interval $t_{j-1} \leq t \leq t_{j+1}$. Let

$$\Lambda_\alpha^j = A_\alpha(t_j)$$

Consider two systems of difference equations as follows:

$$\frac{\varphi^{j-1/2} - \varphi^{j-1}}{\tau} + \Lambda_1^j \frac{\varphi^{j-1/2} + \varphi^{j-1}}{2} = 0$$

$$\frac{\varphi^j - \varphi^{j-1/2}}{\tau} + \Lambda_2^j \frac{\varphi^j + \varphi^{j-1/2}}{2} = 0 \qquad (1.11)$$

$$\frac{\varphi^{j+1/2} - \varphi^j}{\tau} + \Lambda_2^j \frac{\varphi^{j+1/2} + \varphi^j}{2} = 0$$

$$\frac{\varphi^{j+1} - \varphi^{j+1/2}}{\tau} + \Lambda_1^j \frac{\varphi^{j+1} + \varphi^{j+1/2}}{2} = 0 \qquad (1.12)$$

The computation cycle consists of alternating applications of difference schemes (1.11), (1.12). It is easy to show that after the whole cycle of computations using (1.11) and (1.12) we have

$$\varphi^{j+1} = T^j \varphi^{j-1} \qquad (1.13)$$

where

$$T^j = (E + \frac{\tau}{2}\Lambda_1^j)^{-1}(E - \frac{\tau}{2}\Lambda_1^j)(E + \frac{\tau}{2}\Lambda_2^j)^{-1}(E - \frac{\tau}{2}\Lambda_2^j)$$

$$\times (E + \frac{\tau}{2}\Lambda_2^j)^{-1}(E - \frac{\tau}{2}\Lambda_2^j)(E + \frac{\tau}{2}\Lambda_1^j)^{-1}(E - \frac{\tau}{2}\Lambda_1^j)$$

$$= E - 2\tau\Lambda^j + \frac{(2\tau)^2}{2}(\Lambda^j)^2 - \dots$$

If a step operator T^j is comparable with a step operator of the Crank–Nicolson scheme

$$\frac{\varphi^{j+1} - \varphi^{j-1}}{2\tau} + \Lambda^j\frac{\varphi^{j+1} + \varphi^{j-1}}{2} = 0$$

then no matter, are the operators A_α commutative or not, it may be established, that step operators T^j for two-cyclic splitting scheme and that of Crank-Nicolson scheme applied to the doubled time interval coincide with an accuracy of τ^2. This result for the case of two operators $A = A_1 + A_2$, where $A_i \geq 0$, $i = 1, 2$ was obtained independently by G. Strang and G. Marchuk (see [218, 257]). So this technique takes away very strong requirement for the operators to be commutative.

Now let us discuss the question of computational stability of the method. It follows from (1.7) that

$$\|\varphi^{j+1}\| \leq \|T^j\| \cdot \|\varphi^j\|$$

Since, as was shown above

$$\|T^j\| \leq 1$$

for $A_\alpha \geq 0$, then

$$\|\varphi^{j+1}\| \leq \|\varphi^j\| \tag{1.14}$$

It follows immediately that

$$\|\varphi^j\| \leq \|g\| \tag{1.15}$$

If we consider the two-cyclic method, then the estimates of the type (1.14) are valid on each step of the cycle, which means that the two-cyclic method is also absolutely stable.

So, if $A_1(t) \geq 0$ and $A_2(t) \geq 0$ and a solution of problem (1.1) and the elements of matrices $A_1(t)$ and $A_2(t)$ are sufficiently smooth, then the system of difference equations (1.11) and (1.12) is absolutely stable and the scheme (1.13) approximates the original equation (1.1) for $f = 0$ with a second order in τ.

Let us find the solution of inhomogeneous problem using two-cyclic complete spitting. Examine the system of difference equations of the type (1.11) and (1.12) written in more convenient form:

$$(E + \frac{\tau}{2}\Lambda_1^j)\varphi^{j-1/2} = (E - \frac{\tau}{2}\Lambda_1^j)\varphi^{j-1}$$

$$(E + \frac{\tau}{2}\Lambda_2^j)(\varphi^j - \tau f^j) = (E - \frac{\tau}{2}\Lambda_2^j)\varphi^{j-1/2}$$

$$(E + \frac{\tau}{2}\Lambda_2^j)\varphi^{j+1/2} = (E - \frac{\tau}{2}\Lambda_2^j)(\varphi^j + \tau f^j) \qquad (1.16)$$

$$(E + \frac{\tau}{2}\Lambda_1^j)\varphi^{j+1} = (E - \frac{\tau}{2}\Lambda_1^j)\varphi^{j+1/2}$$

where $f^j = f(t_j)$. Resolving this equation for φ^{j+1} we obtain

$$\varphi^{j+1} = T^j\varphi^{j-1} + 2\tau T_1^j T_2^j f^j \qquad (1.17)$$

where

$$T^j = T_1^j T_2^j T_2^j T_1^j \qquad (1.18)$$

$$T_\alpha^j = (E + \frac{\tau}{2}\Lambda_\alpha^j)^{-1}(E - \frac{\tau}{2}\Lambda_\alpha^j) \qquad (1.19)$$

Use an expansion into a power series in small parameter τ to obtain

$$\varphi^{j+1} = \left[E - 2\tau\Lambda^j + \frac{(2\tau)^2}{2}(\Lambda^j)^2\right]\varphi^{j-1} + 2\tau(E - \tau\Lambda^j)f^j + O(\tau^3) \qquad (1.20)$$

which we transform once more into

$$\frac{\varphi^{j+1} - \varphi^{j-1}}{2\tau} + \Lambda^j(E - \tau\Lambda^j)\varphi^{j-1} = (E - \tau\Lambda^j)f^j + O(\tau^2) \qquad (1.21)$$

Exclude φ^{j-1} using expansion into Taylor series in the neighborhood of a point t_{j-1}. We have with an accuracy up to τ^2

$$\varphi^j = \varphi^{j-1} + \left(\frac{\partial\varphi}{\partial t}\right)^{j-1}\tau + O(\tau^2) \qquad (1.22)$$

Exclude $\partial\varphi/\partial t$ using the relationship

$$\left(\frac{\partial\varphi}{\partial t}\right)^{j-1} = -\Lambda^j\varphi^{j-1} + f^j + O(\tau) \qquad (1.23)$$

Substitute (1.23) into (1.22) to obtain

$$\varphi^j = (E - \tau\Lambda^j)\varphi^{j-1} + \tau f^j + O(\tau^2)$$

whence

$$(E - \tau\Lambda^j)\varphi^{j-1} = \varphi^j - \tau f^j + O(\tau^2) \qquad (1.24)$$

Substitute (1.24) into (1.21). As a result we have

$$\frac{\varphi^{j+1} - \varphi^{j-1}}{2\tau} + \Lambda^j\varphi^j = f^j + O(\tau^2) \qquad (1.25)$$

Obviously, equation (1.25) approximates original equation (1.1) on the interval $t_{j-1} \leq t \leq t_{j+1}$ with the second order of approximation in τ. So, we have found second order finite-difference approximation of inhomogeneous evolutionary equation using the two-cyclic method.

The stability of the method may be proved in energetical norm in a quite elementary way. Indeed, let us estimate (1.17) by the norm

$$\|\varphi^{j+1}\| \leq \|T^j\| \, \|\varphi^{j-1}\| + 2\tau\|T_1^j\| \, \|T_2^j\| \, \|f^j\| \qquad (1.26)$$

As was established, $\|T_\alpha^j\| \leq 1$, therefore,

$$\|T^j\| \leq \|T_1^j\| \, \|T_2^j\| \, \|T_2^j\| \, \|T_1^j\| \leq 1$$

Hence

$$\|\varphi^{j+1}\| \leq \|\varphi^{j-1}\| + 2\tau\|f^j\| \qquad (1.27)$$

By recursive relationship (1.27) we obtain

$$\|\varphi^j\| \leq \|g\| + \tau j\|f\| \qquad (1.28)$$

where

$$\|f\| = \max_j \|f^j\|$$

Computational stability of the scheme follows from relationship (1.28) for any finite temporal interval.

The system (1.16) may also be written in the following equivalent form:

$$(E + \frac{\tau}{2}\Lambda_1^j)\varphi^{j-2/3} = (E - \frac{\tau}{2}\Lambda_1^j)\varphi^{j-1}$$

$$(E + \frac{\tau}{2}\Lambda_2^j)\varphi^{j-1/3} = (E - \frac{\tau}{2}\Lambda_2^j)\varphi^{j-2/3}$$

$$\varphi^{j+1/3} = \varphi^{j-1/3} + 2\tau f^j \qquad (1.29)$$

$$(E + \frac{\tau}{2}\Lambda_2^j)\varphi^{j+2/3} = (E - \frac{\tau}{2}\Lambda_2^j)\varphi^{j+1/3}$$

$$(E + \frac{\tau}{2}\Lambda_1^j)\varphi^{j+1} = (E - \frac{\tau}{2}\Lambda_1^j)\varphi^{j+2/3}$$

Excluding unknown variables with fractional indices we arrive at the resolved equation of the type

$$\varphi^{j+1} = T_1^j T_2^j T_2^j T_1^j \varphi^{j-1} + 2\tau T_1^j T_2^j f^j \qquad (1.30)$$

which coincides with (1.17). In some cases it is more convenient to write the equations in the form (1.29) than in the form (1.16).

So, if $A_1(t) \geq 0$ and $A_2(t) \geq 0$ and the solution φ, function f, and the elements of matrices $A_1(t)$ and $A_2(t)$ are sufficiently smooth, the system of difference equations (1.16) is absolutely stable on the interval $0 \leq t \leq T$ and approximates the original equation (1.1) with a second order in τ.

2. Multicomponent Splitting of Problems

We had assumed till now that the original operator A is represented as a sum of two operators of more simple structure. While solving complicated problems of mathematical physics we often need to split the operators into greater number of summands. Consider the case (see [218], [224]) when

$$A = \sum_{\alpha=1}^{n} A_\alpha \qquad (2.1)$$

ere $A_\alpha \geq 0$. Since the case $n = 2$ was discussed in detail in previous ection, we will deal only with the case when $n > 2$. Notice, that it is impossible in a general way to use here the methods developed for the case $n = 2$.

We will try to construct an absolutely stable difference analogue of the problem with a second order of approximation in τ. Keeping in mind the multi-component splitting we assume that

$$\Lambda^j = \sum_{\alpha=1}^{n} \Lambda_\alpha^j \qquad (2.2)$$

where all Λ_α^j are positive semi-definite operators: $\Lambda_\alpha^j \geq 0$. Examine the system of equations

$$(E + \frac{\tau}{2}\Lambda_\alpha^j)\varphi^{j+\alpha/n} = (E - \frac{\tau}{2}\Lambda_\alpha^j)\varphi^{j+(\alpha-1)/n}, \quad \alpha = 1, 2, \ldots, n \qquad (2.3)$$

If $\Lambda_\alpha^j \geq 0$ are commutative and $\Lambda_\alpha^j = A_\alpha^{j+1/2}$ or $\Lambda_\alpha^j = (A_\alpha^{j+1} + A_\alpha^j)/2$ the scheme (2.3) is absolutely stable and has a second order of approximation. This fact may be checked easily with a help of the Fourier method. But, for non-commutative operators Λ_α^j, scheme (2.3) will be the scheme of the first order of approximation in τ and therefore not so interesting for applications, unlike the following scheme of the second order of approximation

$$\begin{aligned} \varphi^{j+\alpha/2n} &= (E - \frac{\tau}{2}\Lambda_\alpha^j)\varphi^{j+(\alpha-1)/2n}, \quad \alpha = 1, \ldots, n \\ (E + \frac{\tau}{2}\Lambda_{2n-\alpha+1}^j)\varphi^{j+\alpha/2n} &= \varphi^{j+(\alpha-1)/2n}, \quad \alpha = n+1, \ldots, 2n \end{aligned} \qquad (2.4)$$

We will try to define a special construction of the full splitting method on the basis of (2.3) which gives the solution to the Cauchy problem for positive semi-definite and non-commutative operators Λ_α^j and is of the second order of approximation. In a certain sense it is a complete solution of a splitting problem.

Notice, that the system of equations (2.3) may be reduced to one equation of the form

$$\varphi^{j+1} = \prod_{\alpha=1}^{n} (E + \frac{\tau}{2}\Lambda_\alpha^j)^{-1}(E - \frac{\tau}{2}\Lambda_\alpha^j)\varphi^j \qquad (2.5)$$

With a help of (4.5) we find the estimate by the norm

$$\|\varphi^{j+1}\| = \prod_{\alpha=1}^{n} \|(E + \frac{\tau}{2}\Lambda_\alpha^j)^{-1}(E - \frac{\tau}{2}\Lambda_\alpha^j)\| \, \|\varphi^j\| \qquad (2.6)$$

By the Kellogg lemma we have

$$\|\varphi^{j+1}\| \le \|\varphi^j\| \le \ldots \le \|g\| \qquad (2.7)$$

If the operators are skew-symmetric, i. e., $(\Lambda_\alpha^j \varphi, \varphi) = 0$, then

$$\|\varphi^{j+1}\| = \|\varphi^j\| = \ldots = \|g\| \qquad (2.8)$$

So, the absolute stability of this scheme has been proved.

In order to define the order of approximation expand into the powers of small parameter τ the following expression (assume that $\frac{\tau}{2}\|\Lambda_\alpha^j\| < 1$):

$$T^j = \prod_{\alpha=1}^n (E + \frac{\tau}{2}\Lambda_\alpha^j)^{-1}(E - \frac{\tau}{2}\Lambda_\alpha^j)$$

Since

$$T^j = \prod_{\alpha=1}^n T_\alpha^j$$

where $T_\alpha^j = (E + \frac{\tau}{2}\Lambda_\alpha^j)^{-1}(E - \frac{\tau}{2}\Lambda_\alpha^j)$, we first expand the operators T_α^j:

$$T_\alpha^j = E - \tau\Lambda_\alpha^j + \frac{\tau^2}{2}(\Lambda_\alpha^j)^2 \ldots \qquad (2.9)$$

As a result we have

$$T^j = E - \tau\Lambda^j + \frac{\tau^2}{2}\left[(\Lambda^j)^2 + \sum_{\alpha=1}^n \sum_{\beta=\alpha+1}^n (\Lambda_\alpha^j\Lambda_\beta^j - \Lambda_\beta^j\Lambda_\alpha^j)\right] + O(\tau^3) \quad (2.10)$$

In the case when the operators Λ_α^j are commutative the expression under the sign of the double sum equals zero and we have

$$T^j = E - \tau\Lambda^j + \frac{\tau^2}{2}(\Lambda^j)^2 + O(\tau^3) \qquad (2.11)$$

which means that scheme (2.3) is of the second order of approximation in τ. If the operators Λ_α^j are non-commutative the scheme happens to be only of the first order of approximation in τ. If we want to construct the scheme of the second order of approximation in τ for non-commutative case, the scheme (2.3) must be substituted by the following:

$$\varphi^j = \prod_{\alpha=1}^n T_\alpha^j\varphi^{j-1}, \quad \varphi^{j+1} = \prod_{\alpha=n}^1 T_\alpha^j\varphi^j \qquad (2.12)$$

It means from an algorithmic point of view that the first system of equations (2.3) is solved on the interval $t_{j-1} \leq t \leq t_j$ for $\alpha = 1, 2, \ldots n$ and then the same system is solved on interval $t_j \leq t \leq t_{j+1}$ but in a reversed sequence ($\alpha = n, n-1, \ldots, 2, 1$):

$$(E + \frac{\tau}{2}\Lambda_\alpha^j)\varphi^{j+\alpha/n-1} = (E - \frac{\tau}{2}\Lambda_\alpha^j)\varphi^{j+(\alpha-1)/n-1}, \ \alpha = 1, \ldots, n$$

$$(2.13)$$

$$(E + \frac{\tau}{2}\Lambda_\alpha^j)\varphi^{j+1-(\alpha-1)/n} = (E - \frac{\tau}{2}\Lambda_\alpha^j)\varphi^{j+1-\alpha/n}, \alpha = n, \ldots, 1$$

It is obvious that for the full cycle (2.13) we have

$$\varphi^{j+1} = T^j \varphi^{j-1}$$

where

$$T^j = \prod_{\alpha=1}^{n} T_\alpha^j \prod_{\alpha=n}^{1} T_\alpha^j = E - 2\tau\Lambda^j + \frac{(2\tau)^2}{2}(\Lambda^j)^2 + O(\tau^3)$$

So, scheme (2.13) has the second order of accuracy in τ on the interval $t_{j-1} \leq t \leq t_{j+1}$.

Notice in conclusion, that the difference scheme (2.13) is absolutely stable for $\Lambda_\alpha^j \geq 0$.

For inhomogeneous equation

$$\frac{d\varphi}{dt} + A\varphi = f, \quad \text{in} \quad \Omega \times \Omega_t, \quad \varphi = g \quad \text{at} \quad t = 0 \qquad (2.14)$$

where $A(t) \geq 0$ and

$$A = \sum_{\alpha=1}^{n} A_\alpha, \quad A_\alpha(t) \geq 0$$

on the interval $t_{j-1} \leq t \leq t_{j+1}$ there is the following splitting scheme:

$$(E + \frac{\tau}{2}\Lambda_1^j)\varphi^{j-(n-1)/n} = (E - \frac{\tau}{2}\Lambda_1^j)\varphi^{j-1}$$

$$\cdots\cdots\cdots\cdots\cdots\cdots\cdots\cdots\cdots\cdots\cdots$$

$$(E + \frac{\tau}{2}\Lambda_n^j)(\varphi^j - \tau f^j) = (E - \frac{\tau}{2}\Lambda_n^j)\varphi^{j-1/n}$$

$$(2.15)$$

$$(E + \frac{\tau}{2}\Lambda_n^j)\varphi^{j+1/n} = (E - \frac{\tau}{2}\Lambda_n^j)(\varphi^j + \tau f^j)$$

$$\cdots\cdots\cdots\cdots\cdots\cdots\cdots\cdots\cdots\cdots\cdots$$

$$(E + \frac{\tau}{2}\Lambda_1^j)\varphi^{j+1} = (E - \frac{\tau}{2}\Lambda_1^j)\varphi^{j+(n-1)/n}$$

where $\Lambda_\alpha^j = A_\alpha(t_j)$. It is easy to check that this scheme has a second order of approximation in τ and is absolutely stable under the assumption of necessary smoothness of φ.

As in the case $n = 2$, n-component system of equations (2.15) may be written in equivalent form:

$$(E + \frac{\tau}{2}\Lambda_\alpha^j)\varphi^{j-(n+1-\alpha)/(n+1)} = (E - \frac{\tau}{2}\Lambda_\alpha^j)\varphi^{j-(n+1-\alpha+1)/(n+1)}$$
$$\alpha = 1, 2$$
$$\varphi^{j+1/(n+1)} = \varphi^{j-1/(n+1)} + 2\tau f^j \qquad (2.16)$$
$$(E + \frac{\tau}{2}\Lambda_{n-\alpha+2}^j)\varphi^{j+\alpha/(n+1)} = (E - \frac{\tau}{2}\Lambda_{n-\alpha+2}^j)\varphi^{j+(\alpha-1)/(n+1)}$$
$$\alpha = 2, 1$$

Now turn to the splitting method for implicit difference approximations. To this end examine the problem

$$\frac{d\varphi}{dt} + A\varphi = 0 \quad \text{in} \quad \Omega \times \Omega_t, \quad \varphi = g \quad \text{at} \quad t = 0 \qquad (2.17)$$

Assume that

$$A = \sum_{\alpha=1}^{n} A_\alpha$$

all $A_\alpha \geq 0$, and A_α does not depend on time. Consider the algorithm of splitting in the following form:

$$\frac{\varphi^{j+1/n} - \varphi^j}{\tau} + A_1\varphi^{j+1/n} = 0$$
$$\dotsb \qquad (2.18)$$
$$\frac{\varphi^{j+1} - \varphi^{j+(n-1)/n}}{\tau} + A_n\varphi^{j+1} = 0$$

Let us show that the algorithm is absolutely stable. Taking an inner product of the equation

$$\frac{\varphi^{j+\alpha/n} - \varphi^{j+(\alpha-1)/n}}{\tau} + A_\alpha\varphi^{j+\alpha/n} = 0 \qquad (2.19)$$

and $\varphi^{j+\alpha/n}$ we obtain the relationship

$$(\varphi^{j+\alpha/n} - \varphi^{j+(\alpha-1)/n}, \varphi^{j+\alpha/n}) + \tau(A_\alpha\varphi^{j+\alpha/n}, \varphi^{j+\alpha/n}) = 0$$

Since the operator A_α is positive semi-definite,

$$\left(\varphi^{j+\alpha/n} - \varphi^{j+(\alpha-1)/n}, \varphi^{j+\alpha/n}\right) \leq 0$$

or

$$\left(\varphi^{j+\alpha/n}, \varphi^{j+\alpha/n}\right) \leq \left(\varphi^{j+\alpha/n}, \varphi^{j+(\alpha-1)/n}\right)$$

But, since

$$\left(\varphi^{j+\alpha/n}, \varphi^{j+(\alpha-1)/n}\right) \leq \frac{1}{2}\left[\left(\varphi^{j+\alpha/n}, \varphi^{j+\alpha/n}\right) + \left(\varphi^{j+(\alpha-1)/n}, \varphi^{j+(\alpha-1)/n}\right)\right]$$

then

$$\left\|\varphi^{j+\alpha/n}\right\|^2 \leq \left\|\varphi^{j+(\alpha-1)/n}\right\|^2, \quad \alpha = 1, 2, ..., n$$

By this recursive inequality we obtain

$$\left\|\varphi^{j+1}\right\| \leq \left\|\varphi^{j}\right\| \qquad (2.20)$$

which means, that with the above assumptions scheme (2.18) is absolutely stable.

It is easy to check that system (2.18) approximates the original problem with the first order of approximation in τ.

Examine the inhomogeneous problem

$$\frac{d\varphi}{dt} + A\varphi = f \quad \text{in} \quad \Omega \times \Omega_t, \quad \varphi = g \quad \text{at} \quad t = 0 \qquad (2.21)$$

and the following splitting scheme for this problem

$$\frac{\varphi^{j+1/n} - \varphi^{j}}{\tau} + A_1\varphi^{j+1/n} = 0$$

$$\cdots\cdots\cdots\cdots\cdots\cdots\cdots\cdots\cdots\cdots\cdots\cdots \qquad (2.22)$$

$$\frac{\varphi^{j+1} - \varphi^{j+(n-1)/n}}{\tau} + A_n\varphi^{j+1} = f^j$$

This splitting scheme approximates the original inhomogeneous equation with the first order of accuracy in τ.

Let us prove stability of scheme (2.22). Take an inner product of each equation and corresponding $\varphi^{j+1/n}, \ldots, \varphi^{j+1}$. Then, like in preceding case we have

$$\left\|\varphi^{j+\alpha/n}\right\|^2 \leq \left\|\varphi^{j+(\alpha-1)/n}\right\|^2, \quad \alpha = 1, 2, ..., n-1 \qquad (2.33)$$

Consider the last equation so obtained from (2.22) in more detail:

$$(\varphi^{j+1}, \varphi^{j+1}) = (\varphi^{j+(n-1)/n}, \varphi^{j+1}) - \tau(A_n\varphi^{j+1}, \varphi^{j+1}) + \tau(f^j, \varphi^{j+1})$$

Taking into account that $A_n \geq 0$ we get

$$(\varphi^{j+1}, \varphi^{j+1}) \leq (\varphi^{j+(n-1)/n}, \varphi^{j+1}) + \tau(f^j, \varphi^{j+1})$$

Since

$$|(\varphi^{j+(n-1)/n}, \varphi^{j+1})| \leq \|\varphi^{j+(n-1)/n}\| \, \|\varphi^{j+1}\|$$
$$|(f^j, \varphi^{j+1})| \leq \|f^j\| \, \|\varphi^{j+1}\|$$

then

$$\|\varphi^{j+1}\|^2 \leq \|\varphi^{j+(n-1)/n}\| \, \|\varphi^{j+1}\| + \tau\|f^j\| \, \|\varphi^{j+1}\|$$

Dividing the relationship by $\|\varphi^{j+1}\|$ we arrive at the inequality

$$\|\varphi^{j+1}\| \leq \|\varphi^{j+(n-1)/n}\| + \tau\|f^j\|$$

Excluding the solutions with fractional indices we have

$$\|\varphi^{j+1}\| \leq \|\varphi^j\| + \tau\|f^j\|$$

With regard to

$$\|\varphi^0\| = \|g\|$$

and excluding intermediate values of the solution we obtain

$$\|\varphi^{j+1}\| \leq \|g\| + \tau j\|f\| \qquad (2.25)$$

where

$$\|f\| = \max_j \|f^j\|$$

Absolute stability of the difference scheme whence follows.

This algorithm may be generalized for the case when the operator A depends on time. In this case an appropriate difference approximation must be taken instead of A on each interval $t_j \leq t \leq t_{j+1}$.

Let us examine an evolutionary problem with operator A depending on time and on solution of the problem:

$$\frac{d\varphi}{dt} + A(t,\varphi)\varphi = 0 \quad \text{in} \quad \Omega \times \Omega_t, \quad \varphi = g \quad \text{at} \quad t = 0 \qquad (2.26)$$

We assume that the operator $A(t, \varphi)$ is non-negative and has the form

$$A(t, \varphi) = \sum_{\alpha=1}^{n} A_\alpha(t, \varphi) \qquad (2.27)$$

The operator $A_\alpha(t, \varphi) \geq 0$ and is sufficiently smooth. Assume further, that the solution φ is also sufficiently smooth function of time. Consider the splitting scheme on the interval $t_{j-1} \leq t \leq t_{j+1}$:

$$\frac{\varphi^{j+1/n-1} - \varphi^{j-1}}{\tau} + A_1^j \frac{\varphi^{j+1/n-1} + \varphi^{j-1}}{2} = 0$$

$$\dotsb$$

$$\frac{\varphi^{j} - \varphi^{j-1/n}}{\tau} + A_n^j \frac{\varphi^{j} + \varphi^{j-1/n}}{2} = 0$$

$$\frac{\varphi^{j+1/n} - \varphi^{j}}{\tau} + A_n^j \frac{\varphi^{j+1/n} + \varphi^{j}}{2} = 0 \qquad (2.28)$$

$$\dotsb$$

$$\frac{\varphi^{j+1} - \varphi^{j+(n-1)/n}}{\tau} + A_1^j \frac{\varphi^{j+1} + \varphi^{j+(n-1)/n}}{2} = 0$$

where

$$A^j = A(t_j, \tilde{\varphi}^j)$$
$$\tilde{\varphi}^j = \varphi^{j-1} - \tau A(t_{j-1}, \varphi^{j-1}) \varphi^{j-1} \qquad (2.29)$$
$$\tau = t_j - t_{j-1}$$

It is easy to show using the methods described above for linear operators depending only on time, that scheme (2.28) with conditions (2.29) has the second order of approximation in τ and is absolutely stable. The splitting method for inhomogeneous quasi-linear equations is defined in a similar way. This opens wide possibilities for applications of the componentwise splitting to the solution of non-stationary quasi-linear problems in hydrodynamics, meteorology, oceanology and other important natural sciences.

3. An Example of Application: Two-Dimensional Problem with Constant Coefficients

Let us examine a special case which will help in understanding the profound essence of the method.

Consider a spatial two-dimensional model problem with constant coefficients

$$\frac{\partial \varphi}{\partial t} + u\frac{\partial \varphi}{\partial x} + v\frac{\partial \varphi}{\partial y} + \sigma\varphi = \mu\Delta\varphi, \quad \varphi = g \quad \text{at} \quad t = 0 \qquad (3.1)$$

Assume $g(x, y)$ as assigned on all the plane (x, y) and decreasing rather fast at infinity. Then a solution may be represented with a help of the Fourier transformation (we assume that all the functions are sufficiently smooth) as

$$\varphi = \int_{-\infty}^{\infty} \int_{-\infty}^{\infty} \Phi(m, n, t)e^{imx+iny} \, dm \, dn \qquad (3.2)$$

Similarly,

$$g = \int_{-\infty}^{\infty} \int_{-\infty}^{\infty} A(m, n)e^{imx+iny} \, dm \, dn \qquad (3.3)$$

Let us multiply the equation and initial data in (3.1) by the factor $e^{-imx-imy}/(4\pi^2)$ and integrate the result over x and y to obtain the problem for Fourier coefficients:

$$\frac{\partial \Phi}{\partial t} + \left[-imu - inv + \sigma + \mu(m^2 + n^2)\right] \Phi = 0, \quad \Phi = A \quad \text{at} \quad t = 0 \qquad (3.4)$$

which has a solution in the form

$$\Phi = A \exp\left\{\left[imu + inv - \sigma - \mu(m^2 + n^2)\right] t\right\} \qquad (3.5)$$

Substituting (3.5) into (3.2) we find the exact solution:

$$\varphi = \int_{-\infty}^{\infty} \int_{-\infty}^{\infty} A(m, n) \exp\left\{im(x + ut) + in(y + vt)\right.$$

$$\left. - \left[\sigma + \mu(m^2 + n^2)\right] t\right\} \, dm \, dn \qquad (3.6)$$

Let us split the problem into two problems on temporal interval $0 \leq t \leq \tau$ and write on this interval two problems:

the first problem

$$\frac{\partial \varphi_1}{\partial t} + u \frac{\partial \varphi_1}{\partial x} + v \frac{\partial \varphi_1}{\partial y} = 0, \quad \varphi_1 = g \quad \text{at} \quad t = 0 \tag{3.7}$$

the second problem:

$$\frac{\partial \varphi_2}{\partial t} - \mu \Delta \varphi_2 + \sigma \varphi_2 = 0 \quad \varphi_2 = \varphi_1(x, y, \tau) \quad \text{at} \quad t = 0 \tag{3.8}$$

Let the function $g(x, y)$ be represented in the form (3.3). Consider similar Fourier-representations for the functions φ_1 and φ_2. Then, like in the above case we arrive at two problems:

for Fourier coefficients Φ_1 of problem (3.7):

$$\frac{\partial \Phi_1}{\partial t} + (imu + inv)\Phi_1 = 0, \quad \Phi_1 = A \quad \text{at} \quad t = 0 \tag{3.9}$$

for Fourier coefficients Φ_2 of problem (3.8):

$$\frac{\partial \Phi_2}{\partial t} + \left[\sigma + \mu(m^2 + n^2) \right] \Phi_2 = 0, \quad \Phi_2 = \Phi_1(\tau) \quad \text{at} \quad t = 0 \tag{3.10}$$

A solution to (3.9) has the form

$$\Phi_1 = A \exp \left\{ (imu + inv)t \right\} \tag{3.11}$$

A solution to (3.10) has the form

$$\Phi_2 = \Phi_1(\tau) \exp \left\{ - \left[\sigma + \mu(m^2 + n^2) \right] t \right\} \tag{3.12}$$

Assuming $t = \tau$ in (3.11) and substituting $\Phi_1(\tau)$ into (3.12) we have

$$\Phi_2(t) = A \exp \left\{ (imu + inv)\tau - \left[\sigma + \mu(m^2 + n^2) \right] t \right\} \tag{3.13}$$

Finally, assume $t = \tau$ in (3.13) to obtain

$$\Phi_2 = A \exp \left\{ \left[imu + inv - \sigma - \mu(m^2 + n^2) \right] \tau \right\}$$

whence

$$\varphi_2 = \int\limits_{-\infty}^{\infty} \int\limits_{-\infty}^{\infty} A(m,n) \exp\{im(x+u\tau)+in(y+v\tau)$$

$$- \left[\sigma + \mu(m^2+n^2)\right]\tau\} \, dm\,dn \quad (3.14)$$

The value of τ was not fixed till now. Let it be arbitrary one. Assume $t = \tau$ in (3.6). Then (3.6) and (3.14) coincide identically (for any τ). It is easy to check that these solutions coincide exactly only at $t = \tau$, i. e., $\varphi_2(x,y,\tau) = \varphi(x,y,\tau)$. If we wish to obtain the value of solution using split problem somewhere inside the interval $0 \le t \le \tau$ it would be only approximate one. So, if we want to find the solution in a discrete row of points $t = t_{j+1}$ knowing a solution at $t = t_j$ we need to solve consequently the split problems that is, the problems

$$\frac{\partial \varphi_1}{\partial t} + u\frac{\partial \varphi_1}{\partial x} + v\frac{\partial \varphi_1}{\partial y} = 0, \quad \varphi_1 = \varphi^j \quad \text{at} \quad t = t_j \quad (3.15)$$

$$\frac{\partial \varphi_2}{\partial t} = \mu\Delta\varphi_2 - \sigma\varphi_2, \quad \varphi_2 = \varphi_1^{j+1} \quad \text{at} \quad t = t_j \quad (3.16)$$

on the interval $t_j \le t \le t_{j+1}$.

This remarkable fact lies in the basis of splitting the problems along the physical processes. Though, since the coefficients u and v in real conditions are not constant, the splitting algorithm does not ensure an exact solution for $t = t_j$. To obtain more precise results one must use appropriate small time intervals thus providing better approximation in the cases when the equation's coefficients change considerably. Two-dimensional case was examined here, but all the layings out are valid for a general three-dimensional case too.

The original problem may be split as follows:

$$\frac{\partial \varphi_1}{\partial t} + u\frac{\partial \varphi_1}{\partial x} = 0, \quad \varphi_1 = g \quad \text{at} \quad t = 0 \quad (3.17)$$

$$\frac{\partial \varphi_2}{\partial t} + v\frac{\partial \varphi_2}{\partial y} = 0, \quad \varphi_2 = \varphi_1(x,y,\tau) \quad \text{at} \quad t = 0 \quad (3.18)$$

$$\frac{\partial \varphi_3}{\partial t} + \sigma\varphi_3 = 0, \quad \varphi_3 = \varphi_2(x,y,\tau) \quad \text{at} \quad t = 0 \quad (3.19)$$

$$\frac{\partial \varphi_4}{\partial t} - \mu\frac{\partial^2 \varphi_4}{\partial x^2} = 0, \quad \varphi_4 = \varphi_3(x,y,\tau) \quad \text{at} \quad t = 0 \quad (3.20)$$

$$\frac{\partial \varphi_5}{\partial t} - \mu\frac{\partial^2 \varphi_5}{\partial y^2} = 0, \quad \varphi_5 = \varphi_4(x,y,\tau) \quad \text{at} \quad t = 0 \quad (3.21)$$

Obtain, as before,

$$\varphi_5(x, y, \tau) = \int\limits_{-\infty}^{\infty} \int\limits_{-\infty}^{\infty} A(m, n) \exp\left\{ im(x + u\tau) + in(y + v\tau) \right.$$

$$\left. - \left[\sigma + \mu(m^2 + n^2) \right] \tau \right\} dm\, dn = \varphi(x, y, \tau) \quad (3.22)$$

Again we have an exact solutions at $t = \tau$, $t = 2\tau$, ... Of course, other variants of splitting are possible.

We see, that there are many variants to split problem (3.1) and all they provide exact solution at assigned temporal points t. An error may appear in real conditions only because the values u and v are not constant. The interval τ is chosen to be sufficiently small in order to minimize possible error of splitting. It is advisable to split general problem (3.1) into two problems or more along physical processes, since each of them in general case is connected with fulfillment of one or another physical balance relationships, and it is important, that these balance relationships would not be disturbed at the stage of construction of difference approximations. As for the solution of difference analogues of the splitting schemes, it may be performed using the methods of componentwise cyclic reduction described in [73].

It is worthy to notice that positive definiteness of elementary operators of problems split along spatial variables ensures the stability while using Crank–Nicolson schemes.

This analysis shows why splitting method brings about such surprisingly exact results while solving problems of transport and diffusion of substances, problems in the theory of climate and so on. Of course, all this is valid for a general three-dimensional case.

All splitting methods considered in the Appendix proved to be extremely effective in the solution of global problems connected with climate changes, general circulation of atmosphere and ocean, environment protection problems.

The reader can find the references corresponding to the Appendix's problems in the author's survey [224].

APPENDIX II

Difference Analogue of Non-Stationary Heat Diffusion Equation in Atmosphere and Ocean

Let us examine the heat diffusion equation[1]

$$c_p \bar{\rho} \frac{dT}{dt} = \frac{\partial}{\partial z} \bar{\nu} \frac{\partial T}{\partial z} + \bar{\mu} \Delta T + F \delta(z) \qquad (1)$$

where T is deviation of air and water particles temperature from certain temperature $\overline{T}$ which is assumed to be "mean" temperature of the troposphere (notice that $\overline{T} = const$, therefore its value is not important in solving equation (1)); F is total radiation flux at the surface of ocean and continent, which is a function of coordinates x, y and time t;

$$\frac{d}{dt} = \frac{\partial}{\partial t} + u \frac{\partial}{\partial x} + v \frac{\partial}{\partial y} + w \frac{\partial}{\partial z}; \quad \Delta = \frac{\partial^2}{\partial x^2} + \frac{\partial^2}{\partial y^2};$$

$\bar{\nu} = c_p \nu \bar{\rho}$; $\bar{\mu} = c_p \mu \bar{\rho}$; ν and μ are coefficients of vertical and horizontal turbulent exchange; c_p is the specific heat; $\bar{\rho}$ is the standard density. While $\bar{\mu} = const$ for the atmosphere and $\bar{\mu} = \bar{\mu}_s = const$ for the ocean is good approximation, $\bar{\nu}$ is more complex function which may be assigned as follows:

$$\bar{\nu}(z) = \begin{cases} bH, & H < z < H_T \\ a + bz, & 0 < z < H \\ c, & -h < z < 0 \\ c_\infty, & -h_T < z \end{cases}$$

; the definition domain $-h_T < z < H_T$ includes the atmosphere with "top boundary" H_T and the ocean with effective depth h_T equal to thermoclyne layer. Here H and $-h$ are respectively top boundary of

---

[1] See the footnote at page 190.

planetary boundary layer of the Earth's atmosphere and friction layer boundary in the ocean. The quantities a, b, c, and c_∞ are constants to be determined as a result of experimental data processing. It should just be noticed that the value c depends heavily on the square of a wind velocity modulus in the planetary boundary layer. All the rest of quantities may be chosen as constants if we have in mind a long range weather forecast.

In the case of atmosphere over a continent the following approximation may be chosen for $\bar{\nu}(z)$:

$$\bar{\nu}(z) = \begin{cases} bH, & H < z < H_T \\ a + bz, & 0 < z < H \\ \bar{\nu}_c, & z < 0 \end{cases}$$

where $\bar{\nu}_c$ is a heat conductivity coefficient for the soil. It is necessary to emphasize that horizontal heat exchange in the soil is negligible, therefore

$$\bar{\mu} = 0 \quad \text{at} \quad z < 0$$

Join boundary conditions to equation (1); in the case of the system atmosphere–ocean

$$\bar{\nu}\frac{\partial T}{\partial z} = 0 \quad \text{at} \quad z = H_T$$

$$\bar{\nu}\frac{\partial T}{\partial z} = 0 \quad \text{at} \quad z = -h_T \tag{2}$$

in the case of the system atmosphere–continent

$$\bar{\nu}\frac{\partial T}{\partial z} = 0 \quad \text{at} \quad z = H_T$$

$$\bar{\nu}\frac{\partial T}{\partial z} = 0 \quad \text{at} \quad z = -h_c \tag{3}$$

Notice, that floating ice in Arctic and Antarctic may be taken into account through corresponding choice of the function F.

Assume the following initial conditions for equation (1):

$$T = T^0 \quad \text{at} \quad t = 0 \tag{4}$$

where T^0 is an assigned function of coordinates.

So, problem (1)–(4) is posed completely with accuracy up to boundary conditions on "lateral" planes which may be introduced easily.

Now, turn to very important question of constructing the difference schemes corresponding to physical peculiarities of problem, such as discontinuous character of turbulent exchange coefficient, presence of δ-like radiation source at $z = 0$, continuity of the function T in all the definition domain along with the flux $\bar{\nu}\frac{\partial T}{\partial z}$ which is discontinuous just at $z = 0$. Allowing for these peculiarities, introduce the main grid of points z_k arranged along decreasing index k: $z_n = H_T$, $z_{n-1}, \ldots, z_2,\ z_1 = H,\ z_0 = 0,\ z_{-1} = -h,\ z_{-2}, \ldots, z_{-m+1},\ z_{-m} = h_T$. For each interval (z_k, z_{k+1}) introduce a medium point

$$z_{k+1/2} = \frac{1}{2}(z_k + z_{k+1})$$

Let us refer to the set of all the points $\{z_{k+1/2}\}$ as an auxiliary system of the grid's points.

Integrate equation (1) in the limits $z_{k-1/2} \leq z \leq z_{k+1/2}$ to obtain

$$\int_{z_{k-1/2}}^{z_{k+1/2}} c_p \frac{dT}{dt} \bar{\rho}\, dz = J_{k+1/2} - J_{k-1/2} + \int_{z_{k-1/2}}^{z_{k+1/2}} \bar{\mu}\Delta T\, dz + \int_{z_{k-1/2}}^{z_{k+1/2}} F\delta(z)dz \quad (5)$$

Examine first relationship (5) for the case when $k \neq 0$, i. e., for internal layers of atmosphere, or ocean, or continent. The integrands in (5) are continuous in this case, therefore, in order to realize the integrals, use the simplest quadrature formula. Then relationship (5) may be approximated as

$$c_p\bar{\rho}\Delta z_k \frac{dT_k}{dt} = J_{k+1/2} - J_{k-1/2} + \Delta z_k \bar{\mu}_k \Delta T_k \quad (k \neq 0)$$

$$J = \bar{\nu}\frac{\partial T}{\partial z} \quad \text{and} \quad J_k = J(z_k) \quad (6)$$

$$\frac{dT_k}{dt} = \frac{\partial T_k}{\partial t} + u_k \frac{\partial T_k}{\partial x} + v_k \frac{\partial T_k}{\partial y} + w_k \frac{\partial T_k}{\partial z}$$

The fourth summand in the right-hand part of the last equality is much smaller than the rest of summands in the problems of climate and general circulation of atmosphere and ocean, therefore we will hereafter neglect this summand and use the approximate equality

$$\frac{dT_k}{dt} = \frac{\partial T_k}{\partial t} + u_k \frac{\partial T_k}{\partial x} + v_k \frac{\partial T_k}{\partial y}$$

Divide both parts of the relationship for a heat flux

$$J(z) = \bar{\nu} \frac{\partial T}{\partial z} \tag{7}$$

by $\bar{\nu}$ to obtain

$$\frac{\partial T}{\partial z} = \frac{J}{\bar{\nu}} \tag{8}$$

Integrate equation (8) in the limits $z_k \le z \le z_{k+1}$ to obtain

$$T_{k+1} - T_k = \int_{z_k}^{z_{k+1}} \frac{J dz}{\bar{\nu}} \tag{9}$$

Since the flux $J(z)$ is continuous function in these limits, the flux J may be taken beyond the integral sign at $z = z_{k+1/2}$, i. e.,

$$T_{k+1} - T_k = J_{k+1/2} \int_{z_k}^{z_{k+1}} \frac{dz}{\bar{\nu}} \tag{10}$$

It follows from (10) that

$$J_{k+1/2} = \frac{T_{k+1} - T_k}{\int_{z_k}^{z_{k+1}} \frac{dz}{\bar{\nu}}} \tag{11}$$

With regard to (11) the boundary conditions of problem may be written in the form

$$J_{n+1} = 0, \quad J_{-m-1/2} = 0 \tag{12}$$

Substitute relationship (11) for the fluxes into (6) to arrive at the difference equations

$$c_p \bar{\rho}_k \Delta z_k \frac{dT_k}{dt} = \frac{\bar{\nu}_{k+1/2}}{\Delta z_{k+1/2}} (T_{k+1} - T_k) - \frac{\bar{\nu}_{k-1/2}}{\Delta z_{k-1/2}} (T_k - T_{k-1})$$

$$+ \bar{\mu}_k \Delta z_k \Delta T_k \quad (k \ne 0) \tag{13}$$

where

$$\bar{\nu}_{k+1/2} = \frac{\Delta z_{k-1/2}}{\int\limits_{z_k}^{z_{k+1}} \frac{dz}{\bar{\nu}}} \tag{14}$$

In order to derive corresponding difference equation for $k = 0$ use again integral relationship (5). Then

$$\int\limits_{z_{-1/2}}^{z_{1/2}} c_p \frac{dT}{dt} \bar{\rho}\, dz = J_{1/2} - J_{-1/2} + \int\limits_{z_{-1/2}}^{z_{1/2}} \bar{\mu}\Delta T\, dz + \int\limits_{z_{-1/2}}^{z_{1/2}} F\delta(z)dz \tag{15}$$

Since the function T is continuous in transition at $z = 0$ and double differentiable in x and y, it is possible to write approximately

$$\int\limits_{z_{-1/2}}^{z_{1/2}} c_p \frac{dT}{dt} \bar{\rho}\, dz = \overline{c_p\rho_0}\Delta z_0 \frac{dT_0}{dt}$$

$$\int\limits_{z_{-1/2}}^{z_{1/2}} \bar{\mu}\Delta T\, dz = \bar{\mu}_0 \Delta z_0 \Delta T_0 \tag{16}$$

$$\int\limits_{z_{-1/2}}^{z_{1/2}} F\delta(z)dz = F$$

where

$$\overline{c_p\rho_0} = \frac{1}{\Delta z_0} \int\limits_{z_{-1/2}}^{z_{1/2}} c_p\bar{\rho}\, dz, \quad \bar{\mu}_0 = \frac{1}{\Delta z_0} \int\limits_{z_{-1/2}}^{z_{1/2}} \bar{\mu}\, dz$$

$$\frac{dT_0}{dt} = \frac{\partial T_0}{\partial t} + u_0 \frac{\partial T_0}{\partial x} + v_0 \frac{\partial T_0}{\partial y}, \quad \Delta z_0 = z_{1/2} - z_{-1/2}$$

$$u_0 = \frac{1}{\overline{c_p\rho_0}} \int\limits_{z_{-1/2}}^{z_{1/2}} c_p\bar{\rho}u\, dz, \quad v_0 = \frac{1}{\overline{c_p\rho_0}} \int\limits_{z_{-1/2}}^{z_{1/2}} c_p\bar{\rho}v\, dz$$

With regard to (16) relationship (15) takes the form

$$\overline{p\rho_0}\Delta z_0 \frac{dT_0}{dt} = \frac{\bar{\nu}_{1/2}}{\Delta z_{1/2}}(T_1 - T_0) - \frac{\bar{\nu}_{-1/2}}{\Delta z_{-1/2}}(T_0 - T_{-1})$$

$$+ \bar{\mu}_0 \Delta z_0 \Delta T_0 + F \tag{17}$$

Combining equations (13)–(17) we arrive at final formulation of difference analogue of the problem:

$$c_p \bar{\rho}_n \Delta z_n \frac{dT_n}{dt} = -\frac{\bar{\nu}_{n-1/2}}{\Delta z_{n-1/2}}(T_n - T_{n-1}) + \bar{\mu}_n \Delta z_n \Delta T_n \quad (k = n)$$

$$c_p \bar{\rho}_k \Delta z_k \frac{dT_k}{dt} = \frac{\bar{\nu}_{k+1/2}}{\Delta z_{k+1/2}}(T_{k+1} - T_k) - \frac{\bar{\nu}_{k-1/2}}{\Delta z_{k-1/2}}(T_k - T_{k-1})$$

$$+ \bar{\mu}_k \Delta z_k \Delta T_k \quad (k = n-1, n-2, \ldots, 1)$$

$$\overline{c_p \rho_0} \Delta z_0 \frac{dT_0}{dt} = \frac{\bar{\nu}_{1/2}}{\Delta z_{1/2}}(T_1 - T_0) - \frac{\bar{\nu}_{-1/2}}{\Delta z_{-1/2}}(T_0 - T_{-1}) + \bar{\mu}_0 \Delta z_0 \Delta T_0 + F \quad (18)$$

$$c_p \bar{\rho}_k \Delta z_k \frac{dT_k}{dt} = \frac{\bar{\nu}_{k+1/2}}{\Delta z_{k+1/2}}(T_{k+1} - T_k) - \frac{\bar{\nu}_{k-1/2}}{\Delta z_{k-1/2}}(T_k - T_{k-1})$$

$$+ \bar{\mu}_k \Delta z_k \Delta T_k \quad (k = -1, -2, \ldots, -m+1)$$

$$c_p \bar{\rho}_{-m} \Delta z_{-m} \frac{dT_{-m}}{dt} = \frac{\bar{\nu}_{-m-1/2}}{\Delta z_{-m-1/2}}(T_{-m+1} - T_{-m}) + \bar{\mu}_{-m} \Delta z_{-m} \Delta T_{-m} \quad (k = -m)$$

Difference analogue (18) may be written in the operator form:

$$\mathbf{B}\frac{d\mathbf{T}}{dt} + \mathbf{A}\mathbf{T} = \mathbf{F} \tag{19}$$

where the vectors $\mathbf{T}$, $\mathbf{F}$ and the matrix operators $\mathbf{B}$, $\mathbf{A}$ are assigned as follows:

$$\mathbf{T} = \begin{bmatrix} T_n \\ T_{n-1} \\ \ldots \\ T_1 \\ T_0 \\ T_{-1} \\ \ldots \\ T_{-m} \end{bmatrix}, \quad \mathbf{F} = \begin{bmatrix} 0 \\ 0 \\ \ldots \\ 0 \\ F \\ 0 \\ \ldots \\ 0 \end{bmatrix}, \quad \mathbf{B} = \begin{bmatrix} c_p \bar{\rho}_n \Delta z_n & & & \\ & c_p \bar{\rho}_{n-1} \Delta z_{n-1} & & \\ & & \ldots\ldots\ldots & 0 \\ & & c_p \bar{\rho}_1 \Delta z_1 & \\ & & \overline{c_p \rho_0} \Delta z_0 & \\ & & c_p \bar{\rho}_{-1} \Delta z_{-1} & \\ & 0 & \ldots\ldots\ldots & \\ & & & c_p \bar{\rho}_{-m} \Delta z_{-m} \end{bmatrix}$$

$$
\mathbf{A} = \begin{bmatrix}
A_n & -a_{n-1} & & & & \\
-a_{n-1} & A_{n-1} + a_{n-2} & -a_{n-2} & & 0 & \\
\cdots & \cdots\cdots\cdots & \cdots\cdots\cdots & \cdots & & \\
& -a_0 & A_0 + a_{-1} & -a_{-1} & & \\
\cdots\cdots & \cdots\cdots\cdots & \cdots\cdots\cdots & & \cdots & \\
& & -a_{-m+1} & A_{-m+1} + a_{-m} & -a_{-m} & \\
& 0 & & -a_{-m} & A_{-m} &
\end{bmatrix}
$$

Here

$$
a_k = \frac{\bar{\nu}_{k+1/2}}{\Delta z_{k+1/2}} \quad (k = n - 1,\, n - 2, \ldots, 1, 0, -1, \ldots, -m)
$$

and the operators A_l are determined by the formula

$$
A_l = a_{l-1} - \bar{\mu}_l \Delta z_l \Delta \quad (l = n,\, n - 1, \ldots, 1, 0, -1, \ldots, -m)
$$

System of equations (18) possesses an important feature: it is balanced. It is enough to sum all equations (18) in k and integrate in (x, y) to verify it. As a result, we arrive at the relationship of complete balance

$$
\sum_k \Delta z_k \int \int c_p \bar{\rho}_k \frac{\partial T}{\partial t} dx dy = \int \int F \, dx dy \tag{20}
$$

The expression in the left-hand part of (20) is a derivative in time of total heat storage Q, i. e., since

$$
Q(t) = \sum_k \Delta z_k \int \int c_p \bar{\rho}_k T \, dx dy
$$

then (20) may be rewritten as

$$
\frac{dQ(t)}{dt} = \int \int F \, dx dy
$$

So, the total increment of a heat in the system atmosphere–ocean depends on total radiation flux F.

The splitting method described in Appendix I allows to formulate absolutely stable algorithm of the second order of accuracy in all independent variables (x, y, z, t) to solve equation (19).

Difference analogues of a system of primitive equations of hydrodynamics of atmosphere and ocean (see [70]) may be formulated in a similar way.

Bibliography

1. А в д о ш и н С. М., Б е л о в В. В., М а с л о в В. П. *Математические аспекты синтеза вычислительных сред.* – М.: МИЭМ, 1984.

2. А г о ш к о в В. И. Оценка скорости сходимости некоторых алгоритмов теории возмущений. – *Препринт ОВМ АН СССР.* – М., 1982. – N. 30.

3. А г о ш к о в В. И. Проекционно-сеточный метод в алгоритмах теории возмущений. – *Препринт ОВМ АН СССР.* – М., 1982. – N. 38.

4. А г о ш к о в В. И. Сопряженные уравнения в алгоритмах возмущений n-го порядка точности. – В: *Сопряженные уравнения и теория возмущений в задачах математической физики.* – М.: ОВМ АН СССР, 1985. – С. 62–85.

5. А г о ш к о в В. И. Алгоритмы возмущений n-го порядка для функционалов от решений нелинейных задач и оценка их скорости сходимости. – В: *Сопряженные уравнения и алгоритмы возмущений в задачах математической физики.* – М.: ОВМ АН СССР, 1989. – С. 3–30.

6. А г о ш к о в В. И. Разрешимость одного класса задач нечувствительного оптимального управления и применение методов возмущений. – В сб.: *Сопряженные уравнения, алгоритмы возмущений и оптимальное управление.* – М.: ИВМ РАН, 1993. – С, 2–13. (Деп. ВИНИТИ, N 453-И93 от 25.02.93).

7. А г о ш к о в В. И., И п а т о в а В. М. О разрешимости одной задачи нечувствительного управления. – В сб.: *Сопряженные уравнения, алгоритмы возмущений и оптимальное управление.* – М.: ИВМ РАН, 1993. – С. 15–27. (Деп. в ВИНИТИ, N. 453-И93 от 25.02.93).

8. А г о ш к о в В. И., М и ш н е в а А. П. О нахождении коэффициента дисперсии в нелинейном параболическом уравнении. – *Препринт ОВМ АН СССР.* – М., 1988. – N. 200.

9. А г о ш к о в В. И., П о п ы к и н А. И., Ш и х о в С. Б. К теории малых возмущений для уравнения переноса. – В: *Сопряженные уравнения и теория возмущений в задачах математической физики.* – М.: ОВМ АН СССР, 1985. – С. 76–84.

10. А л е к с е е в В. В., Г а л к и н Н. А., З а л е с н ы й В. Б. Численное исследование реакции Аравийского моря на вариации ветра// *Океанология,* 1993. – Т. 33. – С. 13–20.

11. А л о я н А. Е., И о р д а н о в Д. Л., П е н е н к о В. В. Численная модель переноса примесей в пограничном слое атмосферы// *Метеорология и гидрология,* – 1981. – N. 8. – С. 32–43.

12. А с а ч е н к о в А. Л. Об одном алгоритме решения обратных задач на основе теории сопряженных уравнений. – *Препринт ОВМ АН СССР.* – М., 1986. – N. 119.

13. Бахвалов Н. С., Панасенко Г. П. *Осреднение процессов в периодических средах. Математические задачи механики композиционных материалов.* – М.: Наука, 1984.

14. Боголюбов Н. Н., Митропольский Ю. А. *Асимптотические методы в теории нелинейных колебаний.* – М.: Физматгиз, 1958.

15. Болтянский В. Г. *Математические методы оптимального управления.* – М.: Наука, 1969.

16. Болтянский В. Г. *Оптимальное управление дискретными системами.* – М.: Наука, 1973.

17. Бочаров Г. А. Исследование асимптотической устойчивости положений равновесия математической модели противовирусного Т-клеточного иммунного ответа на основе теории возмущений. – В: *Сопряженные уравнения и теория возмущений в задачах математической физики.* – М.: ОВМ АН СССР, 1985. – С. 116–126.

18. Бутковский А. Г. *Методы управления системами с распределенными параметрами.* – М.: Наука, 1975.

19. Бухгейм А. Л. Операторные уравнения Вольтерра в шкалах банаховых пространств// *ДАН СССР,* 1978. – Т. 242, N. 2. – С. 272–275.

20. Вайнберг М. М. *Функциональный анализ.* – М.: Просвещение, 1979.

21. Вайнберг М. М., Треногин В. А. *Теория ветвления решений нелинейных уравнений.* – М.: Наука, 1969.

22. Васильев В. Г. Одномерная обратная задача индукционного каротажа. – В: *Некоторые методы и алгоритмы интерпретации геофизических наблюдений*/ Под ред. М.М. Лаврентьева. – М.: Наука, 1967.

23. Васильев Ф. П. *Численные методы решения экстремальных задач.* – М.: Наука, 1988.

24. Васильева А. Б., Бутузов В. Ф. *Асимптотические разложения решений сингулярно возмущенных уравнений.* – М.: Наука, 1973.

25. Вишик М. И., Люстерник Л. А. *Решение некоторых задач о возмущении в случае матриц и самосопряженных и несамосопряженных дифференциальных уравнений. I*// УМН, 1960. – Т. XV, вып. 3. – С. 3–80.

26. Вишик М. И., Люстерник Л. А. Регулярное вырождение и пограничный слой для линейных дифференциальных уравнений с малым параметром// *УМН,* 1957. – Т. XII, вып. 5. – С. 3–122.

27. Вишик М. И., Люстерник Л. А. Асимптотическое поведение решений линейных дифференциальных уравнений с большими или быстро меняющимися коэффициентами и граничными условиями// *УМН,* 1960. – Т. XV, вып. 4. – С. 27–95.

28. Вишик М. И., Люстерник Л. А. Некоторые вопросы возмущений краевых задач для дифференциальных уравнений в частных производных// *ДАН СССР,* 1959. – Т. 129, N. 6. – С. 1203–1206.

29. Вишик М. И., Люстерник Л. А. Возмущение собственных значений и собственных элементов для некоторых несамосопряженных операторов// *ДАН СССР,* 1960. – Т. 130, N. 2. – С. 251–253.

30. В л а д и м и р о в В. С. *Уравнения математической физики.* – М.:
    Наука, 1981.
31. В л а д и м и р о в В. С. Математические задачи односкоростной теории
    переноса частиц// *Тр. МИАН СССР*, 1961. – Вып. 61. – С. 3–157.
32. В л а д и м и р о в В. С. *Обобщенные функции в математической
    физике.* – М.: Наука, 1976.
33. В л а д и м и р о в В. С., В о л о в и ч И. В. Законы сохранения для
    нелинейных уравнений// *ДАН СССР*, 1984. – Т. 279, N 4. – С. 843–847.
34. В л а д и м и р о в В. С., В о л о в и ч И. В. Законы сохранения
    для нелинейных уравнений. – В: *Актуальные проблемы вычислительной
    математики и математического моделирования.* – Новосибирск: Наука,
    1985. – С. 147–162.
35. В о е в о д и н В. В., К у з н е ц о в Ю. А. *Матрицы и вычисления.* –
    М.: Наука, 1984.
36. Г а б а с о в Р., К и р и л л о в а Ф. М. К вопросу о распространении
    принципа максимума Л.С.Понтрягина на дискретные системы// *Автома-
    тика и телемеханика*, 1966. – Т. 27, N. 11.
37. Г а н т м а х е р Ф. Р. *Теория матриц.* – М.: Наука, 1967.
38. Г е л ь ф о н д О. Л. *Исчисление конечных разностей.* – М.: Гостехиздат,
    1952.
39. Д у л и н В. А. *Возмущение критичности реакторов и уточнение
    групповых констант.* – М.: Атомиздат, 1979.
40. Д ы м н и к о в В. П., А л о я н А. Е. Монотонные схемы решения урав-
    нений переноса в задачах прогноза погоды, экологии и теории климата//
    Изв. АН СССР ФАО, 1990. – Т. 26, N. 12. – С. 1237–1246.
41. Е р м а к о в С. М. Об оптимальных несмещенных планах регрессионных
    экспериментов // *Тр. МИАН СССР*, 1970. – Т. III. – С. 252–257.
42. Е р м а к о в С. М. *Метод Монте-Карло и смежные вопросы.*–М.: Наука,
    1975.
43. Е р м а к о в С. М., М а х м у д о в А. А. О планах регрессионных экс-
    периментов, минимизирующих систематическую ошибку // *Завод. лаб.*,
    1977. – N. 7. – С. 854–858.
44. Е р м а к о в С. М., М и р о н е н к о Л. П. Метод существенной выборки
    при моделировании систем связи, описываемых векторными марковскими
    процессами. – В: *Имитационное моделирование систем.* – Новосибирск:
    ВЦ СО АН СССР, 1976. – С. 17–24.
45. Е р м а к о в С. М., П а н к р а т ь е в Ю. Д. Об одном методе
    приближенной линеаризации интегральных уравнений Ляпунова–Шмидта
    // *Вестн. ЛГУ. Математика. Механика. Астрономия*, 1977. – N. 7. –
    С. 44–47.
46. З у е в С. М. *Статистическое оценивание параметров математических
    моделей заболеваний.* – М.: Наука, 1988.
47. И в а н о в В. К. О некорректно поставленных задачах // *Мат. сб.*,
    1963. – Т. 61, N. 2. – С. 211–223.

48. Иорданов Д. Л., Пененко В. В., Алоян А. Е. О вертикальной скорости на верхней границе планетарного пограничного слоя над орографически и термически неоднородной подстилающей поверхностью// *Изв. АН СССР ФАО*, 1979. - Т. 14, N. 11. - С. 1204-1208.

49. Иорданов Д. Л., Пененко В. В., Алоян А. Е. Параметризация стратифицированного планетарного пограничного слоя для численного моделирования атмосферных процессов// *Изв. АН СССР ФАО*, 1978. - Т. 14, N. 8. - С. 815-823.

50. Кадомцев Б. Б. О функции влияния в теории переноса лучистой энергии// *ДАН АН СССР*, 1957. - Т. 113, N. 3. - С. 541-543.

51. Кондратьев К. Я. *Лучистая энергия Солнца.* - Л.: Гидрометеоиздат, 1954.

52. Кондратьев К. Я. *Метеорологические спутники.* - Л.: Гидрометеоиздат, 1963.

53. Красовский Н. Н. *Теория управления движением.* - М.: Наука, 1968.

54. Красовский Н. Н. *Управление динамической системой. Задача о минимуме гарантированного результата.* - М.: Наука, 1985.

55. Крейн С. Г. *Линейные уравнения в банаховом пространстве.* - М.: Наука, 1971.

56. Крейн С. Г. О классах корректности для некоторых задач// *ДАН СССР*, 1957. - Т. 114, N. 6. - С. 1162-1165.

57. Лаврентьев М. М. О постановке некоторых некорректных задач математической физики. - В: *Некоторые вопросы вычислительной и прикладной математики.* - Новосибирск: Наука, 1966. - С. 258-276.

58. Лаврентьев М. М., Васильев В. Г. О постановке некоторых некорректных задач математической физики // *Сиб. мат. журн.*, 1966. - Т. VII, N. 3. - С. 559-576.

59. Ладыженская О. А. *Краевые задачи математической физики.* - М.: Наука, 1973.

60. Ладыженская О. А., Фаддеев Л. Д. О теории возмущения непрерывного спектра// *ДАН СССР*, 1958. - Т. 120, N. 6. - С. 1187-1190.

61. Ландау Л. Д., Лифшиц Е. М. *Квантовая механика.* -М.: Физматгиз, 1963.

62. Летов А. М. *Динамика полета и управление.* - М.: Наука, 1969.

63. Ли Э. Б., Маркус Л. *Основы теории оптимального управления.* - М.: Наука, 1972.

64. Лобарев И. В. Метод возмущений в исследовании простоты собственных значений интегрального уравнения Пайерлса. - В: *Сопряженные уравнения и теория возмущений в задачах математической физики.* - М.: ОВМ АН СССР, 1985. - С. 136-150.

65. Ломов С. А. *Введение в общую теорию сингулярных возмущений.* - М.: Наука, 1981.

66. Лурье К. А. *Оптимальное управление в задачах математической физики.* - М.: Наука, 1977.

67. Люстерник Л. А., Соболев В. И. *Элементы функционального анализа.* – М.: Наука, 1965.

68. Ляпунов А. М. *Собр. соч. в 2 т.* – М.-Л.: Гостехиздат, 1956.

69. Маделунг Э. *Математический аппарат физики.* – М.: Физматгиз, 1960.

70. Марчук Г. И. *Численное решение задачи динамики атмосферы и океана.* – Л.: Гидрометеоиздат, 1974.

71. Марчук Г. И. Применение сопряженных уравнений к решению задач математической физики // *Успехи мех.*, 1981. – Т. 4, вып. 1.– С. 3–27.

72. Марчук Г. И. *Некоторые математические проблемы охраны окружающей среды. Комплексный анализ и его приложения.* – М.: Наука, 1981.

73. Марчук Г. И. *Методы вычислительной математики.* – М.: Наука, 1980.

74. Марчук Г. И. О постановке некоторых обратных задач// *ДАН СССР*, 1964. – Т. 156, N. 3. – С. 503–506.

75. Марчук Г. И. Уравнение для ценности информации с метеорологических спутников и постановка обратных задач// *Космич. исслед.*, 1964. – Т. 2, вып. 3. – С. 462–477.

76. Марчук Г. И. Основные и сопряженные уравнения динамики атмосферы и океана// *Метеорология и гидрология*, 1974. – No. 2. – С. 9–37.

77. Марчук Г. И. *Численные методы расчета ядерных реакторов.* – М.: Атомиздат, 1958.

78. Марчук Г. И. *Сопряженные уравнения и анализ сложных систем.* – М.: Наука, 1992.

79. Марчук Г. И. Методы долгосрочного прогноза погоды на основе решения основных и сопряженных задач// *Метеорология и гидрология*, 1974. – N. 3. – С. 17–34.

80. Марчук Г. И. Окружающая среда и проблемы оптимизации размещения предприятий// *ДАН СССР*, 1976. – Т. 227, N. 5. – С. 1056–1059.

81. Марчук Г. И. *Методы расчета ядерных реакторов.* – М.: Госатомиздат, 1961.

82. Марчук Г. И. Обзор методов расчета ядерных реакторов// *Атом. энергия*, 1961. – Т. 11, N. 4. – С. 356–369.

83. Марчук Г. И. К проблеме математического моделирования экологических ситуаций в акваториях водных бассейнов. – В: *Современные проблемы математической физики и вычислительной математики.* – М.: Наука, 1982. – С. 254–258.

84. Марчук Г. И. Окружающая среда и некоторые проблемы оптимизации. – *Препринт ВЦ СО АН СССР.* – Новосибирск, 1975. – N. 1.

85. Марчук Г. И. К проблеме охраны окружающей среды. –В: *Вычислительные методы в математической физике, геофизике и оптимальном управлении.* – Новосибирск: Наука, 1978. – С. 20–28.

86. Марчук Г. И. Экономические критерии планирования, охраны и восстановления окружающей среды// *ЖВМ и МФ*, 1980. – Т. 20, N. 6. – С. 1365–1372.

87. М а р ч у к Г. И. К проблеме охраны окружающей среды. – *Препринт ВЦ СОАН СССР.* – Новосибирск, 1977. – N. 43.

88. М а р ч у к Г. И. *Математическое моделирование в проблеме окружающей среды.* – М.: Наука, 1982.

89. М а р ч у к Г. И. Окружающая среда и проблемы оптимизации// *Тр. МИАН СССР*, 1984. – Т. 166. – С. 123–129.

90. М а р ч у к Г. И. Моделирование изменений климата и проблема долгосрочного прогноза погоды// *Метеорология и гидрология*, 1979. – N. 7. – С. 25–36.

91. М а р ч у к Г. И., А г о ш к о в В. И. Сопряженные уравнения в нелинейных задачах и их приложения. – В: *Функциональные и численные методы математической физики.* – Киев: Наукова думка, 1988. – С. 138–142.

92. М а р ч у к Г. И., А г о ш к о в В. И. Симметризация нестационарного уравнения переноса и формулировка вариационного принципа. – *Препринт ВЦ СОАН СССР.* – Новосибирск, 1980. – N. 222.

93. М а р ч у к Г. И., А г о ш к о в В. И., Ш у т я е в В. П. Сопряженные уравнения и алгоритмы возмущений в прикладных задачах. – *Вычислительные процессы и системы.* Вып. 4. – М.: Наука, 1986. – С. 5–62.

94. М а р ч у к Г. И., А г о ш к о в В. И., Ш у т я е в В. П. *Сопряженные уравнения и алгоритмы возмущений.* – М.: ОВМ АН СССР, 1986.

95. М а р ч у к Г. И., А г о ш к о в В. И., Ш у т я е в В. П. *Сопряженные уравнения и методы возмущений в нелинейных задачах математической физики.* – М.: Наука, 1993.

96. М а р ч у к Г. И., А л о я н А. Е. Глобальный перенос примеси на сфере// *Изв. АН СССР ФАО*, 1994 (to be published).

97. М а р ч у к Г. И., Б е л ь с к а я Ж. Н. О применении сопряженных уравнений к расчету защиты от излучения// *Вопросы физики защиты реакторов.* – М.: Госатомиздат, 1963. – С. 99–102.

98. М а р ч у к Г. И., Д р о б ы ш е в Ю. П. Некоторые вопросы линейной теории измерений// *Автометрия*, 1967. – N. 3. – С. 24–30.

99. М а р ч у к Г. И., Д ы м н и к о в В. П., К у р б а т к и н Г. П., С а р к и с я н А. С. Программа "Разрезы" и мониторинг Мирового океана// *Метеорология и гидрология*, 1984. – Т. 6, N. 8. – С. 9–17.

100. М а р ч у к Г. И., Е р м а к о в С. М. О некоторых проблемах теории планирования эксперимента. – В: *Математические методы планирования эксперимента.* – Новосибирск: Наука, 1981. – С. 3–18.

101. М а р ч у к Г. И., К а г а н Б. А. *Океанские приливы: Математические модели и численные эксперименты.* – Л.: Гидрометеоиздат, 1977.

102. М а р ч у к Г. И., К у з и н В. И., С к и б а Ю. Н. Проекционно-разностный метод расчета сопряженных функций для модели переноса тепла в системе атмосфера–океан–почва. – В: *Актуальные проблемы вычислительной и прикладной математики.* – Новосибирск: Наука, 1983. – С. 149–154.

103. М а р ч у к Г. И., К у з и н В. И., С к и б а Ю. Н. Применение сопряженных уравнений в численных моделях переноса тепла в системе атмосфера–океан–континент. – В: *Материалы Советско-французского симпозиума по океанографии. Новосибирск, 9–11 июня 1983 г. Ч. 1.* – Новосибирск, 1983. – С. 4–15.

104. М а р ч у к Г. И., К у р б а т к и н Г. П. Физические и математические аспекты анализа и прогноз погоды// *Метеорология и гидрология*, 1977. – N. 11. – С. 25–33.

105. М а р ч у к Г. И., К у р б а т к и н Г. П. Анализ нестационарности полусферного температурного градиента атмосферы// *Изв. АН СССР. ФАО*, 1981. – Т. 17, N. 5. – С. 451–463.

106. М а р ч у к Г. И., Л е б е д е в В. И. *Численные методы в теории переноса нейтронов.* – М.: Атомиздат, 1971.

107. М а р ч у к Г. И., Л ы к о с о в В. Н. Диагностический расчет коэффициентов вертикального перемешивания в верхнем пограничном слое океана. – В: *Математическое моделирование процессов в пограничных слоях атмосферы и океана.* – М.: ОВМ АН СССР, 1989. – С. 4–21.

108. М а р ч у к Г. И., М и х а й л о в Г. А. Решение задач теории переноса излучения методом Монте-Карло. – В: *Теоретические и прикладные проблемы рассеяния света.* – Минск: Наука и техника, 1971. – С. 43–58.

109. М а р ч у к Г. И., М и х а й л о в Г. А., Н а з а р а л и е в М. А. и др. *Метод Монте-Карло в атмосферной оптике.* – Новосибирск: Наука, 1976.

110. М а р ч у к Г. И., О р л о в В. В. К теории сопряженных функций. – В: *Нейтронная физика.* – М.: Госатомиздат, 1961. – С. 30–45.

111. М а р ч у к Г. И., П е н е н к о В. В. Исследование чувствительности дискретных моделей динамики атмосферы и океана// *Изв. АН СССР ФАО*, 1979. – Т. 15, N. 11. – С. 1123–1131.

112. М а р ч у к Г. И., П е н е н к о В. В. Некоторые применения методов оптимизации к проблеме окружающей среды. – В: *Вычислительные методы в прикладной математике.* – Новосибирск: Наука, 1982. – С. 5–22.

113. М а р ч у к Г. И., П е н е н к о В. В., П р о т а с о в А. В. Вариационный принцип в малопараметрической модели динамики атмосферы. – В: *Вариационно-разностные методы в математической физике: Материалы Всесоюзн. конф.* – Новосибирск, 1978. – С. 213–229.

114. М а р ч у к Г. И., С к и б а Ю. Н. Численный расчет сопряженной задачи для модели термического взаимодействия атмосферы с океанами и континентами// *Изв. АН СССР. ФАО*, 1976. – Т. 12, N. 5. – С. 459–469.

115. М а р ч у к Г. И., С к и б а Ю. Н. Об одной модели прогноза осредненных аномалий температуры. – В: *Препринт ВЦ СОАН СССР.* – Новосибирск, 1978. – N. 120.

116. М а р ч у к Г. И., С к и б а Ю. Н. Расчет пространственно-временных функций влияния для среднемесячных аномалий поверхностной температуры воздуха ограниченных регионов. – В: *Динамика атмосферы и океана.* – М.: ОВМ АН СССР, 1990. – С. 35–50.

117. М а р ч у к Г. И., С к и б а Ю. Н., П р о ц е н к о И. Г. Метод расчета эволюции случайных гидродинамических полей на основе сопряженных уравнений// *Изв. АН СССР ФАО*, 1985. – Т. 21, N. 2. – С. 115–122.

118. М а с л о в В. П. *Теория возмущений и асимптотические методы.* – М.: Изд-во МГУ, 1965.

119. М а с л о в В. П. Теория возмущений при переходе от дискретного спектра к непрерывному// *ДАН СССР*, 1956. – Т. 109, N. 1. – С. 267–270.

120. М и х а й л о в Г. А. Использование приближенных решений сопряженной задачи для улучшения алгоритмов метода Монте-Карло// *ЖВМ и МФ*, 1969. – Т. 9, N. 5. – С. 1145–1152.

121. М и х а й л о в Г. А. *Некоторые вопросы теории методов Монте-Карло.* – Новосибирск: Наука, 1974.

122. М и ш н е в а А. П. Отыскание эффективного коэффициента конвективной дисперсии по экспериментальным данным с помощью аппарата сопряженных уравнений и теории возмущений. – В: *Сопряженные уравнения и алгоритмы возмущений.* – М.: ОВМ АН СССР, 1988. – С. 101–111.

123. М и ш н е в а А. П. Прогнозирование эффективного коэффициента конвективной дисперсии при изменении параметров эксперимента. – В: *Сопряженные уравнения и алгоритмы возмущений.* – М.: ОВМ АН СССР, 1988. – С. 112–118.

124. М о и с е е в Н. Н. *Элементы теории оптимальных систем.* – М.: Наука, 1975.

125. М о и с е е в Н. Н. *Асимптотические методы нелинейной механики.* – М.: Наука, 1981.

126. М о н и н А. С., О б у х о в А. М. Основные закономерности турбулентного перемешивания в приземном слое атмосферы// *Тр. Геофиз. Ин-та АН СССР*, 1954. N. 24. – С. 163–187.

127. М о н и н А. С., Я г л о м А. М. *Статистическая гидромеханика. Ч. I.* – М.: Наука, 1965.

128. М у с а э л я н Ш. А. Проблема параметризации процесса передачи лучистой энергии Солнца системе океан – атмосфера и долгосрочный прогноз// *Метеорология и гидрология*, 1974. – N. 10. – С. 9–19.

129. Н а з а р а л и е в М. А. Исследование приближенной функции ценности в расчетах сумеречных эффектов методом Монте-Карло. – В: *Методы Монте-Карло и их применения: Тез. докладов на 3-й Всесоюзн. конф. по методам Монте-Карло.* – Новосибирск: ВЦ СОАН СССР, 1971. – С. 116–117.

130. Назаров В. С., Овсянников А. Н., Спидченко А. Н. К гидрологическому обоснованию регулярных океанографических наблюдений в Мировом океане// *Тр. ГОИН*, 1971. – Вып. 105. – С. 13–36.

131. Н а т а н с о н И. П. *Теория функций вещественной переменной.* – М.: Наука, 1974.

132. Н е к р у т к и н В. В. Прямая и сопряженная схема Неймана–Улама для решения нелинейных интегральных уравнений// *ЖВМ и МФ*, 1974. – Т. 14, N. 6. – С. 1409–1415.

133. О б р а з ц о в Н. Н.   Математическое моделирование и оптимизация в проблемах окружающей среды. – В: *Препринт ОВМ АН СССР*. – М., 1985. – N. 85.

134. О б р а з ц о в Н. Н.   О распространении примеси от точечного источника. – В: *Препринт ОВМ АН СССР*. – М., 1983. – N. 66.

135. О л е й н и к О. А.   Краевые задачи для уравнений с частными производными с малым параметром при старших производных и задача Коши для нелинейных уравнений// *УМН*, 1955. – Т. 10, N. 3. – С. 229–334.

136. П а н к р а т ь е в Ю. Д.   К вопросу о возмущениях линейных функционалов. – В: *Системный анализ и исследование операций*. – Новосибирск: ВЦ СОАН СССР, 1977. – С. 5–14.

137. П е н е н к о В. В.   Вычислительные аспекты моделирования динамики атмосферных процессов и оценки влияния различных факторов на динамику атмосферы. – В: *Некоторые проблемы вычислительной и прикладной математики*. – Новосибирск: Наука, 1975. – С. 61–77.

138. П е н е н к о В. В., А л о я н А. Е.   *Модели и методы для задач охраны окружающей среды*. – Новосибирск: Наука, 1985.

139. П о н т р я г и н Л. С.   Математическая теория оптимальных процессов и дифференциальные игры// *Труды матем. ин-та АН СССР*, 1985. – Т. 169. – С. 119–158.

140. П о н т р я г и н Л. С.   *Избранные труды. Т. 2*. – М.: Наука, 1988.

141. П о н т р я г и н Л. С., Б о л т я н с к и й В. Г., Г а м к р е л и д з е Р. В., М и щ е н к о Е. Ф.   Математическая теория оптимальных процессов. – М.: Наука, 1976.

142. П у п к о В. Я., З р о д н и к о в А. В., Л и х а ч е в Ю. И.   *Метод сопряженных функций в инженерно-физических исследованиях*. – М.: Энергоатомиздат, 1984.

143. С е д у н о в Е. В.   О планах эксперимента, минимизирующих систематическую ошибку. – В: *Математические методы планирования эксперимента*. – Новосибирск: Наука, 1981. – С. 102–140.

144. С о б о л е в С. Л.   *Введение в теорию кубатурных формул*. – М.: Наука, 1971.

145. С т у м б у р Г.   *Применение теории возмущений в физике ядерных реакторов*. – М.: Атомиздат, 1976.

146. *Теория ветвления и нелинейные задачи на собственные значения* /Под ред. Д.Б. Келлера, С. Антмана. – М.: Мир, 1974.

147. Т и х о н о в А. Н.   О решении некорректно поставленных задач и методе регуляризации// *ДАН СССР*, 1963. – Т. 151, N. 3. – С. 501–504.

148. Т и х о н о в А. Н.   О регуляризации некорректно поставленных задач// *ДАН СССР*, 1963. – Т. 153, N. 1. – С. 49–52.

149. Т и х о н о в А. Н., А р с е н и н В. Я.   *Методы решения некорректных задач*. – М.: Наука, 1974.

150. Т р е н о г и н В. А.   Развитие и приложения асимптотического метода Люстерника–Вишика// *УМН*, 1970. – Т. XXV, вып. 4. – С. 123–156.

151. Т р е н о г и н В. А.   *Функциональный анализ*. – М.: Наука, 1980.

152. У с а ч е в Л. Н. Уравнение для ценности нейтронов кинетического реактора и теория возмущений// *Реакторостроение и теория реакторов.* – М.: Изд-во АН СССР, 1955. – С. 251–268.

153. У с а ч е в Л. Н. Теория возмущений для коэффициента воспроизрводства и других отношений чисел различных процессов в реакторе// *Атом. энергия,* 1963. – Т. 15, вып. 6. – С. 472–481.

154. У с а ч е в Л. Н., Б о б к о в Ю. Г. *Теория возмущений и планирование экспериментов в проблеме ядерных данных для реакторов.* –М.: Атомиздат, 1980.

155. Ф а д д е е в Л. Д. О модели Фридрихса в теории возмущений непрерывного спектра// *Тр. МИАН СССР,* 1964. – Т. 73. – С. 292–313.

156. Ф е д о р е н к о Р. П. *Приближенное решение задач оптимального управления.* – М.: Наука, 1978.

157. Ф е д о р о в В. В. Активные регрессионные эксперименты. – В: *Математические методы планирования эксперимента.* – Новосибирск: Наука, 1981. – С. 19–73.

158. Ф е л ь д б а у м А. А. *Основы теории оптимальных автоматических систем.* – М.: Наука, 1966.

159. Ф и л а т о в А. Н. *Асимптотические методы в теории дифференциальных и интегродифференциальных уравнений.* – Ташкент: ФАН, 1974. – 216 с.

160. Ф и л а т о в А. Н., Ш а р о в а Л. В. *Интегральные неравенства и теория нелиненых колебаний.* – М.: Наука, 1976. – 106 с.

161. Ф и л а т о в А. Н., Ш е р ш к о в В. В. *Асимптотические методы в атмосферных моделях.* – Л.: Гидрометеоиздат, 1988. – 270 с.

162. Х и с а м у т д и н о в А. И. Выборка по важности в теории переноса излучений// *ЖВМ и МФ,* 1970. – Т. 10, N. 4. – С. 999–1005.

163. Ч е р н о у с ь к о Ф. Л., Б а н и ч у к В. П. *Вариационные задачи механики и управления.* – М.: Наука, 1973.

164. Ш и х о в С. Б. *Вопросы математической теории реакторов. Линейный анализ.* –М.: Атомиздат, 1972.

165. Ш и ш а т с к и й С. П. Об одном методе приближенного решения некорректной задачи Коши для эволюционного уравнения. – В: *Математические проблемы геофизики. Вып. 3.* – Новосибирск: ВЦ СОАН СССР, 1972. – С. 216–228.

166. Ш у т я е в В. П. Спектральные свойства условно-критической задачи переноса в дискретном приближении и алгоритм теории возмущений. – В: *Сопряженные уравнения и теория возмущений в задачах математической физики.* – М.: ОВМ АН СССР, 1985. – С. 151–169.

167. Ш у т я е в В. П. Свойства сопряженных операторов, возникающих в алгоритмах возмущений для квазилинейной эллиптической задачи. – В: *Сопряженные уравнения и алгоритмы возмущений.* – М.: ОВМ АН СССР, 1988. – С. 119–132.

168. Ш у т я е в В. П. К обоснованию алгоритма возмущений в одной нелинейной гиперболической задаче// *Мат. заметки,* 1991. – Т. 49, вып. 4. – С. 155–156.

169. Э н е е в Т. М. О применении градиентного метода в задачах оптимального управления// *Космические исследования,* 1966. – Т. 4, N. 5.

170. B a l a k r i s h n a n A. V., *Introduction to Optimization Theory in a Hilbert Space*, Berlin, Heidelberg, New York: Springer-Verlag, 1971.

171. B a l a k r i s h n a n A. V., Parameter estimation in stochastic differential systems: Theory and Application, in *Developments in Statistics*, Ed. Paruchuri K. Krishnaiah, New York, San Francisco, London: Academic Press, 1, 1978.

172. B a l a k r i s h n a n A. V., *Stochastic Differential Systems. Filtering and Control*, Lecture Notes in Economics and Math. Systems, Berlin, New York, Heidelberg, Tokyo: Springer–Verlag, 1973.

173. B a l a k r i s h n a n A. V., A note on the Marchuk–Zuev identification problem, in *Vistas in Applied Mathematics. Numerical Analysis. Atmospheric Sciences. Immunology*, New York: Optimization Siftware, Inc. Publications Division, 291, 1986.

174. B a t e s J. R., S e m a z z i F. N., H i g g i n s R. W., Integration of the shallow water equations on the sphere using a vector semi-Lagrangean scheme with a multigrid solver, *Monthly Weather Review*, V. 118, 1615, 1990.

175. B e l l m a n R. *Dinamic Programming*, New Jersey: Princeton Univ. Press, 1957.

176. B e l l m a n R. *Perturbation Techniques in Mathematics, Physics and Engineering*, New York: Holft, 1964.

177. B e l l m a n R., K a l a b a R. E., *Quasilinearization and Nonlinear Boundary-Value Problems*, N.Y.: American Elsevier Publishing Company, Inc., 1965.

178. B e n s o u s s a n A., L i o n s J. L., P a p a n i c o l a u G., *Asymptotic Methods in Periodic Structures*, North Holland, 1978.

179. B r a m v a n L e e r, Towards the ultimate conservative difference scheme: 2. Monotonicity and conservation combined in a second order scheme, *J. Comp. Phys.*, V. 14, 360, 1974.

180. B l o c h C., Sur la théorie des perturbations des états liés, *Nuclear Phys*, 6, 329, 1958.

181. C a c u c i D. G., W e b e r C. F., O b l o w E. M., M a r a b l e J. H., Sensitivity theory for general systems of nonlinear equations, *Nuclear Science and Engineering*, V. 75, 88, 1980.

182. C a r l e m a n I., Applications de la théories des équations intégrales singulières aux équations differentielles de la dynamique, *Arkiv. Mat. Astronom. Fys.*, B. 22, 7, 1, 1932.

183. C i o r a n e s c u D., D o n a t o P., Exact internal controllability in perforated domains, *J. Math. Pures et Appl.*, 68, 185, 1989.

184. C o u r t i e r P., T a l a g r a n d O., Variational assimilation of meteorological observations with the adjoint worticity equation. Part II: Numerical results, *Quart. J. Roy. Meteor. Soc.*, 113, 1329, 1987.

185. D e r b e r J. C., The variational 4-D assimilation of analysis using filtering models as constraints, *Ph. D. Thesis, Univ. of Wisconsin*, Madison, 1985.

186. D e r b e r J. C., Variational four dimensional analysis using the quasi-geostrophic constraint, *Mon. Wea. Rev.*, 115, 998, 1987.

187. D e S z. - N a g y B., Perturbations des tranformations lineaires fermeés,
      Acta Sci. Math. Szeged., V. 14, 125, 1951.
188. E h r l i c h R., H u r w u t z H., Multigroup methods for neutron diffusion
      problems, *Nucleonics*, V. 12, 2, 23, 1954.
189. F r i e d r i c h s K. O., *Perturbation of spectra in Hilbert space, Lectures in
      Applied Math., V. III* Proceedings of the Summer Seminar, Boulder, Colorado,
      1960), American Math. Society, Providence, Rhode Island, 1965.
190. F r i e d r i c h s K. O., Über die Spectralzerlegung eines Integraloperators,
      Math. Ann., V. 115, 249, 1983.
191. F r i e d r i c h s K. O., On the perturbation of continuous spectra, *Comm.
      Pure Appl. Math.*, V. 1, 361, 1984.
192. F r i e d r i c h s K. O., R e j t o P. A., On a perturbation through which
      the discrete spectrum becomes continuous, *Comm. Pure Appl. Math.*, V. 841,
      543, 1962.
193. G a t e s W. L., AMIP: The atmospheric model intercomparison project,
      Bulletin of the American Meteorological Society, v. 73, 12. 1962, 1992.
194. G l a s s t o n e S., E d l u n d M. C., *The Elements of Nuclear Reactor
      Theory*, N.Y. etc., 1952.
195. G l o w i n s k i R., *Numerical Methods for Nonlinear Variational Problems*,
      N.Y.: Springer–Verlag, 1984.
196. G l o w i n s k i R., L i C. M., L i o n s J. L., A numerical approach to the
      exact boundary controllability of the wave equations (F), *Jap. J. of Applied
      Math.*, 1990.
197. K a t o T., On the convergence of the perturbation method, *J. Fac. Sci.*,
      Tokyo Univ., V. 6, 198, 1951.
198. K a t o T., Perturbation of continuous spectra by trace operators, *Proc.
      Japan Acad.*, V. 33, 260, 1957.
199. K a t o T., *Perturbation Theory for Linear Operators*, Berlin: Springer, 1963.
200. K o n t a r e v G. R., The adjoint equation technique applied to meteorological
      problems, in *Technical report No. 21, European Centre for Medium Range
      Weather Forecasts*, Reading, 1980.
201. K u r o d a S. T., Perturbation of continuous spectra by unbounded operators.
      I, II, *J. Math. Soc. Japan*, V. 11, 247, 1959; V. 12, 461, 1960.
202. L a d y z h e n s k a y a O. A., S o l o n n i k o v V. A., U r a l t s e v a
      N. N., *Linear and Quasilinear Equations of Parabolic Type*, Moscow: Nauka,
      1967.
203. L a x P. D., P h i l l i p s R. S., Scattering theory, *Bull. Amer. Math. Soc.*,
      V. 70, 130, 1964.
204. L e D i m e t F. X., A general formalizm of variational analysis, *CIMMS
      Report*, Norman, OK 73091, 22, 1, 1982.
205. L e D i m e t F. X., T a l a g r a n d O., Variational algorithms for analysis
      and assimilation of meteorological observations: theoretical aspects, *Tellus A*,
      38, 97, 1986.

206. L e V e q u e R. J., Time-split methods for partial differential equations, *Report No. STAN-CS-82-904*, Stanford University, 1982.

207. L e v i t u s S., O o r t A. N., Global analysis of oceanographic data, *Bul. Amer. Meteorol. Soc.*, V. 58, 1270, 1977.

208. L e w i n s J., Importance. *The Adjoint Function*, N.Y. etc.: Pergamon Press, 1965.

209. L e w i s J., D e r b e r J., The use of adjoint equations to solve a variational adjustment problem with advective constraints, *Tellus A*, 37, 309, 1985.

210. L i o n s J. - L., Sur les sentinelles des systems distributes. Le cas des conditions initials incompletes, *C. R. Acad. Sci. Paris*, V. 307, Serie I, 819, 1988.

211. L i o n s J. - L., Insensitive controls, in *Computational Mathematics and Applications*, Proceedings of 8-th France–USSR–Italy Joint Symposium, Pavia, October 2–6, 1989, Publicazioni *N* 730, Pavia, 285, 1989.

212. L i o n s J. - L., *Sur le Controle Optimal de Systémes Gouvernes par des Equation aux Derivees Partielles*, Paris: Dunod, Gaithier–Villars, 1968.

213. L i o n s J. - L., *Contrôllabilité Exacte Perturbations et Stabilisation de Systèmes Distribués. T. 1. Contrôllabilité Exacte; T. 2. Perturbations*, Paris: Masson, 1988.

214. L i o n s J. - L., *La Planete Terre*, Conferences à l'Institut d'Espagne, 1990.

215. L i o n s J. - L., M a g e n e s E., *Problémes aux Limites non Homogenes et Applications*, Dunod. Paris, 1968.

216. L i o n s J. - L., T e m a m R., W a n g S., Models for the coupled atmosphere and ocean, *Computational Mechanics Advances*, V. 1, 3, 1993.

217. L o r e n c A. C., Optimal nonlinear objective analysis, *Quart. J. Roy. Meteor. Soc.*, 114, 205, 1988.

218. M a r c h u k G., Some application of splitting-up methods to the solution of mathematical physics problems, *Appl. Math.*, 13, 2, 1968.

219. M a r c h u k G. I., L'etablissment d'un modele de changements de climat et le probleme de la prevision meteorologique a long term, *La Meteorologie*, VI serie, 16, 103, 1979.

220. M a r c h u k G. I., Formulation of the theory of perturbations for complicated models. Part I. The estimation of the climate change, *Geofisica Internacional*, Mexico, V. 15, 2, 103, 1975; Part II. Weather Prediction, Ibid, 169, 1975.

221. M a r c h u k G. I., Mathematical Issues of Industrial Effluent Optimization, *J. of the Meteorological Society of Japan*, V. 60, 481, 1982.

222. M a r c h u k G. I., Perturbation theory and the statement of inverse problems, *Lect. Notes Comput. Sci.*, V. 4, 159, 1973.

223. M a r c h u k G. I., *Methods of Numerical Mathematics*, New York etc.: Springer, 1975.

224. M a r c h u k G. I., Splitting and alternating direction methods, in *Handbook of Numerical Analysis*, V. 1, ed. by P. G. Ciarlet and J.- L. Lions, North–Holland, 197, 1990.

225. M a r c h u k G. I., A g o s h k o v V. I., Conjugate operators and algorithms of perturbation in non-linear problems: 1. Principles of construction of conjugate operators, *Soviet Journal of Numerical Analysis and Mathematical Modelling*, 1, 21, 1988.

226. M a r c h u k G. I., A g o s h k o v V. I., Conjugate operators and algorithms of perturbation in non-linear problems: 2. Perturbation Algorithms, *Soviet Journal of Numerical Analysis and Mathematical Modelling*, 2, 115, 1988.

227. M a r c h u k G. I., A g o s h k o v V. I., On solvalibility and numerical solution of data assimilation problems, *Russ. J. Numer. Anal. Math. Modelling* V. 8, 1, 1, 1993.

228. M a r c h u k G. I., K u s i n V. I., O b r a z t s o v N. N., Numerical modelling of distribution in a water basin, *Preprint, Academy of Science of the USSR. Computing Center*, Novosibirsk, 1979.

229. M a r c h u k G. I., P e n e n k o V. V., Application of optimization method to the problem of mathematical simulation of atmospheric processes and environment, in *Modelling and Optimization of Complex Systems: Proc. of the IFIP – TC7 Work Conf.*, Novosibirsk, 240, 1978.

230. M a r c h u k G. I., S k i b a Yu. N., Numerical calculation of the conjugate problem for a model of the thermal interaction of the atmosphere with the oceans and continents, *Atm. and Ocean. Phys.*, 12. 279, 1976.

231. M a r c h u k G. I., S k i b a Yu. N., Role of adjoint equations in estimating monthly mean air surface temperature anomalies, *Atmosphera*, V. 5, 119, 1992.

232. M a r c h u k G. I., S h u t y a e v V. P., Iterative algorithms for solving a data assimilation problem, *Russ. J. Numer. Anal. Math. Modelling*, 1994 (to be published).

233. M a r c h u k G. I., Z a l e s n y V. B., A numerical technique for geofisical data assimilation problem using Pontryagin's principle and splitting-up method, *Russian J. Num. Anal. Math. Mod.*, V. 8, 4, 1993.

234. M i g n o t F., P u e l J. P., Optimal control in some variational inequalities, *SIAM J. on Control and Opt.*, V. 22, 466, 1984.

235. M i l l m a n M. N., K e l l e r J. B., Perturbation Theory of Nonlinear Boundary Value Problems, *J. of Math. Phys.*, V. 10, 2, 342, 1969.

236. M o s e r J., Stürüngstheorie des kontinuierlichen Spektrums für gewöhnliche Differentialgleichungen zweiter Ordnung, *Math. Ann.*, V. 125, 366, 1953.

237. N a y f e h A. H, *Perturbation Methods*, New York: J. Wiley and Sons, 1973.

238. P e n e n k o V., O b r a z t s o v N. N., A variational initialization method for the fields of the meteorological elements, *Meteorol. Gidrol.*, 11, 1, 1976.

239. P o i n c a r é H., *Les Méthodes Nouvelles de la Mécanique Céleste. Tome 1: Solutions Périodiques. Non Existence des Intégrales Uniforms. Solutions Asymptotiques*, Paris: Gauthier–Villars et'Fils. 8°, 1892.

240. P o n t r y a g i n L.S, B o l t y a n s k i i V. G., G a m k r e l i d z e R. V., M i s c h e n k o E. F., *The Mathematical Theory of Optimal Processes*, New York: Wiley Interscience, 1962.

241. R a h n e m a F., Internal interface perturbation in neutron transport theory, *Nucl. Scien. and Eng.*, V. 86, 76, 1984.
242. R e l l i c h F., Störungthorie des Spektralzerlegung. I–V, *Math. Ann.*, V. 117, 346, 1936; V. 118, 462, 1942.
243. R e l l i c h F., Störungthorie der Spectralzerlegung, *Proc. Internat. Congr. Mathematicians*, New York: Providence, V. 1, 606, 1952.
244. R e l l i c h F., *Perturbation Theory of Eigenvalue Problems*, New York: Gordon and breach Sci. Pub., 1969.
245. R a y l e i g h, L o r d (S t r a t t J. W.), *Theory of Sound*, V. 1–2, London: Mac Millan, 1926.
246. R i e s z F., S z. - N a g y B., *Functional Analysis*, New York: Frederik Ungar, 1955.
247. R o s e n b l u m M., Perturbation of the continuous spectrum and unitary equivalence, *Pacific J. Math.*, V. 7, 997, 1957.
248. S a d o k o v V. P. S h t e y n b o k D. B., Application of adjoint functions to the analysis and forecast of the temperature anomalies, *Soviet Met. and Hydrology*, 8, 6, 1976.
249. S a n c h e z – P a l e n c i a E. *Nonhomogeneous Media and Vibration Theory*, Springer-Verlag, 1980.
250. S a r k i s y a n A. S., I b r a e v R., et all, A note on modelling the world ocean climate, *Ocean Modelling*, 89, 10, 1990.
251. S c h r ö d i n g e r E., Quantisierung als Eigenwertproblem, *Ann. Phys.*, V. 80, 437, 1926.
252. S c h w a r t z L., *Theórie des distributions*, Paris: Hermann, 1966.
253. S m a g o r i n s k y J., General circulation experiments with the primitive equations: 1. The basic experiment, *Mon. Weather Rev.*, V. 91, 99, 1963.
254. S m e d s t a d O. M., O'B r i e n J. J., Variational data assimilation and parameter estimation im an equatorial Pacific Ocean model, *Prog. Oceanolog.*, 26, 179, 1991.
255. S o b o l e v S. L., *Application of Functional Analysis in Mathematical Physics*, Leningrad, 1950 (Transl.: Amer. Math. Soc. Transl. Math. Mono 7, 1963.)
256. S t o k e r L. J, *Water Waves. The Mathematical Theory with Applications*, New York: Interscience Publishers (Pure and Appl. Math. A Series of Texts and Monographs, V. VI), 1957.
257. S t r a n g G., On the construction and comparison of difference schemes, *SIAM J. Numer. Anal.*, 5, 1968.
258. S t r a n g G., *Linear Algebra and its Applications*, N.Y. etc.: Academic Press, 1976.
259. T a l a g r a n d O., C o u r t i e r P., Variational assimilation of meteorological observations with the adjoint vorticity equation. Part I: Theory, *Quart J. Roy. Meteor. Soc.*, 113, 1311, 1987.
260. T i t c h m a r s h E. C., Some theorems on perturbation theory, *J. Analys. Math.*, V. 4, 187, 1954/55.

261. V a n D y k e M. D., *Perturbation Methods in Fluid Mechanics*, New York: Academic Press, 1964.

262. Wallace J.M., Hobbs P.V., *Atmospheric Science (an introductory survey)*, New York, San Francisco, London: Academic Press, 1977.

263. W a n g Z h i, N a v o n I. M., L e D i m e t F. X., Z o u X., The second order adjoint analysis: Theory and Applications, *Meteorology and Atmospheric Physics*, 1992.

264. W e i n b e r g A., W i g n e r E., *The Physical Theory of Neutron Chain Reactors*, Chicago, 1958.

265. W i l k i n s o n J., *The Algebraic Eigenvalue Problem*, Oxford: Claredon Press, 1965.

266. Y o s i d a K., *Functional Analysis*, Berlin: Springer, 1971.

267. Z o u X., N a v o n I. M., L e D i m e t F. X., Incomplete observations and control of gravity waves in variational data assimilation, *Tellus*, 44A, 273, 1992.

268. Z o u X., N a v o n I. M., S e l a J. G., Variational data assimilation with moist threshold processes using the NMC spectral model (Submitted to *Tellus*), 1992.

Index

A

Adiabatic approximation, 351
Adjoint equation, 15, 48
Adjoint function, 40, 193
Adjoint operator, 15, 22, 152, 159
Adjoint problem, 65, 86, 96, 106, 116, 253
Adjointment relationship, 46, 53, 56, 122
Associated system, 153
Average temperature anomaly, 358

B

Baroclinic atmosphere, 351
Barotropic atmosphere, 345
Biorthogonality condition, 17, 24
Boundary conditions, 96, 105
Buoyancy parameter, 315

C

Cauchy problem, 177
Characteristic of measurement, 129
Coefficient of absorbtion,104
Common economic criterion, 296
Commutative operators, 425, 431
Concentration of an aerosol pollutant, 213
Conditionally correct problem, 165, 169
Continuity equation, 303, 345, 361
Control, 390, 393
Controlling family, 170, 171
Correctness set, 169
Crank–Nicolson scheme, 424

D

Data assimilation problem, 411
Data processing, 389
Definition domain of an operator, 10
75, 86, 88, 96, 124, 126, 158, 189
Delta-function, 313
Density of population, 289
Diffusion approximation, 221
Diffusion equation, 95, 230, 242, 441
Duality formulas, 127, 130, 265

E

Eigenfunction, 23, 89
Eigenvalue, 24, 88
Eigenvalue problem, 88
Evolutionary equation, 390

F

Formula of the perturbation theory, 210
Fourier coefficients, 24, 25, 437
Fourier method, 22, 26, 62, 173
Fourier series, 24, 181
Fourier transformation, 250, 437
Fundamental solution, 247, 249, 252
Function of information importance, 32
Functional, 34, 38, 52, 96, 115, 126, 127
Functional's variation, 31, 63, 67

G

Gateáux derivative, 156
Global transport of pollutants, 310
Geophysical data assimilation, 400

Other *Mathematics and Its Applications* titles of interest:

W.L. Miranker: *Numerical Methods for Stiff Equations and Singular Perturbation Problems*. 1980, 220 pp. ISBN 90-277-1107-0

K. Rektorys: *The Method of Discretization in Time and Partial Differential Equations*. 1982, 470 pp. ISBN 90-277-1342-1

L. Ixary: *Numerical Methods and Differential Equations and Applications*. 1984, 360 pp. ISBN 90-277-1597-1

B.S. Razumikhin: *Physical Models and Equilibrium Methods in Programming and Economics*. 1984, 368 pp. ISBN 90-277-1644-7

A. Marciniak: *Numerical Solutions of the N-Body Problem*. 1985, 256 pp.
ISBN 90-277-2058-4

Y. Cherruault: *Mathematical Modelling in Biomedicine*. 1986, 276 pp.
ISBN 90-277-2149-1

C. Cuvelier, A. Segal and A.A. van Steenhoven: *Finite Element Methods and Navier-Stokes Equations*. 1986, 500 pp. ISBN 90-277-2148-3

A. Cuyt (ed.): *Nonlinear Numerical Methods and Rational Approximation*. 1988, 480 pp. ISBN 90-277-2669-8

L. Keviczky, M. Hilger and J. Kolostori: *Mathematics and Control Engineering of Grinding Technology. Ball Mill Grinding*. 1989, 188 pp. ISBN 0-7923-0051-3

N. Bakhvalov and G. Panasenko: *Homogenisation: Averaging Processes in Periodic Media. Mathematical Problems in the Mechanics of Composite Materials.* 1989, 404 pp. ISBN 0-7923-0049-1

R. Spigler (ed.): *Applied and Industrial Mathematics*. Venice-1, 1989. 1991, 388 pp. ISBN 0-7923-0521-3

C.A. Marinov and P. Neittaanmaki: *Mathematical Models in Electrical Circuits. Theory and Applications*. 1991, 160 pp. ISBN 0-7923-1155-8

Z. Zlatev: *Iterative Improvement of Direct Solutions of Large and Sparse Problems*. 1991, 328 pp. ISBN 0-7923-1154-X

M. Vajtersic: *Algorithms for Elliptic Problems*. 1992, 352 pp.
ISBN 0-7923-1918-4

V. Kolmanovskii and A. Myshkis: *Applied Theory of Functional Differential Equations*. 1992, 223 pp. ISBN 0-7923-2013-1

A.D. Egorov, P.I. Sobolevsky and L.A. Yanovich: *Functional Integrals: Approximate Evaluation and Applications*. 1993, 426 pp. ISBN 0-7923-2193-6

G.I. Marchuk: *Adjoint Equations and Analysis of Complex Systems*. 1994
ISBN 0-7923-3013-7

Other *Mathematics and Its Applications* titles of interest:

A. Bakushinsky and A. Goncharsky: *Ill-Posed Problems: Theory and Applications.*
1994, 256 pp. ISBN 0-7923-3073-0